Therapies for Retinal Degeneration

Targeting Common Processes

Drug Discovery Series

Editor-in-chief
David Thurston, *King's College, UK*

Series editors:
David Fox, *Vulpine Science and Learning, UK*
Ana Martinez, *Centro de Investigaciones Biologicas-CSIC, Spain*
David Rotella, *Montclair State University, USA*
Hong Shen, *Roche Innovation Center Shanghai, China*

Editorial advisor:
Ian Storer, *AstraZeneca, UK*

Titles in the Series:
1: Metabolism, Pharmacokinetics and Toxicity of Functional Groups
2: Emerging Drugs and Targets for Alzheimer's Disease; Volume 1
3: Emerging Drugs and Targets for Alzheimer's Disease; Volume 2
4: Accounts in Drug Discovery
5: New Frontiers in Chemical Biology
6: Animal Models for Neurodegenerative Disease
7: Neurodegeneration
8: G Protein-Coupled Receptors
9: Pharmaceutical Process Development
10: Extracellular and Intracellular Signaling
11: New Synthetic Technologies in Medicinal Chemistry
12: New Horizons in Predictive Toxicology
13: Drug Design Strategies: Quantitative Approaches
14: Neglected Diseases and Drug Discovery
15: Biomedical Imaging
16: Pharmaceutical Salts and Cocrystals
17: Polyamine Drug Discovery
18: Proteinases as Drug Targets
19: Kinase Drug Discovery
20: Drug Design Strategies: Computational Techniques and Applications
21: Designing Multi-Target Drugs
22: Nanostructured Biomaterials for Overcoming Biological Barriers
23: Physico-Chemical and Computational Approaches to Drug Discovery
24: Biomarkers for Traumatic Brain Injury
25: Drug Discovery from Natural Products

26: Anti-Inflammatory Drug Discovery
27: New Therapeutic Strategies for Type 2 Diabetes: Small Molecules
28: Drug Discovery for Psychiatric Disorders
29: Organic Chemistry of Drug Degradation
30: Computational Approaches to Nuclear Receptors
31: Traditional Chinese Medicine
32: Successful Strategies for the Discovery of Antiviral Drugs
33: Comprehensive Biomarker Discovery and Validation for Clinical Application
34: Emerging Drugs and Targets for Parkinson's Disease
35: Pain Therapeutics; Current and Future Treatment Paradigms
36: Biotherapeutics: Recent Developments using Chemical and Molecular Biology
37: Inhibitors of Molecular Chaperones as Therapeutic Agents
38: Orphan Drugs and Rare Diseases
39: Ion Channel Drug Discovery
40: Macrocycles in Drug Discovery
41: Human-based Systems for Translational Research
42: Venoms to Drugs: Venom as a Source for the Development of Human Therapeutics
43: Carbohydrates in Drug Design and Discovery
44: Drug Discovery for Schizophrenia
45: Cardiovascular and Metabolic Disease: Scientific Discoveries and New Therapies
46: Green Chemistry Strategies for Drug Discovery
47: Fragment Based Drug Discovery
48: Epigenetics for Drug Discovery
49: New Horizons in Predictive Drug Metabolism and Pharmacokinetics
50: Privileged Scaffolds in Medicinal Chemistry: Design, Synthesis, Evaluation
51: Nanomedicines: Design, Delivery and Detection
52: Synthetic Methods in Drug Discovery: Volume 1
53: Synthetic Methods in Drug Discovery: Volume 2
54: Drug Transporters: Role and Importance in ADME and Drug Development
55: Drug Transporters: Recent Advances and Emerging Technologies
56: Allosterism in Drug Discovery
57: Anti-aging Drugs: From Basic Research to Clinical Practice
58: Antibiotic Drug Discovery: New Targets and Molecular Entities
59: Peptide-based Drug Discovery: Challenges and New Therapeutics
60: Drug Discovery for Leishmaniasis

61: Biophysical Techniques in Drug Discovery
62: Acute Brain Impairment Through Stroke: Drug Discovery and Translational Research
63: Theranostics and Image Guided Drug Delivery
64: Pharmaceutical Formulation: The Science and Technology of Dosage Forms
65: Small-molecule Transcription Factor Inhibitors in Oncology
66: Therapies for Retinal Degeneration: Targeting Common Processes

How to obtain future titles on publication:
A standing order plan is available for this series. A standing order will bring delivery of each new volume immediately on publication.

For further information please contact:
Book Sales Department, Royal Society of Chemistry, Thomas Graham House, Science Park, Milton Road, Cambridge, CB4 0WF, UK
Telephone: +44 (0)1223 420066, Fax: +44 (0)1223 420247,
Email: booksales@rsc.org
Visit our website at www.rsc.org/books

Therapies for Retinal Degeneration
Targeting Common Processes

Edited by

Enrique J. de la Rosa
CSIC Spain, Madrid, Spain
Email: ejdelarosa@cib.csic.es

and

Thomas G. Cotter
University College Cork, Cork, Ireland
Email: t.cotter@ucc.ie

ROYAL SOCIETY OF CHEMISTRY

Drug Discovery Series No. 66

Print ISBN: 978-1-78262-949-8
PDF ISBN: 978-1-78801-366-6
EPUB ISBN: 978-1-78801-536-3
Print ISSN: 2041-3203
Electronic ISSN: 2041-3211

A catalogue record for this book is available from the British Library

The Royal Society of Chemistry is a charity, registered in England and Wales, Number 207890, and a company incorporated in England by Royal Charter (Registered No. RC000524), registered office: Burlington House, Piccadilly, London W1J 0BA, UK, Telephone: +44 (0) 207 4378 6556.

For further information see our web site at www.rsc.org

Printed in the United Kingdom by CPI Group (UK) Ltd, Croydon, CR0 4YY, UK

Preface

The architecture of eye is one of nature's wonders and takes many shapes and forms. From the primitive light-detecting cells of invertebrates to the complex eye of mammals, there are many eye types in between these two ends of the "eye spectrum", and some of these are thought to have evolved in parallel evolution. The eye of many animal types show adaptations to their specific environments. An example of this is seen in birds of prey, which have a significantly higher visual acuity than what we have as humans. Other species such as bees can see and extended spectrum of light into the ultraviolet range, which gives them an advantage in distinguishing different flower patterns. Finally, some animal species have multiple eyes, often of different designs and in different parts of the body. As editors of this book we are privileged to work on one of mother nature's evolutionary wonders.

The eye, like every other organ in the body is prone to disease and malfunctions and this creates both medical and socioeconomic issues that need to be addressed. We have a very poor understanding of these conditions and even poorer therapeutic options. This book seeks to address what future therapeutic avenues to treat degenerative conditions of the retina may open up over the next several years. Since many of these conditions result from cell loss, strategies to either prevent such losses or replace the lost cells are attractive, but technically difficult.

In the opening chapter, Nicolas Cuenca and colleagues describe the course of cell loss during retinal degeneration illustrated with "art-like" confocal images, with the following chapters exploring the cell biology behind this cell loss and how we might control it therapeutically. Mechanistic chapters focused on calcium overload and the role of inflammation in the degenerative process shed some light on the complex biological issues involved.

Drug Discovery Series No. 66
Therapies for Retinal Degeneration: Targeting Common Processes
Edited by Enrique J. de la Rosa and Thomas G. Cotter

Published by the Royal Society of Chemistry, www.rsc.org

Targeting cGMP production (François Paquet-Durand and colleagues), which is often deregulated in retinitis pigmentosa due to a mutation in the cGMP phosphodiesterase, is a rational approach that may yield results. The idea here is that if we can regulate cGMP production we may be able to prevent the loss of photoreceptors which are very sensitive to increased levels of the cyclic nucleotide. Other approaches include the use of biological-based therapeutics, but whatever approach is contemplated, delivery is always going to be an issue. This pivotal issue is explored in detail in chapters by Rocio Herrero-Vanrell and R. V. Rajala and their colleagues, who look at two quite distinct approaches.

Since many degenerative diseases of the retina result from the mutation of one of several different genes, the recent advances of gene therapy to rectify these conditions has several attractions. This may sound easy in theory, but is difficult in practice. This approach is eloquently described by Lolita Petit and V. Kalatzis in their chapter. The two final therapeutic strategies discussed in the book are neuroprotection and stem cell therapy; one prevents cells from dying and the other replaces the dead cells.

Only time will tell which of the above therapeutic strategies will make it to clinic. But with several horses in the metaphorical race there is bound to be at least one winner.

We would like to thank all the authors who have taken time from their busy schedule to contribute a chapter. We would also like to thank the people at the Royal Society of Chemistry for all their encouragement and belief in the subject and us as editors. I hope we did them and the subject justice.

Thomas G. Cotter and Enrique J. de la Rosa

Cork and Madrid

Contents

Drug Discovery Series No. 66
Therapies for Retinal Degeneration: Targeting Common Processes
Edited by Enrique J. de la Rosa and Thomas G. Cotter

Published by the Royal Society of Chemistry, www.rsc.org

CHAPTER 1

The Cellular Course of Retinal Degenerative Conditions

NICOLÁS CUENCA*[a], VICTORIA MANEU[b], LAURA CAMPELLO[a] AND PEDRO LAX[a]

[a]Physiology, Genetics and Microbiology, University of Alicante, Alicante, Spain; [b]Optics, Pharmacology and Anatomy, University of Alicante, Alicante, Spain
*E-mail: cuenca@ua.es

1.1 Introduction

The retina is the light-sensitive tissue that lines the inner surface of the eye. Phototransduction, or the transformation of light signals into electrical impulses, takes place in photoreceptors, thus making the retina the tissue where the processing of visual information begins. The structural and functional complexity of the retina, and its direct exposure to light, makes this tissue one of the most vulnerable to the molecular alterations derived from any sort of pathological injury or disease, leading to blindness.

1.2 Alterations in Retinal Homeostasis

Whatever the origin of the retinal damage, oxidative stress, inflammation and apoptosis pathways are common features and are controlled by a plethora of cell mediators, some of them protective, some of them

Drug Discovery Series No. 66
Therapies for Retinal Degeneration: Targeting Common Processes
Edited by Enrique J. de la Rosa and Thomas G. Cotter

Published by the Royal Society of Chemistry, www.rsc.org

harmful, and several having a dual role depending on the particular conditions present.[1] Therapy for retinal diseases include preventive strategies to neutralize the underlying disease mechanisms and the administration of anti-oxidants, anti-apoptotic and anti-inflammatory compounds, as well as neurotrophic and growth factors, which slow the neurodegeneration of the retina by delaying retinal cell death.[1,2] Due to the tight connection of all these processes and their parallel running, one drug is likely to act in more than one pathway and mechanisms involving more than one effect are common.

1.2.1 Oxidative Stress

Oxidative stress plays an important role in the development and progression of retinal degeneration, with implications in the pathogenesis of age-related macular degeneration (AMD), diabetic retinopathy, retinitis pigmentosa[3] and glaucoma.[1,2,4,5] Photoreceptor cells are continuously exposed to light photons, have a high metabolic rate and are great consumers of oxygen, mainly due to a large accumulation of mitochondria in the ellipsoid and the aerobic metabolism in the membranous disks of photoreceptor outer segments. Hence the retina is a perfect target for reactive oxygen species (ROS) damage. Both overproduction of ROS and/or a reduced ability to neutralize normally produced ROS (due to deficiencies in anti-oxidant enzymes) can bring oxidative stress which, in turn, can induce cell death by different mechanisms: (i) triggering apoptosis by damage in DNA, lipids and proteins; (ii) by oxidizing and inhibiting phosphatases and/or kinases, among other proteins, and altering their downstream signaling pathways. If the levels of oxidative stress are severe the cells undergo necrosis.[2,6] The administration of anti-oxidants has shown good results in degeneration models and has been proven to preserve functional vision for longer in animal models and in patients.[7–11] Supplementation with anti-oxidants such as lutein, zeaxanthin and *meso*-zeaxanthin are recommended to AMD patients at significant risk for visual loss, although its utility for other retinal degenerative processes as well as the therapeutic potential of other anti-oxidants remain to be determined.[12] Impairment of nuclear factor erythroid 2-related factor 2 (Nrf2) signaling increases the vulnerability to oxidative damage and is associated with oxidative stress in several degenerative diseases, such as AMD or diabetic retinopathy (reviewed in Cuenca *et al.*[1]). Activation of the redox-sensitive transcription factor Nrf2 is one of the critical defensive mechanisms against oxidative stress in many species, as it binds to anti-oxidant response elements located in the promoter region of genes encoding many anti-oxidant enzymes and phase II detoxifying enzymes. Nrf2 activators have been tested in a number of retinal pathological models and several studies suggest that Nrf2 activation could be a therapeutic option for oxidative stress-related retinal diseases such as diabetic retinopathy or glaucoma.[13–15]

1.2.2 Neuroinflammation

A common feature of retinal degenerative diseases is neuroinflammation, which is mainly mediated by microglia that become activated in the course of the degeneration and release noxious factors which produce a plethora of inflammatory mediators.[1,3,16–20] Therapy with anti-inflammatory agents seems to be a good strategic point, and several anti-inflammatory compounds have been proven effective against retinal degeneration; among them, natural compounds with anti-oxidant and anti-inflammatory properties such as curcumin or quercetin.[1] Anti-inflammatory drugs as corticosteroids, that suppress multiple pathways of inflammation, have been proposed for the treatment of diabetic macular edema, particularly in cases refractory to laser photocoagulation.[21,22] However, it is necessary to take into consideration the fact that the inflammatory process has a dual effect: it may be pathogenic but also has reparative properties, and some authors have suggested that immunomodulation, rather than an anti-inflammatory approach, may be a more effective therapeutic strategy in retinal pathologies.[23]

The cannabinoid (CB) system has also been suggested as a useful target for the treatment of retinal degenerative diseases due to its implication in different aspects of the degenerative process: inflammation, apoptosis, neurodegeneration or angiogenesis. In this sense, CB1 agonists have showed neuroprotective actions and CB2 agonists neurogenic and anti-inflammatory properties in neurodegenerative diseases.[24,25] Norgestrel, a synthetic form of progesterone, can inhibit apoptosis and inflammation in cells and animal models,[2,26–28] likely through progesterone receptor membrane component 1 (PGRMC1), a key regulator of apoptosis.[29]

1.2.3 Apoptosis, Autophagy, Necroptosis and Necrosis

Cell death is a major determinant of inflammatory disease severity. During inflammation, the pro-survival process of autophagy and the pro-death process of apoptosis interact and influence each other, tilting the balance towards life or death.[30] To date, the whole routes of cell death and survival are not fully known. New unexpected roles for cytokines and other mediators emerge each day, showing more connexions between death and survival mechanisms. In this context, the three major cell death pathways, apoptosis, necrosis and autophagy, are involved in retinal degenerative diseases and are mutually interconnected in such a way that some mediators can influence more than one pathway.[30–34] So, as a therapeutic strategy, it could be essential to act in different steps to avoid the activation of alternative dying routes or to favour the survival side. Whenever a harmful stimulus threatens the cell, its machinery starts working in a way that in the best case will rescue the cell or the tissue from a greater damage, killing injured organelles or cells in a precise, organized way; and in the worst case will end with irreparable tissue damage. The balance between cell death and survival signals will determine the final outcome. Certain genetic mutations, age and environmental factors can trigger specific pathways to induce apoptosis in retinal cells, contributing to the development of many diseases.[1,35] The changes responsible for dystrophic and degenerative retinal diseases, which cause structural and functional damage, may occur at any level of the signal transduction cascade or in any of the morphological components of these differentiated cells. The life-or-death decision seems to be the result of a complex balance between pro- and anti-apoptotic signals at several levels: extracellular, mitochondrial, nuclcar and cytoplasmic.[36,37] The final common pathway of cell death in retinal diseases is apoptosis, which initially affects only certain retinal cells, such as photoreceptors, followed by the apoptosis of all remaining cells in the retina.[1] Hence, the pharmacologic inhibition of cell death through the use of anti-apoptotic agents may prevent disease-associated retinal degeneration. In this sense, several anti-apoptotic agents as tauroursodeoxycholic acid (TUDCA) or pro-insulin have shown good results in retinal neurodegenerative diseases (reviewed in Cuenca *et al.*[1]). Other agents, as the anti-oxidant melatonin, could improve retinal degeneration through the attenuation of apoptosis, but also avoiding the reactive gliosis and microglial activation in rd10 mice, and could have therapeutic interest improving photoreceptors survival in human retinitis pigmentosa.[38] The neuroprotective activity of FAS apoptotic inhibitory molecule 2 (FAIM2), which is an inhibitor of the FAS signaling pathway that is activated by stress in photoreceptors, has been reported recently. Therefore, modulation of the FAS signaling pathway to increase FAIM2 expression may be a potential therapeutic option to prevent photoreceptor death.[39]

Nevertheless, the inhibition of the apoptotic process induced by a noxious stimulus does not completely prevent retinal cell death, thus indicating that many cells can enter other mechanisms of cell death, such as necroptosis. Additional pathways including autophagy and inflammation can also contribute to the loss of retinal cells, as shown in different disease models.[32] In retinal degenerative diseases such as retinitis pigmentosa, both apoptosis and necroptosis can be triggered simultaneously and sometimes by the same death stimuli, but with distinct biochemical and morphological features.[40–43] Necroptosis is the regulated form of necrotic death, is caspase-independent and is triggered by activation of the receptor interacting serine/threonine kinase 3 (RIPK3) and phosphorylation of its pseudokinase substrate mixed lineage kinase-like (MLKL), which translocates to membranes and promotes cell lysis. It is considered a pro-inflammatory form of cell death, because rupturing of the cell releases intracellular contents that can stimulate innate immune cells, so it is believed to be a more potent inducer of inflammation than apoptosis, although additional *in vivo* studies are needed to confirm this statement. The precise mechanisms determining the decision whether a cell will die by apoptosis or necroptosis are not yet fully understood.[40,44–47] To date, no suitable inhibitors of the necroptosis pathway with activity in the CNS of mice or humans have been identified and more research in this field is needed. Blocking necroptotic pathways with synthetic inhibitors or by genetic manipulation relieves neurodegenerative disease both *in vitro* and *in vivo*, which suggests a promising therapeutic strategy for neurodegenerative diseases.[48]

The third major route of cell death, autophagy, also has a main role in sustaining retinal and eye function, and alterations in autophagy and lysosomal pathways are involved in many, if not all, diseases of the eye. Autophagy is evolutionarily a well conserved process from early eukaryotes to mammals. It is generally considered an anti-inflammatory and pro-survival process. It is a lysosome-mediated degradation process that inhibits the accumulation of misfolded proteins and damaged cytoplasmic organelles and removes microorganisms, contributes to the maintenance of metabolic homeostasis and provides energy and recycling macromolecules in response to nutrient and environmental stress.[30,49–52] In CNS diseases, autophagy is a key point in the control of inflammation and production of cytokines by activated microglial cells. Autophagy modulates inflammatory responses through interactions with immune signaling pathways and regulates the secretion of molecular mediators of inflammation in a context-dependent manner.[53,54] Autophagy is also involved in the regulation of proteins relevant for cellular anti-oxidative defence, including Nrf2. In this context, it has been proposed that autophagy is an adaptive response that might confer protection against persistent inflammation in the retina during aging.[55] The decline in lysosomal activity associated with age exacerbates alterations in autophagy, potentially aggravating related conditions. While appropriate autophagy contributes to neuroprotection, inappropriate autophagy could induce cell death and its dysfunction has been related to several neurodegenerative diseases such as Parkinson's or Alzheimer's diseases, among other neuron-affecting pathologies.[54,56,57] In a neurodegenerative process, a mild induction of autophagy should protect cells from damaged proteins and organelles.[57,58] Autophagy is active in the retina from the developmental stage, during which it is primarily implicated in cell death processes. Autophagy plays an important role in recycling photoreceptor outer segments, avoiding lipofuscin accumulation, preventing aggregation and minimizing oxidative and endoplasmic reticulum (ER)-stress, preserving mitochondrial function, attenuating inflammasome activation in the retinal pigment epithelium (RPE) and minimizing ROS levels and sustaining mitochondrial function in retinal ganglion cells (RGCs).[51] Autophagy is involved in AMD pathogenesis, and it has been proposed that autophagy-related genes and proteins can be molecular targets for prevention and therapy of this disease.[59]

ROS prompt apoptosis and autophagy pathways, and stress-specific recruitment of autophagy pathways with cytoprotective function.[60] mTOR, the mammalian target of rapamycin, has a main role in many degenerative diseases as one of the main regulators of autophagy.[3] In this context, work has been published showing that pharmacological treatment with the mTOR inhibitor rapamycin preserved photoreceptor function in metabolic and oxidative stress mouse models that exhibited RPE mitochondrial dysfunction accompanied by activation of the AKT/mTOR pathway.[61] It has also been shown that phosphoinositide 3 kinase (PI3K)/Akt and mTOR/p70S6K pathways play a major role in protection against oxidative stress-induced apoptosis in the ARPE-19 cell line.[62] The mTOR pathway plays an exquisitely complex role in the regulation of retina protein biosynthesis, RNA translation and ER stress-induced apoptosis. mTOR signaling can be suppressed by the inhibition of serine/threonine kinase Akt, which leads to the induction of autophagy.[33] Degradation and inhibition of mTOR attenuates retinal degeneration.[63] Thus, mTOR pathway inhibition has been suggested as a therapeutic strategy for retinal degenerative diseases involving oxidative stress, although it has showed a different response according to the degree of oxidative stress imposed.[2,64] However, clinical trials to assess the effect of rapamycin, an mTOR inhibitor, for the treatment of geographic atrophy reported no clear efficacy and some doubts concerning its safety.[65,66] Although rapamycin's side effects, such as immuno-suppression at early stages of treatment prevented its use in treating neurodegenerative diseases, small molecule enhancers of rapamycin (SMERs) which increase autophagy independently of mTOR pathway have been suggested as potential therapeutic agents to treat neurodegenerative disorders.[3,67]

Other molecules can increase pro-survival signals and have therapeutic potential for retinal degeneration. Calcium channels inhibitors such as amlodipine or amiloride activate the pro-survival kinase Akt, decrease calpain activity, attenuate JNK/c- Jun activation and can improve RGC survival after optic nerve crush, so they could be useful tools for therapeutic interventions in traumatic and degenerative CNS disorders.[68]

Another regulator of autophagy is p53. While the apoptotic activity of p53 is well recognized, several reports have highlighted the functional link between p53 and autophagy. The role of this protein varies depending on its subcellular location.[34,45] Moreover, p53 has been demonstrated to have a neuroprotective role in a *Drosophila* model of tauopathy, by controlling synaptic genes, so it has been proposed that p53-related molecules could have potential therapeutic value in Alzheimer's disease and related neurodegenerative disorders.[69]

Glycogen synthase kinase 3 (GSK3) is a critical central figure in many cellular signaling pathways. It is a multitasking serine/threonine kinase that regulates many cellular functions, such as structure, gene expression, mobility and apoptosis, with more than 100 known substrates to deal with. GSK3 promotes p53-induced apoptosis forming a complex with nuclear p53.[70,71] The inhibition of GSK3 was proved to be one of the mechanisms by which PI3K activation protects neurons from programmed cell death, and it has been proposed that GSK3 be included to the list of potential drug targets for pharmacotherapy of neurodegenerative disorders.[72] There are strong connections between GSK3 and pathophysiology and/or therapeutics of a large number of prevalent diseases that share the common feature of chronic inflammation, as GSK3 is an important positive regulator of the inflammatory process.[73] In this sense, inhibitors of GSK3 may provide a therapeutic strategy to control inflammation[74] and could also be useful therapeutic agents for the treatment of degenerative diseases such as retinitis pigmentosa.[75]

Retinal degenerative diseases entail a non-specific disease-remodeling phenomenon accompanied by common mechanisms of cell stress, harmful and pro-survival signals. The ability of the cell to increase life stimuli and to control death ones tilts the balance towards survival or death. Our help in any of these aspects can have potential therapeutic value to prevent retinal degeneration.

1.3 Morphological Impairments in Retinal Degeneration: Neurons and Circuitries, Microglia, Astrocytes, Müller Cells and Retinal Vasculature

1.3.1 Retinal Morphology

All vertebrate retinas consist of the neural retina (or neuroretina) and the pigment epithelium, which separates the neural retina from the choroid. The neural retina is made up of three nuclear layers, which contain the cell bodies (or somas) of retinal neurons, and two plexiform layers, in which the synaptic contacts are established.[76] The outer nuclear layer (ONL) is formed by the cell bodies of photoreceptors (cones and rods); the inner nuclear layer contains the cell bodies of bipolar, horizontal and amacrine cells; and the ganglion cell layer consists of the cell bodies of ganglion and displaced amacrine cells. Between these three layers of cell bodies there are two plexiform layers where the synaptic contacts take place: the outer plexiform layer (OPL),[77] where the synaptic contacts between the axon terminals of photoreceptors and the dendrites of their two postsynaptic neurons, bipolar and horizontal cells, are established; and the inner plexiform layer (IPL), with synaptic connections between bipolar, amacrine and ganglion cells. In the IPL there is a horizontal flow of information through a large variety of amacrine cells, which together with the bipolar cells integrate the information that reaches the ganglion cells that transmit the information received from bipolar and amacrine cells to the brain *via* the optic nerve.[76,78]

1.3.2 Early Onset of Rod Photoreceptors Degeneration

Animal models of retinitis pigmentosa have an early onset of retinal degeneration. Retinitis pigmentosa-associated genes are primarily expressed in rods. Consequently, this disease is characterized by an initial loss of rods followed by a progressive mutation-independent cone death. In this context, the extensive downstream of neuronal changes that accompany and follow photoreceptor degeneration has been well-established.[1]

One of the first signs of photoreceptor cell impairment is the mislocation of proteins with specific subcellular distribution. Rhodopsin, the rod visual pigment, is localized in the outer segments of rod photoreceptors in healthy retinas. However, during retinal degeneration it is common to observe an abnormal accumulation of this protein in the inner segments, cell bodies and axon terminals of rods (Figure 1.1A and B). A possible explanation for this irregular distribution is that rhodopsin synthesis continues after photoreceptor outer segments are lost due to the degenerative process.[79] Thus, the newly synthesized rhodopsin cannot be translocated to its normal cellular destination, therefore accumulates near the site of synthesis, in the

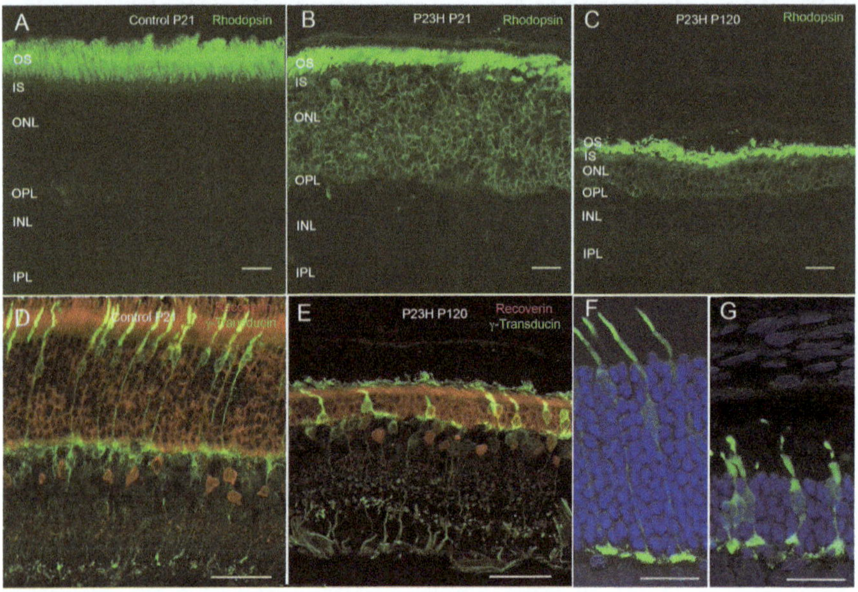

Figure 1.1 Photoreceptor cell changes in retinal degeneration. Vertical retinal sections labeled for rhodopsin (rod outer segments, green), γ-transducin (cone cells, green) and recoverin (cones, rods and some bipolar cells, red), showing the structure of photoreceptors in wild-type animals (A, D, F) and in the P23H rat model of retinitis pigmentosa (B, C, E, G). Nuclei stained with TO-PRO (blue) (F, G). In normal retinas, rhodopsin is located predominantly in the outer segment of rod cells (A). However, in degenerative retinas this protein is mislocated in the inner segments, cell bodies and axon terminals of rods (B). During retinal degeneration there is a notable shortening of outer segments in rods (A–C) followed by the shortening of outer and inner segments as well as axons of cones (D–G). OS: outer segments; IS: inner segments; ONL: outer nuclear layer; OPL: outer plexiform layer; INL: inner nuclear layer; IPL: inner plexiform layer. Scale bars: *A*, *B*, *C*, *D*, *E* = 20 μm; *F*, *G* = 40 μm.

photoreceptor cell soma. The same event can be observed in proteins involved in neurotransmission and synaptic vesicle formation and trafficking, such as synaptophysin and the vesicular glutamate transporter 1 (VGluT1) whose normal distribution is throughout photoreceptor axon terminals (spherules and pedicles) and under degenerative conditions can be found in the cellular bodies.[80] Another prominent feature of photoreceptor degeneration is the progressive outer segment shortening (Figure 1.1A–C), which is also observed in retinal histologic studies in retinitis pigmentosa patients.[81–83] Sprouting of rod processes has also been noted in human retinas with retinitis pigmentosa[83] and in animal models of this disease, *i.e.* rhodopsin transgenic pigs[84] and cats with autosomal dominant retinal dysplasia.[85]

As mentioned earlier, as rod degeneration evolves, cone morphology undergoes progressive changes, including the shortening of outer and inner segments and axons (Figure 1.1D–G).[1,86,87] Additionally, the progressive loss of photoreceptors within the ONL causes the appearance of hypertrophied side branches of Müller cells into the outermost photoreceptor layer. Apical Müller cell processes rearrange in clusters forming firework-like structures. This peculiar disposition of cones and apical Müller cell processes has been described in the S334ter-line-3[88,89] and the P23H[90,91] rat models of retinitis pigmentosa.

The previously described early onset of morphological retinal changes correlate well with altered electroretinogram (ERG) measurements. In concordance, abnormal rod ERG function was detected at an early age in P23H rats[92] and ERG impairment is an early clinical manifestation in autosomal dominant retinitis pigmentosa patients, even without subjective symptoms.[1,93]

1.3.3 Modifications of Second-order Neurons in the Retina

As death of rods and cones progresses, bipolar and horizontal cells, the second-order neurons in the retinal circuitry, become deafferented and display early retraction and loss of dendrites[1,80,94,95] (Figure 1.2). Additionally, in the P23H animal model of retinitis pigmentosa, it has been observed that bipolar and horizontal cells seek out new functional photoreceptors with which to make contacts, thus extending their dendrites (sprouting).[1,94,96] These results resemble closely those seen in the retina of the rd-mutant mice.[97] Therefore, the loss of synaptic connections with photoreceptor cells evokes atrophy and connective re-patterning in their postsynaptic cells.[80,86]

As described earlier, in the context of retinal degeneration bipolar cells undergo dramatic morphological changes accompanying photoreceptor loss. The first obvious changes are the underdevelopment of rod bipolar and horizontal cells dendrites. This is not surprising, since rod bipolar cells receive direct input from rod photoreceptors. These cells are presumably more vulnerable to photoreceptor degeneration at the time of synaptogenesis. As retinal degeneration progresses, rod bipolar cells axonal endings become atrophied, cell bodies lose their normal lamination pattern,

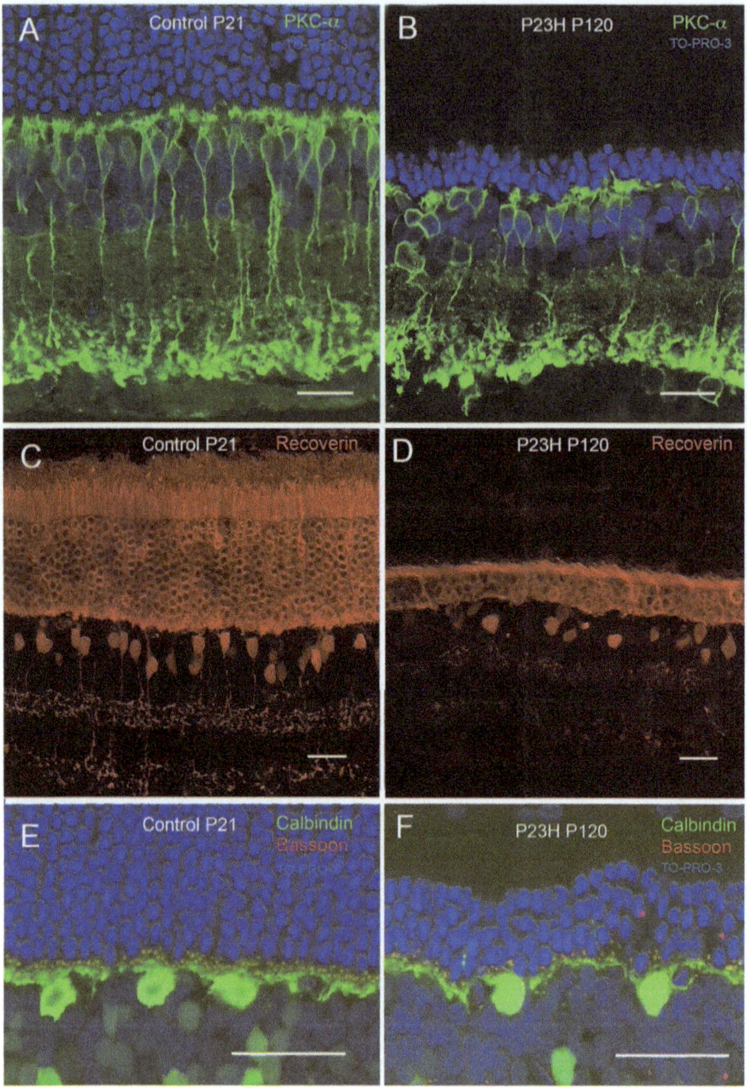

Figure 1.2 Morphological changes in bipolar and horizontal cells in retinal degeneration. Vertical retinal sections immunostained against PKC-α (ON-rod bipolar cells, green), recoverin (photoreceptors and type 8 and type 2 bipolar cells, red), calbindin (green) and bassoon (synaptic ribbons, red) in wild-type animals (A, C, E) and in the P23H rat model of retinitis pigmentosa (B, D, F). Nuclei stained with TO-PRO (blue). As retinal degeneration progresses, the axonal endings of rod bipolar become atrophied, their cell bodies lose their normal lamination pattern and there is a retraction and loss their dendrites (B) compared to control animals (A). Cone bipolar cell morphology is impaired during retinal degeneration (C, D). Horizontal cells also exhibit a decrease in their dendritic arborization, inverted cell bodies and loss of synaptic contacts (D) compared to control animals (C). Scale bars: *A*, *B*, *C*, *D* = 20 μm; *E*, *F* = 40 μm.

become disorganized and, eventually, die (Figure 1.2A and B). The loss of the depolarizing ON rod bipolar cell population is well reflected by the absence of the ERG b-wave in the P23H rat.[86] In a similar way, cone bipolar cells undertake morphological impairment during degeneration (Figure 1.2C and D).

Horizontal cells also show early signs of degeneration as rod photoreceptors are lost. Horizontal cell dendrites connect with cones and the axon terminals contact with rods in the OPL. Thus, morphological changes in horizontal cell axon terminals would be expected. The dendritic arborization of horizontal cells is abnormal and form a weaker network in P23H rat retinas, compared to normal retinas (Figure 1.2E and F). In addition, horizontal cell dendrites appear compromised and seem to become condensed with aberrant rod bipolar cell dendrites.[86] With the progression of retinal degeneration, most of the horizontal cell dendrites are lost, cell bodies become inverted and disorganized and the axon terminals appear to sprout into the ONL as though looking for any remaining degenerate rod spherule. These changes are very similar to those seen in transgenic rd mice by Strettoi *et al.*[97]

1.3.4 Cell Connectivity in the OPL at Late Stages of Retinal Degeneration

The number of retinal neurons decreases progressively as the retinal degeneration proceeds. In concordance, the surviving neurons interconnect in clusters that are uniformly distributed throughout the OPL, establishing new contacts as part of the remodeling process. In this context, the scarce dendrites of rod bipolar cells, which express the metabotropic glutamate receptor mGluR6, partake in these synapses. The remaining dendrites of horizontal cells associated with the presynaptic protein Bassoon are confined within the aforementioned clusters of synaptic contacts, as well as the contacts between cones and rod bipolar cell dendrites. In the absence of rods, rod bipolar cells receive ectopic synapses from cones, as shown in electron microscopic studies performed in different animal models of photoreceptor degeneration, including the P347L pig,[98] rd mouse[98] and RCS rat.[99] Bassoon and synaptophysin, two presynaptic proteins, also downregulate their expression during retinal degeneration (Figure 1.3).

The cartoon in Figure 1.4 summarizes the changes that take place in horizontal and bipolar cells at the OPL level after photoreceptor degeneration.

1.3.5 Modifications in the Inner Plexiform Layer

There is an obvious reduction in the thickness of the IPL after the onset of rod photoreceptor degeneration. This is almost certainly due to the shrinking of the rod bipolar axon terminal in the stratum S5 of the IPL. Cone bipolar cell axons also reduce their size and show morphological impairments. The

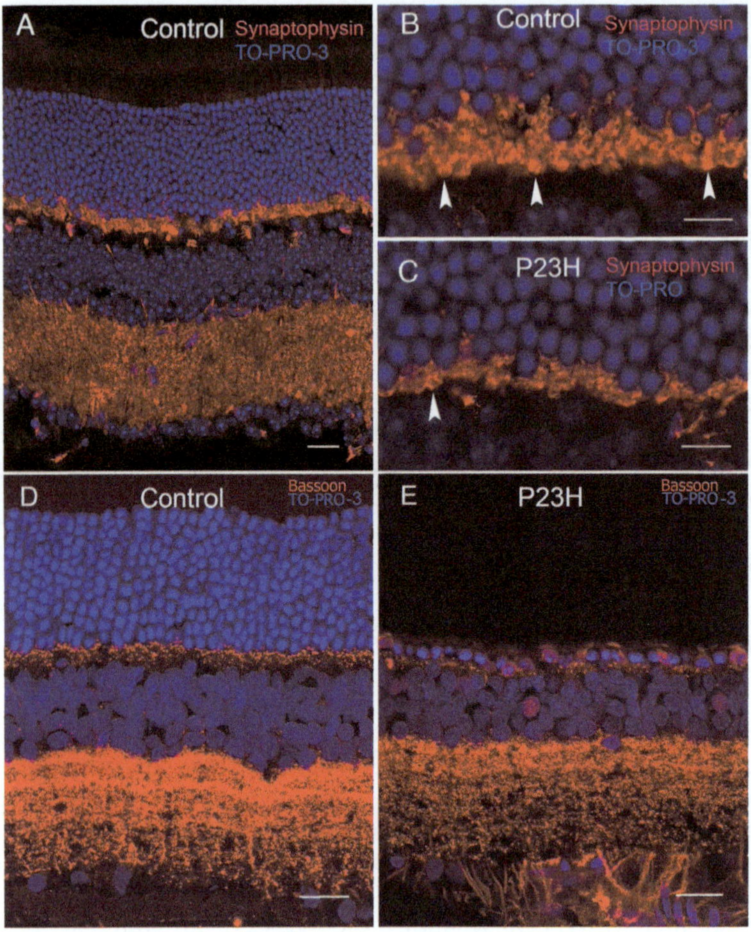

Figure 1.3 Alteration of synaptic connectivity in retinal degeneration. Vertical retinal sections labeled for synaptophysin (axon terminals of photoreceptors, arrowheads, red) (A–C) and Bassoon (photoreceptor synaptic ribbons, red) (D, E) in wild-type animals and in the P23H rat model of retinitis pigmentosa. Nuclei stained with TO-PRO (blue). During the degenerative process the decrease of both synaptophysin and Bassoon immunopositive spots at the outer plexiform layer (OPL) level is evident, which is indicative of loss of synaptic contacts. Scale bars: *A*, *D*, *E* = 20 μm; *B*, *C* = 10 μm.

AII amacrine cells undergo morphological modifications as photoreceptors degenerate. The AII amacrine cells relay rod information from rod bipolar cells to cone bipolar cells and ganglion cells, so the early modification of rod bipolar cells would naturally affect AII amacrine cells.[80] With the progression of retinal degeneration, their lobular appendages get smaller and disappear, and eventually there is complete cell dropout. Like the attenuation of the b-wave of the ERG with the loss of rod bipolar cells, the loss of AII amacrine cells might be reflected in diminution of any scotopic oscillatory potentials.[100]

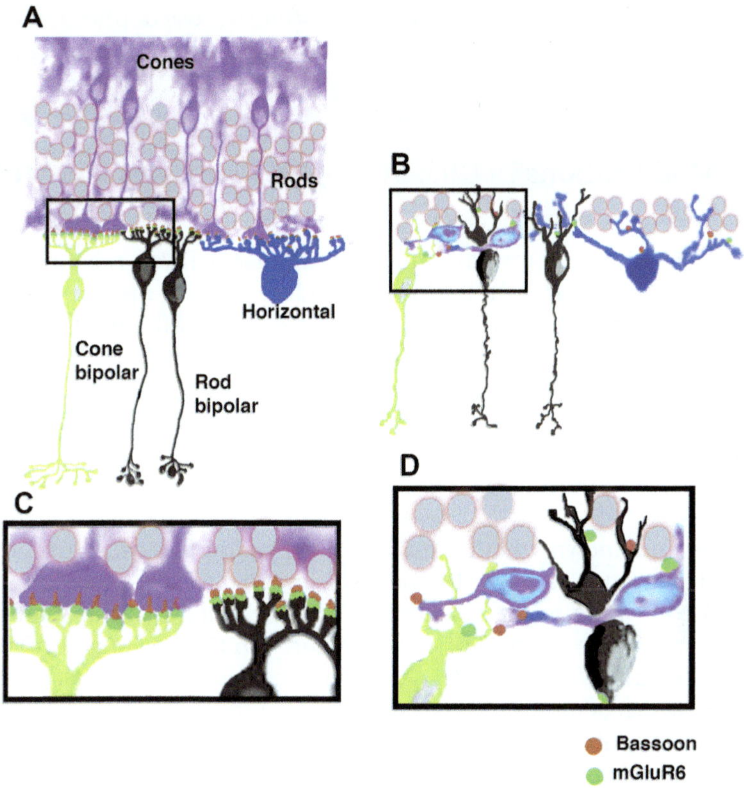

Figure 1.4 Schematic representation of synaptic alterations in the outer plexiform layer in retinal degeneration. Normal synaptic contacts are established between photoreceptors and horizontal and cone and rod bipolar cells (A, C). However, in the context of retinal pathology, there are dramatic changes in the morphology of retinal neurons that lead altered patterns of connectivity (B, D). The retraction of axons and dendrites of bipolar cells entail the mislocalization of mGluR6 and Bassoon synaptic markers from the dendritic tips of bipolar cells to the cell bodies and axon terminal (C, healthy; D, disease). Additionally, there is a marked sprouting process of rod bipolar and horizontal cells that try to establish new synaptic contacts with cones in absence of photoreceptor cells.

The decrease in density of AII amacrine processes, with a simultaneous loss of certain populations of amacrine cells and, eventually, even ganglion cell dendrites, induces the decrease of thickness of this layer. The IPL becomes filled with sprouting Müller cell processes to replace the lost neural synaptic complexes. Simultaneously, the blood vessels become distorted, possibly by Müller and glial cell proliferation. The blood vessels rearrange the previously ordered cell and neuropil layering and, eventually, pull ganglion cell axons into the inner retina. These changes in the IPL due to neuron and Müller cell atrophy and blood vessel attenuation are seen in human specimens of advanced retinitis pigmentosa[101] and have been described in other animal models of retinal dystrophies.[102] In addition, ganglion cells undergo a series of changes after the loss of photoreceptors, and their dysfunction and death has been linked to altered retinal glial cell function.[103]

1.3.6 Modifications in Glia: Microglia, Astrocytes and Müller Cells

Virtually all forms of retinal injury or disease trigger reactive gliosis (reviewed in Cuenca *et al.*[1]). Retinal neurodegenerative diseases are associated with chronic microglial activation and neuroinflammation. In the degenerating retina, endogenous signals activate microglial cells, which in turn can proliferate, migrate, enhance phagocytosis and secrete cytokines, chemokines and neurotoxins (Figure 1.5A and B). These immunological responses and the loss of limiting control mechanisms may contribute significantly to retinal tissue damage and pro-apoptotic events in retinal dystrophies.[1,104–106]

It remains unknown whether microglia activation is a cause or a consequence of neuronal damage. In the CNS, microglia have protective functions and contribute to the secretion of trophic factors, anti-oxidants and cytokines and to the removal of cellular debris. In the first stages of retinal neurodegeneration, microglia trigger repair mechanisms such as glial scar formation, which demarcates the lesion area and separates the injured tissue from its surroundings, promoting neuronal survival. But microglia chronically activate release noxious factors that mediate neuroinflammation and toxicity.[1,104,105,107] Microglia produce a plethora of inflammatory mediators, including cytokines, chemokines, trophic factors and small molecules, that promote and perpetuate the inflammatory response, potentially leading to neurodegeneration and photoreceptor death.[1,3,16–18] In this context, it has been published that cytokines play roles as mediators and modulators of diverse forms of neurodegeneration in the CNS, including the retina.[19,20] Due to the dual role of microglia activation in the progression of neurodegenerative diseases, both protecting and harmful, doubts remain regarding the optimal therapeutic intervention to halt or reduce the degeneration. On one hand, the selective inhibition of the overactive microglial activity and the preservation of their trophic and homeostatic functions seems a promising treatment for degenerative diseases. In this sense, minocycline, a semi-synthetic tetracycline analog, is an inhibitor of microglial activation, it counter-regulates microgliosis and is proven to have potent anti-inflammatory and neuroprotective effects in mouse models of retinal degeneration, indicating a promising concept for the treatment of retinal pathologies.[108–111] Clinical trials are under development to evaluate minocycline safety and efficacy on AMD patients (ClinicalTrials.gov identifier: NCT02564978). However, the inhibition of the protective effect of microglial activation could have harmful consequences because reactive microglia also participate in regenerative

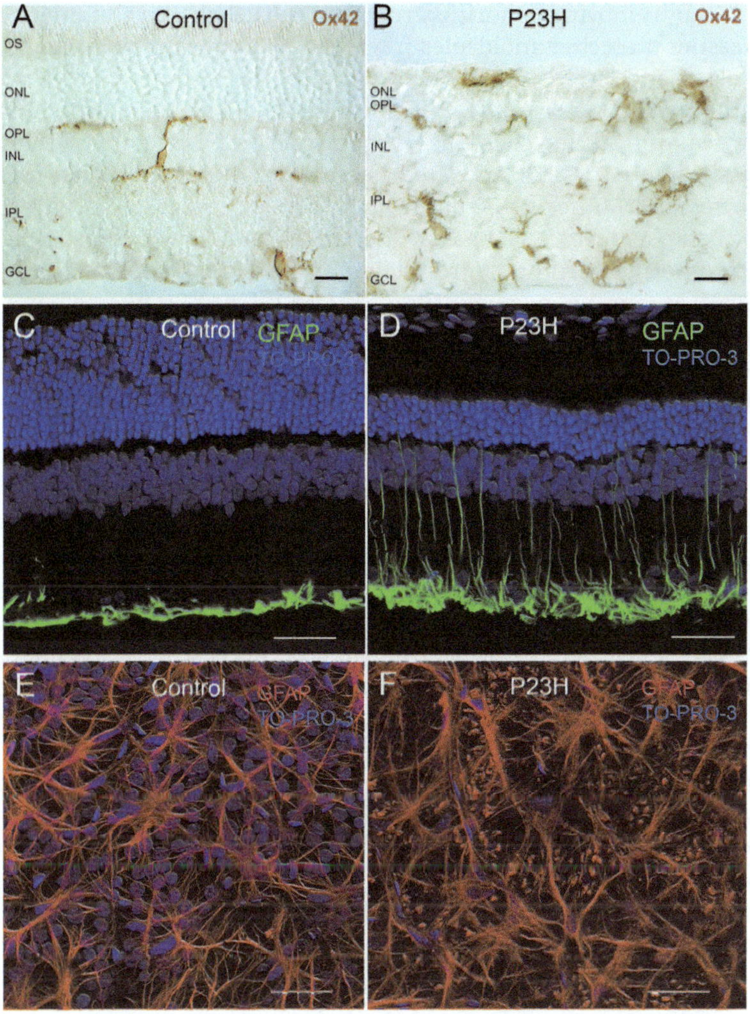

Figure 1.5 Activation of microglia, astrocytes and Müller cells during the degenerative process. Images of healthy (A, C, E) and P23H (B, D, F) rat retina labeled with antibodies against Ox42 (A, B), a marker of microglia, or glial fibrillary acidic protein (GFAP), a marker of reactive gliosis in Müller cells (green; C, D) and astrocytes (red; E, F). There is an increase in the number of microglial cells, which migrate throughout the retina and adopt the typical amoeboid shape of activated microglia, in the P23H rat retina (B) compared to control (A). The number of activated Müller cells is also evident in the diseased retinas (D). Additionally, activated astrocytes (F) become less ramified and hypertrophic than in control rats. OS: outer segments; ONL: outer nuclear layer; OPL: outer plexiform layer; INL: inner nuclear layer; IPL: inner plexiform layer; GCL: ganglion cell layer. Scale bars: *A*, *B* = 20 μm; *C*, *D*, *E*, *F* = 40 μm.

processes by removing dendritic structures, cell debris and glutamate, as well as releasing protective molecules (reviewed by Cuenca *et al.*,[1] Langmann[104] and Polazzi and Monti[112]).

In the early stages of degeneration, macroglial cells, which include astrocytes and Müller cells, also become activated as part of the gliosis process (Figure 1.5C–F). Activated microglia induce macroglia, including astrocytes and Müller cells, and RPE to secrete cytotoxic factors such as tumor necrosis factor (TNF)-α or interleukin (IL)-1β, which act synergistically, promoting chronic neuroinflammation and collaborate to neuronal cell death and retinal degeneration.[113,114] While reactive gliosis has a direct neuroprotective effect on the retina, its chronicity exacerbates disease progression, increasing vascular permeability, infiltration of toxic compounds and neovascularization.[115,116]

The fractalkine (CX3CL1)/CX3CR1 signaling pathway has a relevant role in the control of retinal inflammation and hence in the pathogenesis of retinal inflammatory and neurodegenerative diseases of the retina.[117–119] This pathway protects neurons by regulating retinal microglial activation and migration. Fractalkine modulates TNF-α and IL-6 secretion and neurotoxicity induced by microglial activation[120–123] and can reduce neurotoxicity and microglial activation in neurodegenerative models.[124] In rd10 mice, a synthetic progesterone analog (norgestrel) upregulates fractalkine-CX3CR1 signaling and has dual actions as a neuroprotective and anti-inflammatory agent in the retina, becoming a promising treatment for retinitis pigmentosa.[25,125]

Current therapeutic approaches used to protect photoreceptors and slow down degeneration include the administration of neurotrophic factors. Some of them, such as methyl 3,4-dihydroxybenzoate (MDHB), can exert their protective actions at least in part by reducing reactive gliosis,[126] apart from other anti-angiogenic, neuroprotective and anti-inflammatory activities,[1,127] and can have a role in retinal therapy.[128]

1.3.7 Alterations of Retinal Vascularization

Due to the high metabolic activity of the retina and its great oxygen demand, the supply of blood and the ability to regulate blood flow is a key point for retinal health. All degenerative diseases affect retinal vascularization.[1] In pathological conditions, astrocytes and Müller cells are involved in the development of new vascularization by releasing angiogenic factors, such as the vascular endothelial growth factor (VEGF), in response to pathogenic stimuli.[8,129–131] Anti-VEGF factors such as bevacizumab have been proposed and tested as therapeutic tools in degenerative diseases, and clinical trials with aflibercept, bevacizumab and ranibizumab have shown good results when treating patients with retinal degeneration.[22,132] Unfortunately, to date, anti-VEGF factors cannot rescue the normal retinal vasculature and have serious drawbacks, for example in the treatment of infants affected by retinopathy of prematurity.[133–135] For patients with a suboptimal response to anti-VEGF therapy, a multimodality approach has been proposed, as the addition of the corticoid dexamethasone can induce better results.[136] It has been proposed that revascularization by using proangiogenic cells can be an option for therapeutic revascularization in the treatment of retinopathy of prematurity.[137]

Following photoreceptor degeneration, the inner retina becomes hyperoxic. In consequence, there is a reduction of VEGF expression and a loss of the deep capillary plexus.[1,138] The disruption of the deep capillary plexus alters the normal nutrient and oxygen supply to retinal cells, and thereby accelerates the progress of retinal degeneration. Retinal vasculature remodeling has been evidenced in retinitis pigmentosa models.[1,138] In conclusion, the loss of photoreceptors in retinitis pigmentosa induces changes in retinal vasculature, which in turn may induce the activation of astrocytes and Müller cells, which can induce inner retinal remodeling,[1,86,139–141] including ganglion cell degeneration.[101,142,143]

1.4 Physiological Degeneration of the Retina

Independently of the etiology of the damage, retinal neurodegenerative diseases result in morphological and functional impairment of retinal cells that lead to well-established physiological changes affecting vision. Reliable assessment of visual capacity can provide key information about the mechanisms and progression of the degenerative events associated with retinal diseases.

1.4.1 Electroretinogram

The recording of electrical responses of the eye to light stimulus constitutes the ERG. The negative a-wave mainly represents light-evoked hyperpolarization of photoreceptors in the outer retina,[144,145] whereas the positive b-wave indirectly represents the activity of the on-bipolar cells.[146–150] The oscillatory potentials are rapid oscillatory responses superimposed on the ascending b-wave of the ERG.[151,152] Although the specific cellular origins of the oscillatory potentials have yet to be established, early oscillatory potentials have been associated with the activity of photoreceptors and bipolar cells in the outer retina, whereas later oscillatory potentials are attributed to activity in amacrine and ganglion cells in the inner retina.[100,153,154]

Even though considerable variability exists in the onset and evolution of retinal diseases, ERG recordings can be used to distinguish common abnormalities. The amplitude and implicit time of the ERG components reflect morphological alterations occurring during degenerative processes, including the progressive loss of photoreceptors and synaptic dysfunction in both the OPL and IPL. In patients with retinitis pigmentosa, ERGs characteristically show reduced a- and b-wave amplitudes, as well as delayed rod and/or cone b-wave implicit times.[155,156] In contrast, the ERGs of patients with cone dystrophy characteristically exhibit normal, albeit slower, rod b-waves and reduced or absent cone ERG responses.[157] In addition, ERG recordings provide a means to assess the progression of retinal degeneration in animal models[86,94,158] and to evaluate the therapeutic effects of neuroprotective agents on them.[159–163] In this context, it has been shown that the number of rows of photoreceptor nuclei positively correlates with the amplitude of the scotopic b-waves recorded in P23H rats,[160,162] a model of retinitis pigmentosa. Furthermore, the thickness of the ONL was found to be proportional to the scotopic a- and b-wave amplitudes in a rotenone rat model of Parkinson's disease.[164]

The analysis of oscillatory potentials has been shown to be very useful in the study of retinal diseases.[100,165–167] Oscillatory potentials are a good indicator of the functional integrity of the microcirculation of the inner retina and are drastically affected by acute disturbances occurring in areas supplied by central retinal vessels.[167,168] Thereby, abnormal oscillatory potential responses reveal a pathological microcirculation in the inner retina.[100] However, oscillatory potentials have also been found to be aberrant in outer retinal diseases. In retinitis pigmentosa, characterized by a primary degeneration of the outer retina, diminished or delayed oscillatory potentials have been detected.[169,170]

Ganglion cells have very little contribution to the scotopic ERG responses to bright stimuli. However, the dark-adapted ERG response to very weak light flashes (scotopic threshold response; STR) depends directly upon ganglion cell function.[171,172] The negative component of these scotopic responses (nSTR) likely reflects primarily amacrine cells function, while the positive component (pSTR) has been postulated to reflect primarily retinal ganglion cell function.[173,174] Thus, STRs are reduced or extinguished after substantial retinal ganglion cell loss associated with glaucomatous damages[171,172,175,176] or excitotoxic insults.[177]

The main limitation of the global or full-field ERG is that the recordings do not permit efficient mapping of retinal responsiveness for the detection of small dysfunctional areas. In this sense, the multifocal (mf)ERG allows concurrent testing of ERG activity in a large number of retinal locations, which enables functional mapping of the retina.[178,179] The mfERG is mainly used in the clinic to localize damage spatially, in such a way that variations in the topographic array of signals are more important than the absolute signal size. The mfERG is particularly valuable in cases in which the fundus appears normal and it is difficult to distinguish between diseases.[180,181] Thereby, the mfERG has been shown to be useful to evaluate maculopathies, Best disease, Stargardt disease, diabetic retinopathy and AMD.[182]

1.4.2 Visual Evoked Potentials

VEPs consist of electrophysiological signals recorded from the electroencephalographic activity in the visual cortex in response to visual stimulation.[183,184] Since the visual cortex is activated primarily by visual inputs, VEPs depend on the functional integrity of each element of the visual pathway, including the retina, optic nerve, optic radiations and occipital cortex.[184] In addition to detecting disorders of the anterior visual pathway, posterior visual pathway dysfunctions can be assessed by analyzing the VEPs.[185] Thus, VEPs can be valuable in the diagnosis of optic neuropathies and non-organic visual loss, as well as in the assessment of visual function in infants or children.[186] Moreover, VEP results can be predictive of visual recovery in traumatic optic neuropathy[186] and VEP evaluation is used to detect subclinical demyelinating lesions in multiple sclerosis.[187]

The multifocal (mf)VEPs provide local VEP responses from the visual field. Thus, mfVEPs allow the identification of spatially localized damage and pathologies that may be missed with a traditional single VEP, and facilitates investigations on structural–functional correlations.[180,188,189] The mfVEP is used to study visual field defects caused by ganglion cell or optic nerve damage[190–193] and has been considered a powerful tool for the identification of small, peripheral lesions in glaucoma, for example.[194–197] Additionally, mfVEP can be applied as a valuable method to detect visual pathway involvement in multiple sclerosis.[198]

1.4.3 Visual Acuity

Visual acuity is the most frequently performed measure of visual function in clinical practice. Visual acuity measures the ability of the visual system to discriminate between two stimuli separated in space, with a high contrast in relation to the background.[199,200] Thus, visual acuity represents a practical tool for tracing the course of ocular disorders and therapy.[201]

More recently, contrast sensitivity testing has been proposed as a valuable tool for psychophysical assessment of visual function. Contrast sensitivity measures how much contrast a person requires to see a target[202] and concerns many aspects of vision, including motion detection, visual field, pattern recognition, adaptation to darkness and visual acuity.[203] Low-contrast sensitivity is not specific to any particular disease, as many visual disturbances have similar effects on the contrast sensitivity function. Nevertheless, the testing of contrast sensitivity is a valuable tool for detecting eye disease and assessing the efficacy of treatment.[203] Contrast sensitivity tests have been useful for evaluating intraocular lenses, cataract surgery and ophthalmic pathologies such as glaucoma, optic neuritis, diabetic retinopathy or AMD.[203–209] Conversely, the correlation between impaired contrast sensitivity and degenerative dementias has been demonstrated;[210,211] hence contrast sensitivity tests may be a useful biomarker for neurodegenerative diseases.

Visual function in animals has been evaluated using different approaches. The optomotor test enables the generation of a psychophysical threshold in a reduced period of time, and does not involve the failure of older animals to learn a task.[212–214] Pigmented animal responses are stronger and easier to recognize than those of albino mice or rats, which do not show clear responses to the optomotor test.[215] The optomotor test has been used successfully as a visual test in different animal models of retinal degeneration.[209,216]

1.5 Retinal Remodeling

Morphological and functional changes associated with retinal degeneration affect neuronal cells, glia and retinal vasculature, a process that ends in a generalized retinal remodeling that ultimately leads to blindness (Figure 1.6).[1,86,94,97,99,217] The remodeling process is independent of the initial death stimulus affecting rod cells and has been observed in many models of retinitis pigmentosa. In several studies performed in our lab we have observed that none of these changes in the inner retina or the subsequent synaptic remodeling are exclusive features of retinitis pigmentosa. Instead, this set of changes has also been found in other retinal diseases, such as

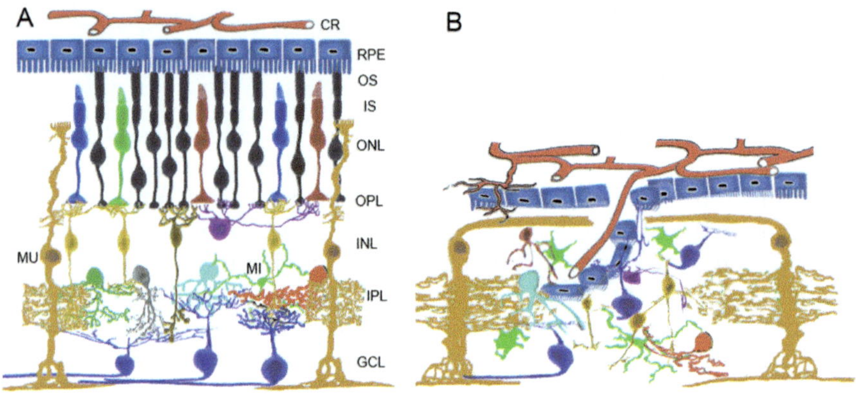

Figure 1.6 Schematic representation of retinal remodeling in healthy (A) and late (B) stages of retinal degeneration. Advanced stages of retinal remodeling are characterized by degeneration and death of photoreceptor cells, reduction in cell density in the inner nuclear layer (INL) and remodeling of neurites in both outer and inner plexiform layers (OPL and IPL, respectively). At this stage, gliosis is more intense, with hypertrophy of Müller cells (MU). The retina is restructured, and neuronal cells migrate; amacrine and bipolar cells translocate into the inner plexiform and ganglion cell layers. A deep synaptic remodeling arises in all postsynaptic neurons, which form microneuromas. In later stages of retinal degeneration, death of neuronal cells progress, hypertrophy of Müller cells remains and activation of microglial cells (MI) augments. Deterioration of the retinal blood barrier become evident at this stage, the retinal pigment epithelium (RPE) and Brunch's membrane degenerate, and choroidal vessels enter the retina. At these stages of retinal degeneration there is a lack of visual capacity because of the absence of photoreceptor cells. CR: choroid; OS: outer segments; IS: inner segments; ONL: outer nuclear layer.

metabolic disorders,[218,219] animal models of ocular hypertension[220,221] and even in human organotypic cultures,[222] where retinal remodeling had also occurred at the OPL level. Several reviews about retinal remodeling are available.[1,101,139,141]

All the aforementioned mechanisms of photoreceptor cell death trigger a vast repertoire of cellular responses of other retinal cells. Understanding these responses may be extremely helpful in improving and orienting current therapies.

Development of treatments that provide novel ways of replacing photoreceptors assumes that the inner retina remains intact, and that ganglion cells (the output neurons of the retina) remain capable of transmit the electrical signals to higher cortical areas. However, animal models of retinal degeneration and retinitis pigmentosa patients show significant alterations and plastic neuronal changes in the inner retina. Preserving the inner retina degeneration after photoreceptor loss is crucial for the optimal success of new therapeutic strategies such as electronic retinal implants (bionic eye), optogenetics, gene therapy or stem cell transplantation.

Acknowledgements

This work was supported by grants from the Spanish Ministry of Economy and Competitiveness-FEDER (BFU2015-67139-R), Instituto de Salud Carlos III RETICS-FEDER RD16/0008/0016, Organización Nacional de Ciegos Españoles (ONCE), FUNDALUCE and Generalitat Valenciana, PROMETEO/2016/158.

References

1. N. Cuenca, L. Fernandez-Sanchez, L. Campello, V. Maneu, P. De la Villa, P. Lax and I. Pinilla, *Prog. Retinal Eye Res.*, 2014, **43**, 17.
2. F. Doonan, G. Groeger and T. G. Cotter, *Exp. Cell Res.*, 2012, **318**, 1278.
3. S. Ghavami, S. Shojaei, B. Yeganeh, S. R. Ande, J. R. Jangamreddy, M. Mehrpour, J. Christoffersson, W. Chaabane, A. R. Moghadam, H. H. Kashani, M. Hashemi, A. A. Owji and M. J. Los, *Prog. Neurobiol.*, 2014, **112**, 24.
4. T. Ueta, T. Inoue, T. Furukawa, Y. Tamaki, Y. Nakagawa, H. Imai and Y. Yanagi, *J. Biol. Chem.*, 2012, **287**, 7675.
5. L. Lu, B. C. Oveson, Y. J. Jo, T. W. Lauer, S. Usui, K. Komeima, B. Xie and P. A. Campochiaro, *Antioxid. Redox Signaling*, 2009, **11**, 715.
6. J. F. Curtin, M. Donovan and T. G. Cotter, *J. Immunol. Methods*, 2002, **265**, 49.
7. K. Komeima, B. S. Rogers, L. Lu and P. A. Campochiaro, *Proc. Natl. Acad. Sci. U. S. A.*, 2006, **103**, 11300.
8. N. J. Coorey, W. Shen, S. H. Chung, L. Zhu and M. C. Gillies, *Clin. Exp. Optom.*, 2012, **95**, 266.
9. D. M. van Reyk, M. C. Gillies and M. J. Davies, *Redox Rep.*, 2003, **8**, 187–192.

10. B. Falsini, M. Piccardi, G. Iarossi, A. Fadda, E. Merendino and P. Valentini, *Ophthalmology*, 2003, **110**, 51, discussion 61.
11. K. Neelam, C. J. Goenadi, K. Lun, C. C. Yip and K. G. Au Eong, *Br. J. Ophthalmol.*, 2017, **101**, 551.
12. P. S. Bernstein, B. Li, P. P. Vachali, A. Gorusupudi, R. Shyam, B. S. Henriksen and J. M. Nolan, *Prog. Retinal Eye Res.*, 2016, **50**, 34.
13. Y. Nakagami, *Oxid. Med. Cell. Longevity*, 2016, **2016**, 7469326.
14. N. Himori, K. Yamamoto, K. Maruyama, M. Ryu, K. Taguchi, M. Yamamoto and T. Nakazawa, *J. Neurochem.*, 2013, **127**, 669.
15. Z. Xu, Y. Wei, J. Gong, H. Cho, J. K. Park, E. R. Sung, H. Huang, L. Wu, C. Eberhart, J. T. Handa, Y. Du, T. S. Kern, R. Thimmulappa, A. J. Barber, S. Biswal and E. J. Duh, *Diabetologia*, 2014, **57**, 204.
16. L. Carniglia, D. Ramirez, D. Durand, J. Saba, J. Turati, C. Caruso, T. N. Scimonelli and M. Lasaga, *Mediators Inflammation*, 2017, **2017**, 5048616.
17. M. E. Lull and M. L. Block, *Neurotherapeutics*, 2010, **7**, 354.
18. A. L. Wang, A. C. Yu, L. T. Lau, C. Lee, M. Wu le, X. Zhu and M. O. Tso, *Neurochem. Int.*, 2005, **47**, 152.
19. S. M. Allan and N. J. Rothwell, *Nat. Rev. Neurosci.*, 2001, **2**, 734.
20. M. M. Al-Gayyar and N. M. Elsherbiny, *Eur. Cytokine Network*, 2013, **24**, 27.
21. T. Y. Wong, C. M. Cheung, M. Larsen, S. Sharma and R. Simo, *Nat. Rev. Dis. Primers*, 2016, **2**, 16012.
22. C. C. Wykoff, *Curr. Opin. Ophthalmol.*, 2017, **28**, 213.
23. R. Russo, G. P. Varano, A. Adornetto, C. Nucci, M. T. Corasaniti, G. Bagetta and L. A. Morrone, *Eur. J. Pharmacol.*, 2016, **787**, 134.
24. T. Behl, I. Kaur and A. Kotwani, *Diabetes Metab. Res. Rev.*, 2016, **32**, 251.
25. S. L. Roche, A. C. Wyse-Jackson, V. Gomez-Vicente, P. Lax, A. M. Ruiz-Lopez, A. M. Byrne, N. Cuenca and T. G. Cotter, *PLoS One*, 2016, **11**, e0165197.
26. C. Choksuchat, S. Zhao, T. D. Deutch, T. D. Kimble and D. F. Archer, *Contraception*, 2009, **79**, 139.
27. F. Doonan, C. O'Driscoll, P. Kenna and T. G. Cotter, *J. Neurochem.*, 2011, **118**, 915.
28. I. Roglio, R. Bianchi, S. Gotti, S. Scurati, S. Giatti, M. Pesaresi, D. Caruso, G. C. Panzica and R. C. Melcangi, *Neuroscience*, 2008, **155**, 673.
29. A. C. Jackson, S. L. Roche, A. M. Byrne, A. M. Ruiz-Lopez and T. G. Cotter, *J. Neurochem.*, 2016, **136**, 63.
30. J. S. Messer, *Cell. Mol. Life Sci.*, 2017, **74**, 1281.
31. Y. Murakami, S. Notomi, T. Hisatomi, T. Nakazawa, T. Ishibashi, J. W. Miller and D. G. Vavvas, *Prog. Retinal Eye Res.*, 2013, **37**, 114.
32. N. D. Chinskey, C. G. Besirli and D. N. Zacks, *Curr. Opin. Ophthalmol.*, 2014, **25**, 228.
33. M. V. Jain, A. M. Paczulla, T. Klonisch, F. N. Dimgba, S. B. Rao, K. Roberg, F. Schweizer, C. Lengerke, P. Davoodpour, V. R. Palicharla, S. Maddika and M. Los, *J. Cell. Mol. Med.*, 2013, **17**, 12.
34. V. Nikoletopoulou, M. Markaki, K. Palikaras and N. Tavernarakis, *Biochim. Biophys. Acta*, 2013, **1833**, 3448.

35. C. C. Chien, C. J. Huang, L. T. Tien, Y. C. Cheng, C. Y. Ke and Y. J. Lee, *Invest. Ophthalmol. Visual Sci.*, 2017, **58**, 3107.
36. C. Y. Kuan, K. A. Roth, R. A. Flavell and P. Rakic, *Trends Neurosci.*, 2000, **23**, 291.
37. A. Strasser, L. O'Connor and V. M. Dixit, *Annu. Rev. Biochem.*, 2000, **69**, 217.
38. X. J. Xu, S. M. Wang, Y. Jin, Y. T. Hu, K. Feng and Z. Z. Ma, *J. Pineal Res.*, 2017, **63**, e12428.
39. M. Pawar, B. Busov, A. Chandrasekhar, J. Yao, D. N. Zacks and C. G. Besirli, *Cell Death Differ.*, 2017, **24**, 1799.
40. A. Linkermann and D. R. Green, *N. Engl. J. Med.*, 2014, **370**, 455.
41. G. Trichonas, Y. Murakami, A. Thanos, Y. Morizane, M. Kayama, C. M. Debouck, T. Hisatomi, J. W. Miller and D. G. Vavvas, *Proc. Natl. Acad. Sci. U. S. A.*, 2010, **107**, 21695.
42. Y. Murakami, H. Matsumoto, M. Roh, J. Suzuki, T. Hisatomi, Y. Ikeda, J. W. Miller and D. G. Vavvas, *Proc. Natl. Acad. Sci. U. S. A.*, 2012, **109**, 14598.
43. M. Los, M. Mozoluk, D. Ferrari, A. Stepczynska, C. Stroh, A. Renz, Z. Herceg, Z. Q. Wang and K. Schulze-Osthoff, *Mol. Biol. Cell*, 2002, **13**, 978.
44. M. Pasparakis and P. Vandenabeele, *Nature*, 2015, **517**, 311.
45. A. Ranjan and T. Iwakuma, *Int. J. Mol. Sci.*, 2016, **17**, E2068.
46. K. Newton, D. L. Dugger, A. Maltzman, J. M. Greve, M. Hedehus, B. Martin-McNulty, R. A. Carano, T. C. Cao, N. van Bruggen, L. Bernstein, W. P. Lee, X. Wu, J. DeVoss, J. Zhang, S. Jeet, I. Peng, B. S. McKenzie, M. Roose-Girma, P. Caplazi, L. Diehl, J. D. Webster and D. Vucic, *Cell Death Differ.*, 2016, **23**, 1565.
47. R. Weinlich, A. Oberst, H. M. Beere and D. R. Green, *Nat. Rev. Mol. Cell Biol.*, 2017, **18**, 127.
48. S. Zhang, M. B. Tang, H. Y. Luo, C. H. Shi and Y. M. Xu, *Cell Death Dis.*, 2017, **8**, e2905.
49. S. Elmore, *Toxicol. Pathol.*, 2007, **35**, 495.
50. B. Levine and G. Kroemer, *Cell*, 2008, **132**, 27.
51. P. Boya, L. Esteban-Martinez, A. Serrano-Puebla, R. Gomez-Sintes and B. Villarejo-Zori, *Prog. Retinal Eye Res.*, 2016, **55**, 206.
52. N. Mizushima and M. Komatsu, *Cell*, 2011, **147**, 728.
53. N. T. Pun, A. Subedi, M. J. Kim and P. H. Park, *PLoS One*, 2015, **10**, e0124636.
54. J. Song, Y. Oh and J. E. Lee, *Exp. Neurobiol.*, 2015, **24**, 117.
55. A. I. Arroba, L. Rodriguez-de la Rosa, S. Murillo-Cuesta, L. Vaquero-Villanueva, J. M. Hurle, I. Varela-Nieto and A. M. Valverde, *Dis. Models & Mech.*, 2016, **9**, 965.
56. A. Rosello, G. Warnes and U. C. Meier, *Clin. Exp. Immunol.*, 2012, **168**, 52.
57. Y. Meng, Y. Yong, G. Yang, H. Ding, Z. Fan, Y. Tang, J. Luo and Z. J. Ke, *J. Neurochem.*, 2013, **126**, 805.
58. Y. C. Kim and K. L. Guan, *J. Clin. Invest.*, 2015, **125**, 25.
59. K. Kaarniranta, P. Tokarz, A. Koskela, J. Paterno and J. Blasiak, *Cell Biol. Toxicol.*, 2017, **33**, 113.

60. M. Dewaele, W. Martinet, N. Rubio, T. Verfaillie, P. A. de Witte, J. Piette and P. Agostinis, *J. Cell. Mol. Med.*, 2011, **15**, 1402.
61. C. Zhao, D. Yasumura, X. Li, M. Matthes, M. Lloyd, G. Nielsen, K. Ahern, M. Snyder, D. Bok, J. L. Dunaief, M. M. LaVail and D. Vollrath, *J. Clin. Invest.*, 2011, **121**, 369.
62. Z. Faghiri and N. G. Bazan, *Exp. Eye Res.*, 2010, **90**, 718.
63. B. Fan, Y. J. Sun, S. Y. Liu, L. Che and G. Y. Li, *Int. J. Mol. Sci.*, 2017, **18**, E201.
64. F. G. Holz, S. Schmitz-Valckenberg and M. Fleckenstein, *J. Clin. Invest.*, 2014, **124**, 1430.
65. W. T. Wong, S. Dresner, F. Forooghian, T. Glaser, L. Doss, M. Zhou, D. Cunningham, K. Shimel, M. Harrington, K. Hammel, C. A. Cukras, F. L. Ferris and E. Y. Chew, *Invest. Ophthalmol. Visual Sci.*, 2013, **54**, 2941.
66. P. A. Petrou, D. Cunningham, K. Shimel, M. Harrington, K. Hammel, C. A. Cukras, F. L. Ferris, E. Y. Chew and W. T. Wong, *Invest. Ophthalmol. Visual Sci.*, 2014, **56**, 330.
67. R. A. Floto, S. Sarkar, E. O. Perlstein, B. Kampmann, S. L. Schreiber and D. C. Rubinsztein, *Autophagy*, 2007, **3**, 620.
68. V. T. Ribas, J. C. Koch, U. Michel, M. Bahr and P. Lingor, *Mol. Neurobiol.*, 2017, **54**, 72.
69. P. Merlo, B. Frost, S. Peng, Y. J. Yang, P. J. Park and M. Feany, *Proc. Natl. Acad. Sci. U. S. A.*, 2014, **111**, 18055.
70. C. A. Grimes and R. S. Jope, *Prog. Neurobiol.*, 2001, **65**, 391.
71. E. Beurel, S. F. Grieco and R. S. Jope, *Pharmacol. Ther.*, 2015, **148**, 114.
72. M. Hetman, J. E. Cavanaugh, D. Kimelman and Z. Xia, *J. Neurosci.*, 2000, **20**, 2567.
73. A. M. Orellana, A. R. Vasconcelos, J. A. Leite, L. de Sa Lima, D. Z. Andreotti, C. D. Munhoz, E. M. Kawamoto and C. Scavone, *Aging (Albany NY)*, 2015, **7**, 1094.
74. R. S. Jope, C. J. Yuskaitis and E. Beurel, *Neurochem. Res.*, 2007, **32**, 577.
75. M. Marchena, B. Villarejo-Zori, J. Zaldivar-Diez, V. Palomo, C. Gil, C. Hernandez-Sanchez, A. Martinez and E. J. de la Rosa, *J. Enzyme Inhib. Med. Chem.*, 2017, **32**, 522.
76. H. Kolb and D. Marshak, *Doc. Ophthalmol.*, 2003, **106**, 67.
77. D. A. Copland, C. J. Calder, B. J. Raveney, L. B. Nicholson, J. Phillips, H. Cherwinski, M. Jenmalm, J. D. Sedgwick and A. D. Dick, *Am. J. Pathol.*, 2007, **171**, 580.
78. H. Kolb, in *Principles and Practices of Clinical Electrophysiology of Vision*, ed. J. R. Heckenlively and B. A. Geoffrey, Mosby Year Book Inc., 1991, p. 25.
79. D. P. Edward, K. Lim, S. Sawaguchi and M. O. Tso, *Graefes Arch. Clin. Exp. Ophthalmol.*, 1993, **231**, 289.
80. R. Barhoum, G. Martinez-Navarrete, S. Corrochano, F. Germain, L. Fernandez-Sanchez, E. J. de la Rosa, P. de la Villa and N. Cuenca, *Neuroscience*, 2008, **155**, 698.
81. E. L. Berson, B. Rosner and E. Simonoff, *Am. J. Ophthalmol.*, 1980, **89**, 763.

82. H. Kolb and P. Gouras, *Invest. Ophthalmol.*, 1974, **13**, 487.
83. A. H. Milam, Z. Y. Li and R. N. Fariss, *Prog. Retinal Eye Res.*, 1998, **17**, 175.
84. Z. Y. Li, F. Wong, J. H. Chang, D. E. Possin, Y. Hao, R. M. Petters and A. H. Milam, *Invest. Ophthalmol. Visual Sci.*, 1998, **39**, 808.
85. N. H. Chong, R. A. Alexander, K. C. Barnett, A. C. Bird and P. J. Luthert, *Exp. Eye Res.*, 1999, **68**, 51.
86. N. Cuenca, I. Pinilla, Y. Sauve, B. Lu, S. Wang and R. D. Lund, *Neuroscience*, 2004, **127**, 301.
87. B. Lu, C. W. Morgans, S. Girman, R. Lund and S. Wang, *Vis. Neurosci.*, 2013, **30**, 77.
88. Y. Ji, W. Q. Yu, Y. S. Eom, F. Bruce, C. M. Craft, N. M. Grzywacz and E. J. Lee, *Invest. Ophthalmol. Visual Sci.*, 2014, **56**, 352.
89. E. J. Lee, Y. Ji, C. L. Zhu and N. M. Grzywacz, *Glia*, 2011, **59**, 1107.
90. L. Fernandez-Sanchez, P. Lax, L. Campello, I. Pinilla and N. Cuenca, *Front. Cell. Neurosci.*, 2015, **9**, 484.
91. D. Garcia-Ayuso, A. Ortin-Martinez, M. Jimenez-Lopez, C. Galindo-Romero, N. Cuenca, I. Pinilla, M. Vidal-Sanz, M. Agudo-Barriuso and M. P. Villegas-Perez, *Invest. Ophthalmol. Visual Sci.*, 2013, **54**, 5888.
92. S. Machida, M. Kondo, J. a. Jamison, N. W. Khan, L. T. Kononen, T. Sugawara, R. a. Bush and P. a. Sieving, *Invest. Ophthalmol. Visual Sci.*, 2000, **41**, 3200.
93. E. L. Berson, *Int. Ophthalmol.*, 1981, **4**, 7.
94. N. Cuenca, I. Pinilla, Y. Sauve and R. Lund, *Eur. J. Neurosci.*, 2005, **22**, 1057.
95. G. Martinez-Navarrete, M. J. Seiler, R. B. Aramant, L. Fernandez-Sanchez, I. Pinilla and N. Cuenca, *Exp. Eye Res.*, 2011, **92**, 227.
96. R. K. Sullivan, E. Woldemussie and D. V. Pow, *Invest. Ophthalmol. Visual Sci.*, 2007, **48**, 2782.
97. E. Strettoi, V. Porciatti, B. Falsini, V. Pignatelli and C. Rossi, *J. Neurosci.*, 2002, **22**, 5492.
98. Y. W. Peng, Y. Hao, R. M. Petters and F. Wong, *Nat. Neurosci.*, 2000, **3**, 1121.
99. Y.-W. Peng, T. Senda, Y. Hao, K. Matsuno and F. Wong, *Neuroscience*, 2003, **119**, 813.
100. L. Wachtmeister, *Prog. Retinal Eye Res.*, 1998, **17**, 485.
101. B. W. Jones, C. B. Watt, J. M. Frederick, W. Baehr, C. K. Chen, E. M. Levine, A. H. Milam, M. M. Lavail and R. E. Marc, *J. Comp. Neurol.*, 2003, **464**, 1.
102. A. J. Eisenfeld, M. M. LaVail and J. H. LaVail, *J. Comp. Neurol.*, 1984, **223**, 22.
103. B. I. Gallego, J. J. Salazar, R. de Hoz, B. Rojas, A. I. Ramirez, M. Salinas-Navarro, A. Ortin-Martinez, F. J. Valiente-Soriano, M. Aviles-Trigueros, M. P. Villegas-Perez, M. Vidal-Sanz, A. Trivino and J. M. Ramirez, *J. Neuroinflammation*, 2012, **9**, 92.
104. T. Langmann, *J. Leukocyte Biol.*, 2007, **81**, 1345.

105. M. Karlstetter, S. Ebert and T. Langmann, *Immunobiology*, 2010, **215**, 685.
106. N. Gupta, K. E. Brown and A. H. Milam, *Exp. Eye Res.*, 2003, **76**, 463.
107. A. Noailles, V. Maneu, L. Campello, P. Lax and N. Cuenca, *Cell Death Dis.*, 2018, **9**, 350.
108. L. Zhao, W. Ma, R. N. Fariss and W. T. Wong, *Am. J. Pathol.*, 2011, **179**, 1265.
109. R. Scholz, M. Sobotka, A. Caramoy, T. Stempfl, C. Moehle and T. Langmann, *J. Neuroinflammation*, 2015, **12**, 209.
110. M. P. Bordone, M. F. Gonzalez Fleitas, L. A. Pasquini, A. Bosco, P. H. Sande, R. E. Rosenstein and D. Dorfman, *J. Neurochem.*, 2017, **142**, 323.
111. M. W. Stewart, H. W. Flynn Jr, S. G. Schwartz and I. U. Scott, *Expert Opin. Drug Delivery*, 2016, **13**, 1277.
112. E. Polazzi and B. Monti, *Prog. Neurobiol.*, 2010, **92**, 293.
113. R. de Hoz, B. Rojas, A. I. Ramirez, J. J. Salazar, B. I. Gallego, A. Trivino and J. M. Ramirez, *BioMed Res. Int.*, 2016, **2016**, 2954721.
114. R. Natoli, N. Fernando, M. Madigan, J. A. Chu-Tan, K. Valter, J. Provis and M. Rutar, *Mol. Neurodegener.*, 2017, **12**, 31.
115. J. S. Penn, A. Madan, R. B. Caldwell, M. Bartoli, R. W. Caldwell and M. E. Hartnett, *Prog. Retinal Eye Res.*, 2008, **27**, 331.
116. A. Bringmann, T. Pannicke, J. Grosche, M. Francke, P. Wiedemann, S. N. Skatchkov, N. N. Osborne and A. Reichenbach, *Prog. Retinal Eye Res.*, 2006, **25**, 397.
117. S. M. Cardona, A. S. Mendiola, Y. C. Yang, S. L. Adkins, V. Torres and A. E. Cardona, *ASN Neuro*, 2015, **7**, 1.
118. M. Zhang, G. Xu, W. Liu, Y. Ni and W. Zhou, *PLoS One*, 2012, **7**, e35446.
119. M. Zieger, P. K. Ahnelt and P. Uhrin, *PLoS One*, 2014, **9**, e106562.
120. A. E. Cardona, E. P. Pioro, M. E. Sasse, V. Kostenko, S. M. Cardona, I. M. Dijkstra, D. Huang, G. Kidd, S. Dombrowski, R. Dutta, J. C. Lee, D. N. Cook, S. Jung, S. A. Lira, D. R. Littman and R. M. Ransohoff, *Nat. Neurosci.*, 2006, **9**, 917.
121. S. H. Cho, B. Sun, Y. Zhou, T. M. Kauppinen, B. Halabisky, P. Wes, R. M. Ransohoff and L. Gan, *J. Biol. Chem.*, 2011, **286**, 32713.
122. T. Mizuno, J. Kawanokuchi, K. Numata and A. Suzumura, *Brain Res.*, 2003, **979**, 65.
123. V. Zujovic, J. Benavides, X. Vige, C. Carter and V. Taupin, *Glia*, 2000, **29**, 305.
124. M. M. Pabon, A. D. Bachstetter, C. E. Hudson, C. Gemma and P. C. Bickford, *J. Neuroinflammation*, 2011, **8**, 9.
125. S. L. Roche, A. C. Wyse-Jackson, A. M. Ruiz-Lopez, A. M. Byrne and T. G. Cotter, *Sci. Rep.*, 2017, **7**, 43067.
126. J. Zhang, D. Xu, H. Ouyang, S. Hu, A. Li, H. Luo and Y. Xu, *Exp. Eye Res.*, 2017, **162**, 86.
127. M. Sacchetti, F. Mantelli, M. L. Rocco, A. Micera, L. Brandolini, L. Focareta, C. Pisano, L. Aloe and A. Lambiase, *Curr. Eye Res.*, 2017, **42**, 1064.

128. P. Baranov, H. Lin, K. McCabe, D. Gale, S. Cai, B. Lieppman, D. Morrow, P. Lei, J. Liao and M. Young, *J. Ocul. Pharmacol. Ther.*, 2017, **33**, 412.
129. T. Chan-Ling, D. S. McLeod, S. Hughes, L. Baxter, Y. Chu, T. Hasegawa and G. A. Lutty, *Invest. Ophthalmol. Visual Sci.*, 2004, **45**, 2020.
130. Y. Kubota and T. Suda, *Trends Cardiovasc. Med.*, 2009, **19**, 38.
131. J. Kur, E. A. Newman and T. Chan-Ling, *Prog. Retinal Eye Res.*, 2012, **31**, 377.
132. S. B. Bressler, D. Liu, A. R. Glassman, B. A. Blodi, A. A. Castellarin, L. M. Jampol, P. L. Kaufman, M. Melia, H. Singh, J. A. Wells and Diabetic Retinopathy Clinical Research, *JAMA Ophthalmol.*, 2017, **135**, 558.
133. S. Ittiara, M. P. Blair, M. J. Shapiro and S. J. Lichtenstein, *J. AAPOS*, 2013, **17**, 323.
134. R. K. Wong, S. Hubschman and I. Tsui, *Retina*, 2015, **35**, 675.
135. H. Wang, *Eye Brain*, 2016, **8**, 81.
136. Z. Khan, R. K. Kuriakose, M. Khan, E. K. Chin and D. R. Almeida, *Ophthalmic Surg. Lasers Imaging Retina*, 2017, **48**, 160.
137. T. L. P. Trinh, S. Li Calzi, L. C. Shaw, M. C. Yoder and M. B. Grant, *Eye Brain*, 2016, **8**, 113.
138. M. E. Pennesi, S. Nishikawa, M. T. Matthes, D. Yasumura and M. M. LaVail, *Exp. Eye Res.*, 2008, **87**, 561.
139. R. E. Marc, B. W. Jones, C. B. Watt and E. Strettoi, *Prog. Retinal Eye Res.*, 2003, **22**, 607.
140. B. W. Jones and R. E. Marc, *Exp. Eye Res.*, 2005, **81**, 123.
141. B. W. Jones, M. Kondo, H. Terasaki, Y. Lin, M. McCall and R. E. Marc, *Jpn. J. Ophthalmol.*, 2012, **56**, 289.
142. D. Garcia-Ayuso, M. Salinas-Navarro, M. Agudo, N. Cuenca, I. Pinilla, M. Vidal-Sanz and M. P. Villegas-Perez, *Exp. Eye Res.*, 2010, **91**, 800.
143. B. Kolomiets, E. Dubus, M. Simonutti, S. Rosolen, J. a. Sahel and S. Picaud, *Neurobiol. Dis.*, 2010, **38**, 47.
144. K. T. Brown, *Vis. Res.*, 1968, **8**, 633.
145. K. T. Brown and T. N. Wiesel, *J. Physiol.*, 1961, **158**, 257.
146. J. G. Robson and L. J. Frishman, *Vis. Neurosci.*, 1995, **12**, 837.
147. D. G. Green and N. V. Kapousta-Bruneau, *Vis. Neurosci.*, 1999, **16**, 727.
148. B. Lei and I. Perlman, *Vis. Neurosci.*, 1999, **16**, 743.
149. R. A. Shiells and G. Falk, *Vis. Neurosci.*, 1999, **16**, 503.
150. R. A. Stockton and M. M. Slaughter, *J. Gen. Physiol.*, 1989, **93**, 101.
151. W. A. Cobb and H. B. Morton, *J. Physiol.*, 1954, **123**, 36.
152. R. Granit and A. Munsterhjelm, *J. Physiol.*, 1937, **88**, 436.
153. S. Korol, P. M. Leuenberger, U. Englert and J. Babel, *Brain Res.*, 1975, **97**, 235.
154. C. J. Dong, P. Agey and W. A. Hare, *Vis. Neurosci.*, 2004, **21**, 533.
155. E. L. Berson, *Jpn. J. Ophthalmol.*, 1987, **31**, 327.
156. I. Pinilla, R. D. Lund and Y. Sauve, *Neurosci. Lett.*, 2005, **382**, 16.
157. U. Kellner and M. H. Foerster, *Ger. J. Ophthalmol.*, 1993, **2**, 170.
158. I. Pinilla, N. Cuenca, Y. Sauve, S. Wang and R. D. Lund, *Exp. Eye Res.*, 2007, **85**, 381.

159. L. Fernandez-Sanchez, P. Lax, G. Esquiva, J. Martin-Nieto, I. Pinilla and N. Cuenca, *PLoS One*, 2012, **7**, e43074.
160. L. Fernandez-Sanchez, P. Lax, C. Isiegas, E. Ayuso, J. M. Ruiz, P. de la Villa, F. Bosch, E. J. de la Rosa and N. Cuenca, *Hum. Gene Ther.*, 2012, **23**, 1290.
161. L. Fernandez-Sanchez, P. Lax, I. Pinilla, J. Martin-Nieto and N. Cuenca, *Invest. Ophthalmol. Visual Sci.*, 2011, **52**, 4998.
162. P. Lax, G. Esquiva, C. Altavilla and N. Cuenca, *Exp. Eye Res.*, 2014, **120**, 175.
163. P. Lax, B. B. Otalora, G. Esquiva, L. Rol Mde, J. A. Madrid and N. Cuenca, *J. Pineal Res.*, 2011, **50**, 183.
164. J. Esteve-Rudd, L. Fernandez-Sanchez, P. Lax, E. De Juan, J. Martin-Nieto and N. Cuenca, *Neurobiol. Dis.*, 2011, **44**, 102.
165. D. Yonemura, K. Tsuzuki and T. Aoki, *Acta Ophthalmol.*, 1962, (suppl. 70), 115.
166. P. Algvere, *Acta Ophthalmol (copenh).*, 1968, **46**, 993.
167. L. Wachtmeister, *Prog. Brain Res.*, 2001, **131**, 465.
168. P. Speros and J. Price, *Surv. Ophthalmol.*, 1981, **25**, 237.
169. A. V. Cideciyan and S. G. Jacobson, *Invest. Ophthalmol. Visual Sci.*, 1993, **34**, 3253.
170. I. Pinilla, N. Cuenca, G. Martinez-Navarrete, R. D. Lund and Y. Sauve, *Vis. Res.*, 2009, **49**, 2067.
171. L. J. Frishman, F. F. Shen, L. Du, J. G. Robson, R. S. Harwerth, E. L. Smith 3rd, L. Carter-Dawson and M. L. Crawford, *Invest. Ophthalmol. Visual Sci.*, 1996, **37**, 125.
172. B. V. Bui and B. Fortune, *J. Physiol.*, 2004, **555**, 153.
173. A. Moshiri, E. Gonzalez, K. Tagawa, H. Maeda, M. Wang, L. J. Frishman and S. W. Wang, *Dev. Biol.*, 2008, **316**, 214.
174. S. M. Saszik, J. G. Robson and L. J. Frishman, *J. Physiol.*, 2002, **543**, 899.
175. V. Chrysostomou and J. G. Crowston, *Invest. Ophthalmol. Visual Sci.*, 2013, **54**, 4691.
176. T. I. Tsai, B. V. Bui and A. J. Vingrys, *Invest. Ophthalmol. Visual Sci.*, 2014, **55**, 1067.
177. V. Gomez-Vicente, P. Lax, L. Fernandez-Sanchez, N. Rondon, G. Esquiva, F. Germain, P. de la Villa and N. Cuenca, *PLoS One*, 2015, **10**, e0137826.
178. E. E. Sutter and D. Tran, *Vis. Res.*, 1992, **32**, 433.
179. M. A. Bearse Jr and E. E. Sutter, *J. Opt. Soc. Am. A. opt. Image sci. Vis.*, 1996, **13**, 634.
180. D. C. Hood, J. G. Odel, C. S. Chen and B. J. Winn, *J. Neuro Ophthalmol.*, 2003, **23**, 225.
181. U. Kretschmann, M. W. Seeliger, K. Ruether, T. Usui, E. Apfelstedt-Sylla and E. Zrenner, *Br. J. Ophthalmol.*, 1998, **82**, 267.
182. E. M. Schonbach, V. Chaikitmongkol, R. Annam, E. C. McDonnell, Y. Wolfson, E. Fletcher and H. P. N. Scholl, *Ophthalmic Res.*, 2017, **58**, 117.
183. S. Sokol, *Surv. Ophthalmol.*, 1976, **21**, 18.
184. J. V. Odom, M. Bach, M. Brigell, G. E. Holder, D. L. McCulloch, A. P. Tormene and Vaegan, *Doc. Ophthalmol.*, 2010, **120**, 111.
185. G. E. Holder, *Prog. Retinal Eye Res.*, 2001, **20**, 531.

186. B. Young, E. Eggenberger and D. Kaufman, *Curr. Opin. Ophthalmol.*, 2012, **23**, 497.
187. P. Fuhr, A. Borggrefe-Chappuis, C. Schindler and L. Kappos, *Brain*, 2001, **124**, 2162.
188. H. A. Baseler, E. E. Sutter, S. A. Klein and T. Carney, *Electroencephalogr. Clin. Neurophysiol.*, 1994, **90**, 65.
189. D. Creel, in *Webvision: The Organization of the Retina and Visual System*, ed. H. Kolb, E. Fernandez and R. Nelson, Salt Lake City (UT), 1995.
190. D. C. Hood, X. Zhang, V. C. Greenstein, S. Kangovi, J. G. Odel, J. M. Liebmann and R. Ritch, *Invest. Ophthalmol. Visual Sci.*, 2000, **41**, 1580.
191. A. I. Klistorner, S. L. Graham, J. R. Grigg and F. A. Billson, *Invest. Ophthalmol. Visual Sci.*, 1998, **39**, 937.
192. Y. Betsuin, Y. Mashima, H. Ohde, R. Inoue and Y. Oguchi, *Curr. Eye Res.*, 2001, **22**, 54.
193. G. E. Holder, R. P. Gale, J. F. Acheson and A. G. Robson, *Curr. Opin. Neurol.*, 2009, **22**, 3.
194. D. C. Hood and V. C. Greenstein, *Prog. Retinal Eye Res.*, 2003, **22**, 201.
195. A. Klistorner, S. L. Graham, A. Martins, J. R. Grigg, H. Arvind, R. S. Kumar, A. C. James and F. A. Billson, *Ophthalmology*, 2007, **114**, 1613.
196. B. Fortune, S. Demirel, X. Zhang, D. C. Hood, E. Patterson, A. Jamil, S. L. Mansberger, G. A. Cioffi and C. A. Johnson, *Invest. Ophthalmol. Visual Sci.*, 2007, **48**, 1173.
197. A. Klistorner and S. L. Graham, *Ophthalmology*, 2000, **107**, 2283.
198. G. Pihl-Jensen, M. F. Schmidt and J. L. Frederiksen, *Clin. Neurophysiol.*, 2017, **128**, 1234.
199. G. Westheimer, *Annu. Rev. Psychol.*, 1965, **16**, 359.
200. C. Kniestedt and R. L. Stamper, *Ophthalmol. Clin. North Am.*, 2003, **16**, 155, v.
201. G. Westheimer, *Prog. Retinal Eye Res.*, 2009, **28**, 178.
202. C. Owsley, *Ophthalmol. Clin. North Am.*, 2003, **16**, 171.
203. J. Richman, G. L. Spaeth and B. Wirostko, *J. Cataract Refractive Surg.*, 2013, **39**, 1100.
204. S. C. Howes, T. Caelli and P. Mitchell, *Aust. J. Ophthalmol.*, 1982, **10**, 173.
205. J. E. Ross, A. J. Bron and D. D. Clarke, *Br. J. Ophthalmol.*, 1984, **68**, 821.
206. A. P. Ginsburg, *Curr. Opin. Ophthalmol.*, 2006, **17**, 19.
207. G. C. Woo, *Am. J. Optom. Physiol. Opt.*, 1985, **62**, 648.
208. I. L. Bailey, *Optom. Vis. Sci.*, 1993, **70**, 299.
209. Y. Umino and E. Solessio, *Invest. Ophthalmol. Visual Sci.*, 2013, **54**, 1536.
210. F. K. Cormack, M. Tovee and C. Ballard, *Int. J. Geriatr. Psychiatr.*, 2000, **15**, 614.
211. A. Ridder, M. L. Muller, V. Kotagal, K. A. Frey, R. L. Albin and N. I. Bohnen, *Parkinsonism Relat. Disord.*, 2017, **34**, 15.
212. R. M. Douglas, N. M. Alam, B. D. Silver, T. J. McGill, W. W. Tschetter and G. T. Prusky, *Vis. Neurosci.*, 2005, **22**, 677.
213. G. T. Prusky, N. M. Alam, S. Beekman and R. M. Douglas, *Invest. Ophthalmol. Visual Sci.*, 2004, **45**, 4611.
214. T. Darland and J. E. Dowling, *Proc. Natl. Acad. Sci. U. S. A.*, 2001, **98**, 11691.

215. J. Abdeljalil, M. Hamid, O. Abdel-Mouttalib, R. Stephane, R. Raymond, A. Johan, S. Jose, C. Pierre and P. Serge, *Vis. Res.*, 2005, **45**, 1439.
216. P. Barabas, W. Huang, H. Chen, C. L. Koehler, G. Howell, S. W. John, N. Tian, R. C. Renteria and D. Krizaj, *Invest. Ophthalmol. Visual Sci.*, 2011, **52**, 6766.
217. C. F. de Souza, M. Kalloniatis, P. J. Polkinghorne, C. N. McGhee and M. L. Acosta, *Exp. Eye Res.*, 2012, **97**, 73.
218. L. Rodriguez-de la Rosa, L. Fernandez-Sanchez, F. Germain, S. Murillo-Cuesta, I. Varela-Nieto, P. de la Villa and N. Cuenca, *Neurobiol. Dis.*, 2012, **46**, 476.
219. Y. Umino, N. Cuenca, D. Everhart, L. Fernandez-Sanchez, R. B. Barlow and E. Solessio, *Invest. Ophthalmol. Visual Sci.*, 2012, **53**, 915.
220. N. Cuenca, I. Pinilla, L. Fernandez-Sanchez, M. Salinas-Navarro, L. Alarcon-Martinez, M. Aviles-Trigueros, P. de la Villa, J. Miralles de Imperial, M. P. Villegas-Perez and M. Vidal-Sanz, *Exp. Eye Res.*, 2010, **91**, 273.
221. L. Fernandez-Sanchez, L. P. de Sevilla Muller, N. C. Brecha and N. Cuenca, *Invest. Ophthalmol. Visual Sci.*, 2014, **55**, 6059.
222. I. Fernandez-Bueno, L. Fernandez-Sanchez, M. J. Gayoso, M. T. Garcia-Gutierrez, J. C. Pastor and N. Cuenca, *Exp. Eye Res.*, 2012, **104**, 26.

CHAPTER 2

Immune Response of the Retina

MEI CHEN* AND HEPING XU

Centre for Experimental Medicine, School of Medicine, Dentistry and Biomedical Sciences, Queen's University Belfast, University Road, Belfast, BT7 1NN, UK
*E-mail: m.chen@qub.ac.uk

2.1 Introduction

Retinal degenerative diseases such as age-related macular degeneration (AMD), diabetic retinopathy, and glaucomatous retinopathy are the major causes of blindness in developed countries. The aetiology of different types of retinal degenerative diseases differs, although inflammation is known to play an important role in the development of retinal pathologies in all conditions. Reactive immune response to age-related oxidative insults or degenerative neurons not only cleans up noxious molecules, dead cells, or debris, but also contributes to the disease's progression. In this chapter, we discuss how the immune system responds to retinal endogenous insults in the ageing eye as well as in eyes with retinal degenerative conditions.

2.2 The Immune System and Inflammation

2.2.1 The Immune System

Immunity is the ability of the host to resist a particular infection or toxin. This is achieved by a sophisticated immune system that comprises a highly organised network of organs, tissues, cells and soluble molecules.

Drug Discovery Series No. 66
Therapies for Retinal Degeneration: Targeting Common Processes
Edited by Enrique J. de la Rosa and Thomas G. Cotter

Published by the Royal Society of Chemistry, www.rsc.org

There are two arms of the immune system: the innate (natural or native) and the acquired (adaptive) immune systems. The innate immune system is the first line of defence. The response is rapid, non-specific and independent of previous exposure to the antigens released by foreign invaders or necrotic cells. In contrast, the acquired immune response is relatively slow, but is highly specific to a particular pathogen and has immunological memory.

The innate immune system consists of physiological barriers (*e.g.* the skin and the mucous membranes), phagocytic and cytotoxic cells (*e.g.* macrophages, neutrophils, eosinophils, basophils and natural killer cells), soluble molecules (*e.g.* complement) that normally present in the body fluids, or cytokines often released by cells responding to the attack. The acquired immune system includes T- and B-lymphocytes, which have specific receptors that recognise foreign antigens. Upon engaging with pathogens, the antigen-specific T-helper cells and B-cells may release molecules such as cytokines and antibodies, respectively. Some cells, such as macrophages and dendritic cells are central to both innate and acquired immune systems.

Myeloid cells, including granulocytes (neutrophils, basophils, eosinophils and mast cells) and mononuclear cells (monocytes, macrophages and dendritic cells) are important components of the innate immune system. One major function of myeloid cells is to phagocytose invading organisms, damaged host cells and tissue debris. During acute inflammation they are recruited from the circulation to the site of infection or injury. Depending on the type of insult (*e.g.* bacterial, virus, fungal infection, injury, *etc.*), different innate immune cells may be recruited. Neutrophils have relatively short life and are the most abundant type of white blood cells in the circulation; they are attracted to the site of inflammation by chemotaxis at the early stages of infection or tissue damage. Other granulocytes (*i.e.* basophils, eosinophils and mast cells) participate in the immune response to allergens by releasing granules, and sequentially induce inflammation. Monocytes are recruited from the circulation to the injury site by chemotaxis and then differentiate to macrophages. Macrophages are central to both innate and acquired immune responses. As part of the innate immune system, they phagocytose invading organisms, damaged host cells and tissue debris. They also release various cytokines which can activate other lymphocytes or surrounding tissue cells. Macrophages may serve as antigen-presenting cells and participate in the acquired immune response. Dendritic cells are present in the skin, conjunctiva and the inner layers of the respiratory system and guts. Dendritic cells endocytose foreign organisms and process the antigens, which then be presented to T- and B-lymphocytes *via* human leukocyte antigen molecules. Dendritic cells are the most potent antigen-presenting cells.

Lymphocytes (T-cells and B-cells) belong to the acquired immune system. T-cells proliferate and release cytokines upon recognition and uptake of the antigen presented by dendritic cells and macrophages. In general, $CD8^+$ T-cells (also called cytotoxic T-cells) are activated upon recognising antigens presented by the major histocompatibility complex (MHC)-I molecules, whereas $CD4^+$ T-cells (also known as T-helper cells) are activated when they are presented with antigens by the MHC-II molecule. Depending on the antigens and other cytokines in the microenvironment, the $CD4^+$ T-cells can be differentiated into different T-helper cells, such as Th1, Th2 and Th17 cells. These T-helper cells release specific cytokines such as interferon (IFN)γ (by Th1 cells), interleukin (IL)-4 (by Th2 cells) and IL-17 (by Th17 cells) and participate in adaptive immune response. B-cells are activated through the B-cell receptor, which can be T-cell-dependent or -independent. Once activated, B-cells proliferate and mature into plasma cells and produce antibodies, which form an important part of the adaptive immunity.

The complement system includes a series of complement proteins, from C1 to C9, and soluble- or membrane-form regulators such as C4b-binding protein, factor H (CFH), factor I (CFI), and decoy-accelerating factors such as CD55 and CD59. Complement proteins are synthesised predominantly by hepatocytes in the liver and are released into the circulation in a latent form. Circulating complement proteins remain in an inactive form, and their activation is controlled by various inhibitors. These inhibitors not only inhibit the activation of various enzymes, but also limit the response of the complement system once activated. The complement system can be activated by at least three pathways: the classical, the alternative and the lectin pathways (Figure 2.1).[1] The conversion of C3 is the central axis of the complement

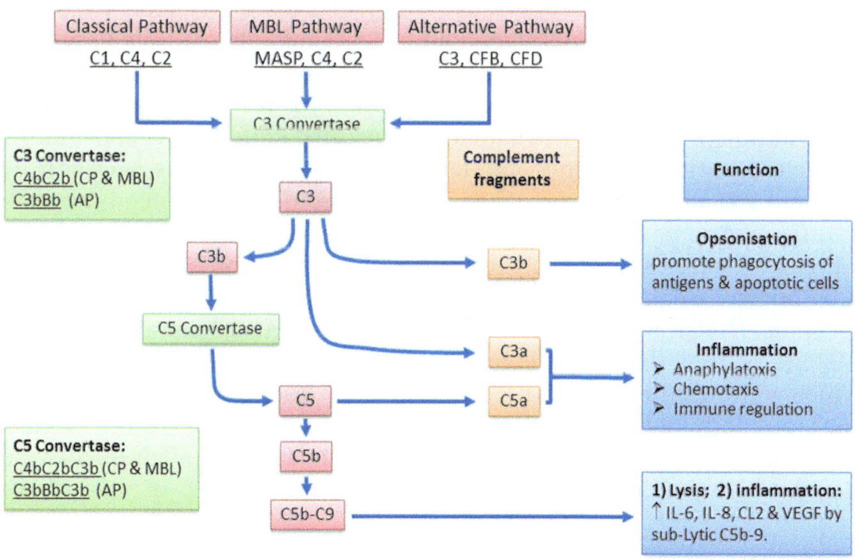

Figure 2.1 Complement activation and immune regulation. The complement system can be activated by the classical pathway (CP), mannose-binding lectin (MBL) pathway and the alternative pathway (AP): all lead to the cleavage of C3 and C5 and the formation of C5b-9.[1] Figure adapted from ref. 1, with permission from Elsevier. Published under a Creative Commons License (DOI: 10.1016/j.ejphar.2016.03.001).

activation pathways (Figure 2.1).[1] Once C3 is activated by C3bBb in the alternative pathway or C4bC2a in the classical and lectin pathway, it forms C3b3bBb or C4b3b2b to break down C5. The activation of C5 results in further sequential cleavage of C6, C7, C8 and C9, and ultimately, the formation of the C5b-9 (membrane attack complex; MAC), leading to cell lysis and death.[1] In addition to MAC, complement activation generates various complement fragments, such as C3a, C5a and C4a, which are actively involved in various immune responses.[1]

2.2.2 Inflammation

Classical inflammation is the response of the immune system to exogenous or endogenous insults. Medzhitov[2] defines a classical inflammatory response into four components: inducers, sensors, mediators and effectors. The inducers are foreign organisms such as bacteria and viruses and related pathogens, called pathogen-associated molecular patterns (PAMPs), as well as endogenous noxious factors generated through a pathophysiological process, known as damage-associated molecular patterns (DAMPs). The sensors include a range of receptors of the innate immune system (and other cell types), they are collectively named pattern recognition receptors (PRRs). The mediators include various cytokines and chemokines such as tumour necrosis factor (TNF)α, IL-1β, IL-6, chemokine (C–C motif) ligand (CCL)2 and histamine, released by sensor-expressing cells. The mediators may recruit circulating immune cells to the site of inflammation. They may also activate tissue cells and induce the autonomous inflammatory response. The effectors are infiltrating active immune cells that are important for the eradication of pathogens, although they may also damage tissue cells. Both sensors and effectors can release various cytokines and chemokines. Some cytokines, such as IL-1β, IL-6 and TNFα have systemic effects when secreted in sufficient amounts.[2]

Inflammasome activation is a special type of inflammatory response, which can occur in both tissue cells and immune cells. Upon detecting a pathogen or danger molecule, a special group of PRRs, such as Nod-like receptor (NLR)1, 3, or 4, or absent-in-melanoma (AIM)2 are able to assemble and oligomerise into a common structure called an "inflammasome", which activates the caspase-1 cascade, thereby leading to the production of the active form of inflammatory cytokines such as IL-1β and IL-18.[3,4] Inflammasome activation is known to be involved in various inflammatory and degenerative conditions.[4]

2.3 Immune Defence System of the Retina

2.3.1 Immune Privilege of the Eye

Although the systemic immune system is very powerful and effective in protecting the host from exogenous and endogenous insults, it has little impact on the retina under basal homeostatic conditions due to its "immune privilege" state.

The phrase "immune privilege" was first introduced in 1940s by Sir Peter Medawar. He found that the eye lacks an immune response against allografts placed into the anterior chamber of the eye and he described this phenomenon as immune privilege.[5] We now understand that not only the eye, but many other tissues/organs, such as brain, placenta and liver are privileged, one way or another, from the systemic immune attack.[6]

Several mechanisms are known to be involved in ocular immune privilege. First, the retina is segregated by physical barriers from the systemic immune system. There are two components of blood–retina barriers (BRBs): the inner BRB and outer BRB. The inner BRB is formed by tight junctions between the endothelial cells of the retinal blood vessels. The outer BRB is formed by tight junctions between the retinal pigment epithelial (RPE) cells, which filter blood from the fenestrated, leaky choroidal vessels. Circulating immune cells and molecules are not able to pass the intact BRBs into retinal parenchyma. In addition, the BRBs restrict and sequester retinal antigens within the intraocular compartment, avoiding T-cell activation, a phenomenon called immunologic ignorance.[7] The retina has a lack of lymphatic vessels. Therefore, retinal endogenous DAMPs generated during ageing or chronic stress, such as diabetes or glaucoma, cannot be detected by circulating cells.

In addition to the physical barriers, the retinal microenvironment also contributes to immunological privilege. For example, retinal neurons and RPE cells express various immune modulators, such as CD200, CX3CL1, Fas ligand (FasL), TNF-related apoptosis-inducing ligand (TRAIL), CD55, CD46, *etc.*[8–13] These immunoregulatory molecules suppress T-cell activation, induce the death of infiltrating immune cells, and suppress complement activation. The ocular fluid contains a number of immunoinhibitory molecules, such as transforming growth factor (TGF)β2, neuropeptides, and vasoactive intestinal peptide.[14,15] These mediators and molecules form an "immune-suppressive microenvironment" inside the eye. Table 2.1 lists the major known contributors, and it is likely that many more will be discovered in the future.[16]

2.3.2 Retinal Resident Immune Cells

Although the retina is considered an immune privileged tissue, immune response does occur, and inflammatory or autoimmune retinal disorders are the major causes of blindness. The retina has its own defence system, including resident immune cells and the complement system. There are three known types of resident immune cells in the retina: microglia, perivascular macrophages and dendritic cells.

2.3.2.1 Microglia

Microglia are the resident macrophages of the central nervous system, including the brain and the retina. In the retina, microglia reside in the inner neuronal layers, including the ganglion layer, inner plexiform layer and outer plexiform layer.[7,17] Microglia play an essential role in retinal immune

Table 2.1 Mediators involved in immune privilege in the eye. Data taken from ref. 16.[a]

Type of factor	Example	T-cell function	Macrophage activity	Antigen presentation	Cell killing/ apoptosis
Neuropeptide	VIP, PACAP, CGRP,	Inhibits	Inhibits		
	melanocortin, substance P		Promotes		
Neurotransmitter	GABA, glycine	Inhibits			
Cytokine	TGF-β, IL-10		Promotes[b]		
Chemokine	MCP-1, SDF-1				
Membrane ligand	FasL, CD200, TRAIL, APRIL				Promotes
Membrane receptor	TLRs	?Inhibits	Promotes		
Cellular enzymes	IDO, L-arginase	Inhibits		Inhibits	Promotes
MHC protein	Qa-1			Inhibits	
Prostaglandins	PGE2, LTAb4	Inhibits		Inhibits	
Cannabinoids		Inhibits		Inhibits?	Promotes
Free radicals	Peroxynitrate		Promotes	Inhibits	Promotes

[a]APRIL, a proliferation-inducing ligand; CGRP, calcitonin gene-related peptide; GABA, γ-amino-butyric acid, LTA, lipoteichoic acid; MCP, monocyte chemoattractant protein; PACAP, pituitary adenylate cyclase activating polypeptide; PGE, prostaglandin E; TGF, transforming growth factor; TLR, Toll-like receptor; TRAIL, TNF-related apoptosis-inducing ligand; SDF, stromal cell-derived factor; VIP, vasoactive intestinal peptide.
[b]Scavenger activity?

defence. The morphology of microglia in the retina often reflects their functional stage. The resting microglia have a ramified morphology with a small cell body and long branched dendrites. They express lower levels of costimulatory molecules and possess relatively low phagocytic activity under normal physiological conditions. Although resting microglia have a relatively stable cell body, their dendrites are highly motile and are constantly patrolling the microenvironment.[18] They can uptake and scavenge damaged neurons and synapses through cell surface PRRs such as the Toll-like receptors (TLRs) – a process called immune surveillance.[8] Microglial activation is tightly controlled by a network of regulatory molecules such as TGFβ, IL-10, CX3CL1 and CD200 expressed by retinal neurons.[19] In addition to their immune surveillance role, microglia also play a role in the physiology and survival of retinal neurons through secreting several polypeptide neurotrophic factors, including brain-derived neurotrophic factor, ciliary neurotrophic factor, glial-derived neurotrophic factor, nerve growth factor, neurotrophin-3 and fibroblast growth factor.[19]

Microglia activation is a common response to many ocular insults such as infection and neuronal injury. Upon engaging with danger signals, microglia may convert from a resting surveillance state to an active state. Activated microglia are morphologically and functionally distinct from resting microglia. Morphologically, the cell body increases in size and dendrites shorten and enlarge, and fully activated microglia often adapt an amoeboid shape. This is often accompanied by the expression and upregulation of several molecules such as F4/80, complement receptor 3, CD68, major histocompatibility complex (MHC)II and *Griffonia simplicifolia* isolectin B4.[19] These molecules are now considered microglial activation markers and are routinely used in immunohistochemistry or immunofluorescence-staining studies.[20] Functionally, the activated microglia can proliferate and migrate to the damaged area, where they phagocytose dead cells or debris and produce various bioactive molecules.[19] We have reported that one microglial cell can contain multiple phagocytic bodies under acute oxidative stress conditions,[21] suggesting that active microglia have an extremely high phagocytosis capacity. Apart from phagocytosis, active microglia secret nitric oxide and various pro-inflammatory cytokines such as IL-1β, TNFα, IL-6 and CCL2, which are toxic to retinal neurons, leading to further retinal damage.

It is important to note that under chronic stress conditions such as ageing and diabetes, microglia often undergo a low level of intermediate activation, characterised by shorter dendrites and large soma (cell bodies) compared to the resting cells.[7] The low level of intermediate microglial activation is a common phenomenon of many retinal degenerative diseases.[19,22]

2.3.2.2 Perivascular Macrophages

Perivascular macrophages are a subset of macrophages residing along retinal blood vessels. They have a distinct morphology and phenotype.[7,16] Although perivascular macrophages share a few common makers with microglia, such as CD11b, MHCII and F4/80, the expression levels of CD45 are generally higher in the former. In addition, perivascular macrophages express CD14, the receptor for lipopolysaccharide (LPS), whereas microglial cells do not.[23,24] As important immune cells that safeguard the retina around blood vessels, they are often activated when high levels of immune activation is evident in the circulation. For example, in the mouse model of interphotoreceptor retinoid binding protein peptide-induced experimental autoimmune uveoretinitis (EAU), activation of retinal vascular endothelial cells occurs 2–3 days prior to the onset of retinal inflammation.[25] Activation of perivascular macrophages can also be observed at the early stages of EAU, particularly around retinal venules (Figure 2.2A). As inflammation progresses, massive immune cell infiltration often occurs around retinal venules (Figure 2.2B), suggesting that perivascular macrophages are actively involved in retinal inflammation during retinal autoimmunity.

2.3.2.3 Dendritic Cells

A small population of dendritic cells have been observed in the normal retina by a few groups.[7,26–28] We have shown that mouse retinal dendritic cells are MHC-II^{+}33D1^{+}, and are strategically located around the optic disc and

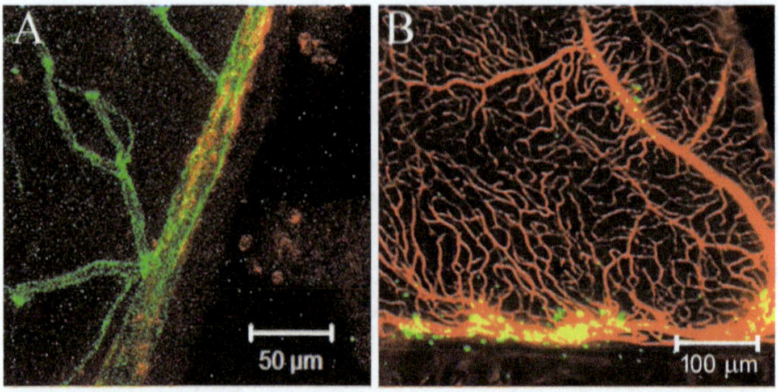

Figure 2.2 Retinal perivascular macrophages in experimental autoimmune uveoretinitis (EAU). (A) Retinal flatmount from a day-10 EAU mouse was stained for collagen VI (green) and major histocompatibility complex (MHC)-II (red). (B) Calcein AM labelled CD4 T-cells were adaptively transferred into day-12 EAU mice through the tail vein. 24 h later, animals received 100 μL Evans blue injection, and were then sacrificed. Retinal tissues were collected for flatmount investigation. Red: Evans blue-stained retinal blood vessels; green: calcein AM-labelled CD4 T-cells.

peripheral retinal margin area.[28] The specific location might suggest that these dendritic cells have gatekeeper functions. Other studies have shown that retinal dendritic cells are activated in retinal inflammation[29] and upon optic nerve damage.[30]

2.3.3 The Complement System in the Retina

Compelling evidence suggests that the complement system plays an important role in retinal pathophysiology. Although circulating complement proteins are not able to travel to the neuronal retinal under normal physiological conditions, we and others have shown that a complement regulatory system exists in the retina. Retinal cells can produce various complement proteins and regulators. For example, the mRNAs or proteins of complement components C1qa/b, Cr1, C2, C3, C4, CFB, CFD, C5 and C7, complement regulatory factors Serping-1, CD46, CD55, CFH, CFI and CD59, as well as complement receptors CR1, C3aR and C5aR were detected in neural retina of human and mouse.[1,31,32] We have shown that microglia and RPE cells are the major cellular sources of complement expression in the retina.[32] Furthermore, we and others have shown that complement expression in RPEs can be regulated by inflammatory cytokines and chemokines such as TNFα, IFN-γ, IL-27[1,33,34] or the supernatants of macrophages.[35] The results suggest that retinal complement expression and activation is regulated by a local ocular microenvironment and uncontrolled or dysregulated retinal complement activation may contribute to retinal degeneration.[1]

2.4 Immune Response in Retinal Degenerative Conditions

Common retinal degenerative diseases include inherited retinal degeneration and age-associated retinal diseases, such as AMD, glaucoma and diabetic retinopathy. Vision deficits in inherited retinal degeneration are due to mainly photoreceptor death, although the mechanisms vary.[36] AMD is due to damage in the macula, leading to the loss of central vision. The pathologies in AMD are known to be related to dysregulation of the innate immune response, including the complement system, macrophages and microglia.[7,17,36,37] Visual impairment in glaucoma is due to progressive ganglion cell death resulting from elevated intraocular pressure.[38] Uncontrolled or dysregulated inflammation is known to be involved in glaucomatous retinal degeneration.[39] Diabetic retinopathy is a major microvascular complication of long-term diabetes. Patients may suffer from ischaemic retinal damage at the early stages and the development of neovascular membrane at the advanced stages of diabetic retinopathy. It is well appreciated that local inflammatory response plays a central role in the pathogenesis of diabetic retinopathy.[40,41]

2.4.1 Parainflammation in the Ageing Retina

Inflammation is the body's protective response to tissue damage. The damage can be severe, such as during infection or tissue injury. Or, the damage can be caused by low levels of chronic stress that are subclinical and not noticeable, such as hypertension, high blood glucose and the presence of excessive free radicals whereby tissue cells suffer from low-level noxious insults. Depending on the level and extent of tissue damage, the inflammatory response can be overt and acute or unobserved. In response to infection or tissue injury, acute inflammation in the form of rapid and massive immune cell infiltration may occur, which may lead to full recovery. The response may turn into chronic inflammation if the pathogens or damage cannot be fully removed in time. Chronic inflammation often leads to tissue destruction or malfunction. Under chronic stress conditions such as during ageing, a low level of immune response may be initiated. Medzhitov named this as "para-inflammation" in a review article in 2008.[42] The physiological purpose of the parainflammatory response is to remove the tissue stress and maintain homeostasis and functionality. The response is between the basal homeostatic state and a classic inflammation, and may not be detectable using classical inflammatory biomarkers.[17,42] It is now recognised that dysregulated or malfunction of the parainflammatory response many contribute to various age-related degenerative diseases.[7,17,42–44]

As we age the retina undergoes significant changes including a decreased number of neurons, accumulation of lipofuscin in the RPE cells, increased thickness of Bruch's membrane, the accumulation of basal laminar deposits and the formation of drusen.[17] Importantly, oxidised proteins accumulate and are considered to be the major factor in age-related retinal cell death. Other endogenous molecules such as advanced glycation endproducts (AGEs) and advanced lipoxidation endproducts (ALEs) are formed during the ageing process and accumulate in the retina. In response to these noxious molecules, a parainflammatory response, characterised by low levels of microglial activation, microglia subretinal accumulation and increased complement expression is initiated to maintain retinal homeostasis. This has been intensively reviewed.[17]

2.4.2 Inherited Retinal Degeneration

Inherited retinal degeneration (IRD) such as retinitis pigmentosa is caused by specific gene mutations. So far, more than 200 genes have been identified to be related to inherited retinal degeneration (https://sph.uth.edu/RETNET/). Photoreceptor cell death is the major cause of visual loss in IRD. Although inflammation is not the initial trigger of IRD, microglial activation secondary to photoreceptor death may further participate in photoreceptor degeneration. Activated microglia accumulate in the damaged area and display enhanced proliferating, migrating and phagocytic functions. The large amoeboid-shaped microglia are predominantly found in the region of rod cell death in retinitis pigmentosa and other animal models of photoreceptor degeneration.[36,45] The activation of microglia is considered an early feature of retinal degeneration. In addition, NLRP3 inflammasome activation is known to be involved in bystander cone photoreceptor cell death in a P23H rhodopsin model of retinal degeneration.[46] Experimental studies have shown that control microglial activation can reduce photoreceptor degeneration in various models, suggesting that targeting microglia activation may be a good therapeutic approach for retinal degeneration.[45] Interestingly, little information is available about the role of the complement system in disease progression of inherited retinal degeneration.

2.4.3 Age-related Macular Degeneration

AMD is the leading cause of blindness in developed countries.[47] Although the initial trigger of this disease is not clear, we now know that ageing, family history of AMD, smoking, sunlight exposure and polymorphisms of various immune related genes are risk factors for AMD.[7,17,47] Genome-wide association studies have identified that the Y402H sequence variant in CFH significantly increased the incidence of AMD.[48–51] Other studies have shown that polymorphisms in complement components C2, CFB, C3 and SERPING1 are associated with AMD.[1,36] Various complement components, including C3, C5b-9, CFB and CFH have been detected in drusen deposits and in AMD lesions.[31,52] Complement components C3a, c5a and C4a have been reported to be significantly higher in the serum of AMD patients compared to age-matched controls.[1,53,54] Collectively, these data suggest that the complement system, in particular the alternative pathway, may be dysregulated and play an important role in the development of AMD.

Microglia undergo low levels of activation during ageing. Our group has found that microglia migrate and accumulate in the subretinal space in ageing mice.[20] The expression levels of TLR3/4, CD11c, 33D1 and MHCII[17] and numbers of microglia[21] were increased in the ageing mouse retina. Bloated microglia phagocytosing photoreceptor cells were found in the retina from patients with geographic atrophy within the retinal nuclear layers.[55] This suggests that microglial activation is an early and long-lasting chronic feature in AMD pathogenesis. Monocyte-derived macrophages may also play a pathogenic role in AMD.[37] These cells infiltrate and accumulate in AMD lesions and express various markers including IBA1, CCR2, CD14, CD163, *etc.* and produce high levels of inflammatory cytokines, including IL-1β, IL-6 and CCL2.[37] Aberrant inflammasome activation has also been associated with AMD.[56] NLRP3 inflammasome components are expressed in the RPE cells of AMD eyes.[57,58] Alu RNA-induced NLRP3 inflammasome activation is involved in RPE death and may contribute to geographic atrophy in AMD.[59] *In vitro* studies have suggested that NLRP3 in RPE cells can be activated by various AMD relevant molecules, such as complement proteins, drusen components, peroxidised lipids, or lysosomal destabilisation and overexpression of VEGF.[60–63] Active microglia may also promote RPE inflammasome activation as supernatant from LPS-treated microglia can increase pro-inflammatory gene expression including IL-1β and caspase-1 in RPE cells.[64]

Our present knowledge regarding adaptive immunity in AMD is limited. Various autoantibodies against the retina and RPEs were detected in the serum of AMD patients;[65] infiltrating lymphocytes have been observed in AMD lesions;[66] and elevated IL-17 in the sera[67] and ocular tissues of AMD[68] have been reported; all of which suggest that adaptive immunity may be involved in the pathogenesis of AMD. However, exactly how the adaptive immune system is activated in AMD is not yet known. The role of the retinal autoantibodies or T-cells in AMD lesion development remains to be elucidated.

2.4.4 Diabetic Retinopathy

Diabetic retinopathy is a major microvascular complication of long-term diabetes. Various studies have shown that diabetes has a significant impact on the immune system, particularly the innate immune system.[69] Circulating monocytes and neutrophils from type 1 and type 2 diabetes express higher levels of adhesion molecules and produce more pro-inflammatory cytokines.[70] They also express higher levels of CCR5 and are preferentially trapped in blood vessels of the retina.[71] The clinical symptoms of diabetic retinopathy are often linked to the disrupted immune privilege of the retina. The breakdown of BRBs is considered as the consequence of uncontrolled vascular endothelial inflammation (*e.g.* leukostasis). The increased endothelia–leukocytes interaction may result in increased inflammatory cytokine production and endothelial death.[72] Once the integrity of BRB is compromised, serum proteins, including complement, may leak into the parenchyma of the retina and further damage blood vessels and neurons.[73] The leaking serum proteins may also interfere with the immune-suppressive microenvironment of the retina.

Uncontrolled or dysregulated complement activation is known to be involved in diabetic retinopathy.[74] C5 polymorphism is associated with proliferative diabetic retinopathy (PDR) in type 2 diabetes patients.[75] Higher level of systemic complement activation is reported to be positively correlated with progression of diabetic retinopathy.[76] Uncontrolled complement activation may also occur locally in the retina. High levels of C3d and MAC was detected in the choriocapillaries of diabetic retinopathy eyes, and MAC deposition was detectable in the retinal vessels of patients with >9 years type 2 diabetes.[77,78] In addition, increased levels of C3, C5a and CFI have been detected in the vitreous of PDR eyes.[79,80] Experimental studies have shown that complement activation may contribute to retinal pericyte death and vascular damage in diabetes.[73,81] Taken together, these data suggest that uncontrolled complement activation is involved in the development and progression of diabetic retinopathy, although how the complement system is activated in the diabetic eye is not known.

In animal models of diabetic retinopathy, reactive microglia are known to participate at all stages of diabetic retinopathy.[82] During the early stages of diabetic retinopathy, noxious metabolites such as AGEs and ALEs pose a low level of threat to the retina, which may induce a parainflammatory response to maintain homeostasis, and patients may maintain good visual function. During this time, microglia increase moderately in number and size in all retina layers.[82] Amoeboid microglia clusters can be detected around soft exudates and the optic nerve region in the pre-PDR stage. In the advanced stages of diabetic retinopathy, *i.e.* PDR, microglia are found in the dilated diseased blood vessels.

Inflammasome activation also occurs in diabetic retinopathy. A recent study by a Finnish group[83] has found that the levels of IL-18 and caspase-1 were significantly increased in the vitreous of diabetic retinopathy patients, especially in PDR eyes. The levels of NLRP3 were found to be higher in PDR eyes with tractional retinal detachment than in PDR eyes with fully attached retina, suggesting that the inflammasome may be related to the severity of PDR.

2.4.5 Glaucoma

Glaucoma is a heterogeneous disorder related to retinal ganglion cell death resulting from increased intraocular pressure.[84] Neurodegenerative insults and glial activation initiate an immune response to restore tissue homeostasis and facilitate tissue repair and remodelling.[85] Phagocytic microglia have been found in the optic nerve head in human glaucoma eyes[86] and in the retina of animal models of glaucoma.[36] Active microglia in glaucoma are known to express higher levels of metalloproteinases (MMP), including MMP1, MMP2, MMP3 and MMP14, although the beneficial and detrimental roles of microglial activation in the development and progression of glaucomatous retinopathy remain to be elucidated.

Complement activation is known to be involved in glaucomatous retinopathy, in particular, the classical pathway of the complement system. Many components of the classical pathway of the complement system were upregulated in glaucomatous retina.[87] Animal studies have shown that C3 is upregulated at the early stage of glaucoma and complement activation is an early immune response in glaucoma.[88] Furthermore, mice with C1q or C5 deficiency were protected from glaucoma-induced retinal damage,[89,90] suggesting that complement activation in glaucoma contributes to ganglion cell death.

Despite the extensive studies of the role of inflammation in glaucomatous retinopathy, many questions remain to be answered. For example, is inflammation in glaucoma secondary to ganglion death or a primary cause of retinal degeneration? What immune pathways are involved in microglial activation and complement expression in glaucoma? What is the role of inflammasome in glaucomatous retinopathy? Understanding these questions will be essential for developing effective immune therapy for glaucoma patients.

2.5 Conclusions

Although the retina is an immune privilege tissue, it is not privileged from inflammation-mediated pathology. The retina has its own immune defence system. When the BRB is intact, the retinal defence system can protect the neuronal retina from various insults that may exist in the ageing eye, or eyes with diabetes, glaucoma, or retinitis pigmentosa. If the insults persist for sustained period of time, the immune response may turn into a chronic state and become dysregulated due to genetic or epigenetic modification. Such chronic inflammation may further damage the retina and contribute to the development of retinal degeneration. In addition to microglial cells, the complement system and inflammasome activation are known to be key inflammatory pathways involved in various forms of retinal degeneration, and warrant further investigation.

References

1. H. Xu and M. Chen, *Eur. J. Pharmacol.*, 2016, **787**, 94.
2. R. Medzhitov, *Cell*, 2010, **140**, 771.
3. E. Latz, T. S. Xiao and A. Stutz, *Nat. Rev. Immunol.*, 2013, **13**, 397.
4. H. Guo, J. B. Callaway and J. P. Ting, *Nat. Med.*, 2015, **21**, 677.
5. P. B. Medawar, *Br. J. Exp. Pathol.*, 1948, **29**, 58.

6. J. V. Forrester, H. Xu, T. Lambe and R. Cornall, *Mucosal Immunol.*, 2008, **1**, 372.
7. M. Chen and H. Xu, *J. Leukocyte Biol.*, 2015, **98**, 713.
8. A. D. Dick, D. Carter, M. Robertson, C. Broderick, E. Hughes and J. V. Forrester, *et al.*, *J. Leukocyte Biol.*, 2003, **74**, 161.
9. C. Combadiere, C. Feumi, W. Raoul, N. Keller, M. Rodero and A. Pezard, *et al.*, *J. Clin. Invest.*, 2007, **117**, 2920.
10. T. S. Griffithand and T. A. Ferguson, *Immunol. Today*, 1997, **18**, 240.
11. T. A. Ferguson and T. S. Griffith, *Chem. Immunol. Allergy*, 2007, **92**, 140.
12. J. H. Sohn, H. J. Kaplan, H. J. Suk, P. S. Bora and N. S. Bora, *Invest. Ophthalmol. Visual Sci.*, 2000, **41**, 4195.
13. J. H. Sohn, P. S. Bora, P. Jha, T. H. Tezel, H. J. Kaplan and N. S. Bora, *Chem. Immunol. Allergy*, 2007, **92**, 105.
14. A. W. Taylor, *Neuroimmunomodulation*, 1996, **3**, 195.
15. A. W. Taylor and D. Lee, *Adv. Exp. Med. Biol.*, 2010, **681**, 143.
16. J. V. Forrester, H. Xu, L. Kuffova, A. D. Dick and P. G. McMenamin, *Immunol. Rev.*, 2010, **234**, 282.
17. H. Xu, M. Chen and J. V. Forrester, *Prog. Retinal Eye Res.*, 2009, **28**, 348.
18. A. Nimmerjahn, F. Kirchhoff and F. Helmchen, *Science*, 2005, **308**, 1314.
19. T. Langmann, *J. Leukocyte Biol.*, 2007, **81**, 1345.
20. H. Xu, M. Chen, A. Manivannan, N. Lois and J. V. Forrester, *Aging Cell*, 2008, **7**, 58.
21. M. Chen, J. R. Hombrebueno, C. Luo, R. Penalva, J. Zhao and L. Colhoun, *et al.*, *PLoS One*, 2013, **8**, e61381.
22. E. Schuetz and S. Thanos, *Curr. Drug Targets*, 2004, **5**, 619.
23. A. D. Dick, A. L. Ford, J. V. Forrester and J. D. Sedgwick, *Br. J. Ophthalmol.*, 1995, **79**, 834.
24. H. G. Fischer and G. Reichmann, *J. Immunol.*, 2001, **166**, 2717.
25. H. Xu, J. V. Forrester, J. Liversidge and I. J. Crane, *Invest. Ophthalmol. Visual Sci.*, 2003, **44**, 226.
26. J. Zhang, G. S. Wu, S. Ishimoto, G. Pararajasegaram and N. A. Rao, *Invest. Ophthalmol. Visual Sci.*, 1997, **38**, 1848.
27. D. S. Gregerson and J. Yang, *Invest. Ophthalmol. Visual Sci.*, 2003, **44**, 3083.
28. H. Xu, R. Dawson, J. V. Forrester and J. Liversidge, *Invest. Ophthalmol. Visual Sci.*, 2007, **48**, 1701.
29. N. D. Heuss, U. Lehmann, C. C. Norbury, S. W. McPherson and D. S. Gregerson, *J. Immunol.*, 2012, **188**, 1191.
30. U. Lehmann, N. D. Heuss, S. W. McPherson, H. Roehrich and D. S. Gregerson, *Neurobiol. Dis.*, 2010, **40**, 177.
31. D. H. Anderson, M. J. Radeke, N. B. Gallo, E. A. Chapin, P. T. Johnson and C. R. Curletti, *et al.*, *Prog. Retinal Eye Res.*, 2010, **29**, 95.
32. C. Luo, M. Chen and H. Xu, *Mol. Vision*, 2011, **17**, 1588.
33. M. Chen, J. V. Forrester and H. Xu, *Exp. Eye Res.*, 2007, **84**, 635.
34. M. Chen, E. Muckersie, M. Robertson, J. V. Forrester and H. Xu, *Exp. Eye Res.*, 2008, **87**, 543.

35. C. Luo, J. Zhao, A. Madden, M. Chen and H. Xu, *Exp. Eye Res.*, 2013, **112C**, 93.
36. M. Karlstetter, R. Scholz, M. Rutar, W. T. Wong, J. M. Provis and T. Langmann, *Prog. Retinal Eye Res.*, 2015, **45**, 30.
37. X. Guillonneau, C. M. Eandi, M. Paques, J. A. Sahel, P. Sapieha and F. Sennlaub, *Prog. Retinal Eye Res.*, 2017, **61**, 98.
38. H. A. Quigley and A. T. Broman, *Br. J. Ophthalmol.*, 2006, **90**, 262.
39. G. Tezel and M. B. Wax, *Curr. Opin. Ophthalmol.*, 2004, **15**, 80.
40. J. Tang and T. S. Kern, *Prog. Retinal Eye Res.*, 2011, **30**, 343.
41. A. W. Stitt, T. M. Curtis, M. Chen, R. J. Medina, G. J. McKay and A. Jenkins, *et al.*, *Prog. Retinal Eye Res.*, 2016, **51**, 156.
42. R. Medzhitov, *Nature*, 2008, **454**, 428.
43. P. L. McGeer and E. G. McGeer, *Ann. N. Y. Acad. Sci.*, 2004, **1035**, 104.
44. Y. Shen and S. Meri, *Prog. Neurobiol.*, 2003, **70**, 463.
45. R. Scholz, A. Caramoy, M. B. Bhuckory, K. Rashid, M. Chen and H. Xu, *et al.*, *J. Neuroinflammation*, 2015, **12**, 201.
46. I. A. Viringipurampeer, A. L. Metcalfe, A. E. Bashar, O. Sivak, A. Yanai and Z. Mohammadi, *et al.*, *Hum. Mol. Genet.*, 2016, **25**, 1501.
47. L. S. Lim, P. Mitchell, J. M. Seddon, F. G. Holz and T. Y. Wong, *Lancet*, 5 May, 2012, **379**(9827), 1728–1738.
48. A. O. Edwards, R. Ritter, 3rd, K. J. Abel, A. Manning, C. Panhuysen and L. A. Farrer, *Science*, 2005, **308**, 421.
49. J. L. Haines, M. A. Hauser, S. Schmidt, W. K. Scott, L. M. Olson and P. Gallins, *et al.*, *Science*, 2005, **308**, 419.
50. R. J. Klein, C. Zeiss, E. Y. Chew, J. Y. Tsai, R. S. Sackler and C. Haynes, *et al.*, *Science*, 2005, **308**, 385.
51. G. S. Hageman, D. H. Anderson, L. V. Johnson, L. S. Hancox, A. J. Taiber and L. I. Hardisty, *et al.*, *Proc. Natl. Acad. Sci. U. S. A.*, 2005, **102**, 7227.
52. D. H. Anderson, R. F. Mullins, G. S. Hageman and L. V. Johnson, *Am. J. Ophthalmol.*, 2002, **134**, 411.
53. J. Lechner, M. Chen, R. E. Hogg, L. Toth, G. Silvestri and U. Chakravarthy, *et al.*, *Immun. Ageing*, 2016, **13**, 4.
54. R. Reynolds, M. E. Hartnett, J. P. Atkinson, P. C. Giclas, B. Rosner and J. M. Seddon, *Invest. Ophthalmol. Visual Sci.*, 2009, **50**, 5818.
55. N. Gupta, K. E. Brown and A. H. Milam, *Exp. Eye Res.*, 2003, **76**, 463.
56. J. Ambati, J. P. Atkinson and B. D. Gelfand, *Nat. Rev. Immunol.*, 2013, **13**, 438.
57. W. A. Tseng, T. Thein, K. Kinnunen, K. Lashkari, M. S. Gregory and P. A. D'Amore, *et al.*, *Invest. Ophthalmol. Visual Sci.*, 2013, **54**, 110.
58. Y. Wang, J. W. Hanus, M. S. Abu-Asab, D. Shen, A. Ogilvy and J. Ou, *et al.*, *Int. J. Mol. Sci.*, 2016, **17**, E73.
59. V. Tarallo, Y. Hirano, B. D. Gelfand, S. Dridi, N. Kerur and Y. Kim, *et al.*, *Cell*, 2012, **149**, 847.
60. A. G. Marneros, *Cell Rep.*, 2013, **4**, 945.

61. S. L. Doyle, M. Campbell, E. Ozaki, R. G. Salomon, A. Mori and P. F. Kenna, *et al.*, *Nat. Med.*, 2012, **18**, 791.
62. A. Kauppinen, H. Niskanen, T. Suuronen, K. Kinnunen, A. Salminen and K. Kaarniranta, *Immunol. Lett.*, 2012, **147**, 29.
63. P. Prager, M. Hollborn, A. Steffen, P. Wiedemann, L. Kohen and A. Bringmann, *PLoS One*, 2016, **11**, e0165653.
64. C. Nebel, A. Aslanidis, K. Rashid and T. Langmann, *Biochem. Biophys. Res. Commun.*, 2017, **484**, 681.
65. S. Camelo, *Autoimmune Dis.*, 2014, **2014**, 532487.
66. C. Tsutsumi-Miyahara, K. H. Sonoda, K. Egashira, M. Ishibashi, H. Qiao and T. Oshima, *et al.*, *Br. J. Ophthalmol.*, 2004, **88**, 1217.
67. B. Liu, L. Wei, C. Meyerle, J. Tuo, H. N. Sen and Z. Li, *et al.*, *J. Transl. Med.*, 2011, **9**, 1.
68. C. C. Chan and D. Ardeljan, *Adv. Exp. Med. Biol.*, 2014, **801**, 193.
69. H. Xu and M. Chen, *Vision Res.*, 2017, **139**, 39.
70. K. Josefsen, H. Nielsen, S. Lorentzen, P. Damsbo and K. Buschard, *Clin. Exp. Immunol.*, 1994, **98**, 489.
71. A. M. Serra, J. Waddell, A. Manivannan, H. Xu, M. Cotter and J. V. Forrester, *Am. J. Pathol.*, 2012, **181**, 719.
72. R. Chibber, B. M. Ben-Mahmud, S. Chibber and E. M. Kohner, *Curr. Diabetes Rev.*, 2007, **3**, 3.
73. Y. Li, D. Smith, Q. Li, N. Sheibani, S. Huang and T. Kern, *et al.*, *Invest. Ophthalmol. Visual Sci.*, 2012, **53**, 5520.
74. Y. Yang, H. L. Lu, J. Zhang, H. Y. Yu, H. W. Wang and M. X. Zhang, *et al.*, *Int. J. Obes (Lond).*, 2006, **30**, 439.
75. D. Xu, H. Yi, S. Yu, X. Li, Y. Qiao and W. Deng, *PLoS One*, 2016, **11**, e0149704.
76. P. Geng, Y. Ding, L. Qiu and Y. Lu, *Diabetes Care*, 2015, **38**, 868.
77. V. B. Gerl, J. Bohl, S. Pitz, B. Stoffelns, N. Pfeiffer and S. Bhakdi, *Invest. Ophthalmol. Visual Sci.*, 2002, **43**, 1104.
78. J. Zhang, C. Gerhardinger and M. Lorenzi, *Diabetes*, 2002, **51**, 3499.
79. B. B. Gao, X. Chen, N. Timothy, L. P. Aiello and E. P. Feener, *J. Proteome Res.*, 2008, **7**, 2516.
80. H. Wang, L. Feng, J. Hu, C. Xie and F. Wang, *Exp. Eye Res.*, 2013, **108**, 110.
81. M. Nozaki, B. J. Raisler, E. Sakurai, J. V. Sarma, S. R. Barnum and J. D. Lambris, *et al.*, *Proc. Natl. Acad. Sci. U. S. A.*, 2006, **103**, 2328.
82. H. Y. Zeng, W. R. Green and M. O. Tso, *Arch. Ophthalmol.*, 2008, **126**, 227.
83. S. Loukovaara, N. Piippo, K. Kinnunen, M. Hytti, K. Kaarniranta and A. Kauppinen, *Acta Ophthalmol.*, 2017, **95**, 803.
84. R. W. Nickells, G. R. Howell, I. Soto and S. W. John, *Annu. Rev. Neurosci.*, 2012, **35**, 153.
85. G. Tezel, *Exp. Eye Res.*, 2011, **93**, 178.
86. A. H. Neufeld, *Arch. Ophthalmol.*, 1999, **117**, 1050.
87. G. Tezel, X. Yang, C. Luo, A. D. Kain, D. W. Powell and M. H. Kuehn, *et al.*, *Invest. Ophthalmol. Visual Sci.*, 2010, **51**, 5071.

88. J. M. Harder, C. E. Braine, P. A. Williams, X. Zhu, K. H. MacNicoll and G. L. Sousa, *et al.*, *Proc. Natl. Acad. Sci. U. S. A.*, 2017, **114**, E3839.
89. G. R. Howell, I. Soto, M. Ryan, L. C. Graham, R. S. Smith and S. W. John, *J. Neuroinflammation*, 2013, **10**, 76.
90. G. R. Howell, D. G. Macalinao, G. L. Sousa, M. Walden, I. Soto and S. C. Kneeland, *et al.*, *J. Clin. Invest.*, 2011, **121**, 1429.

CHAPTER 3

Modulation of Calcium Overload and Calpain Activity

FRANÇOIS PAQUET-DURAND[a], PER EKSTRÖM[b] AND VALERIA MARIGO*[c]

[a]Institute for Ophthalmic Research, University of Tübingen, Germany; [b]Ophthalmology, Department of Clinical Sciences Lund, Faculty of Medicine, Lund University, Sweden; [c]Department of Life Sciences, University of Modena and Reggio Emilia, via G. Campi 287, 41125 Modena, Italy
*E-mail: valeria.marigo@unimore.it

3.1 Overview

The most recent advances in the characterization of molecular mechanisms leading to cell demise during retinal degeneration have allowed the identification of several therapeutic targets proven to play key roles in the progression of the disease. Despite the great heterogeneity of genetic or environmental insults causing retinal degeneration, some common cell death pathways have been identified. Concomitantly, new molecular tools and delivery systems have facilitated the development of novel therapeutic interventions aimed at correcting the genetic defects. Gene therapy approaches based on gene replacement are already in clinical trials for recessive genetic forms of the disease and demonstrated to be effective, although recent studies have questioned whether the benefit is persistent after treatment.[1–5]

Drug Discovery Series No. 66
Therapies for Retinal Degeneration: Targeting Common Processes
Edited by Enrique J. de la Rosa and Thomas G. Cotter

Published by the Royal Society of Chemistry, www.rsc.org

New strategies based on zinc finger and CRISPR/Cas9 nucleases showed interesting potentials for the treatment of dominant hereditary retinal dystrophies but, at the moment, show limited efficiency.[6–10] All these elegant approaches are tailored for a limited number of patients that bear mutations in a specific gene or even a specific mutation in one gene. Gene therapy thus requires the development of expensive personalized treatments that can benefit few patients.

Neuroprotection systems targeting molecular processes shared by otherwise distinct genetic insults can benefit a larger cohort of patients. Neuroprotection can be achieved through different approaches based either on very general trophic factors or on specific molecules interfering with death-inducing factors. Trophic factors delivered to the diseased retina can have pleiotropic effects by targeting several intracellular and/or extracellular pathways. The positive aspect of using trophic factors is that one molecule impacts different points of the degeneration cascade and the effect can be amplified. However, side effects are less controllable. Synthesis or expression of molecules precisely interfering with one of the events of the cell death cascade may be more specific, but may activate secondary cell death pathways, thus limiting a full neuroprotection. Another general aspect of neuroprotection is that such treatments can be stopped at will, if any untoward effects are noted.

Plenty of studies have analyzed rod photoreceptors during the progression of the disease and identified intracellular changes and activation of molecules that are amenable to drug development.[11,12] One important change found was unbalanced cGMP, which has been detected in several animal models and in patients with retinal degeneration.[13–21] Molecules developed to counteract high levels of cGMP are discussed in Chapter 6 of this book. One of the consequences of high intracellular cGMP is the opening of the cyclic nucleotide-gated (CNG) channels with the resulting influx of cations, including calcium ions (Ca^{2+}). Changes in intracellular Ca^{2+} homeostasis occur also as a result of stress in the endoplasmic reticulum, as happens when mutations cause protein misfolding. Here we discuss the consequences of high Ca^{2+} levels and the resulting activation of calpains, cysteine proteases described in Section 3.3, leading to cell death, as depicted in Figure 3.1.

3.2 Targeting Calcium Overload

Ca^{2+} plays fundamental roles in the regulation of different functions in photoreceptors. Ca^{2+} ebbs and flows from the extracellular environment and within intracellular organelles, such as mitochondria and endoplasmic reticulum, generates large fluctuations of free cytosolic ions. The cell maintains cytosolic Ca^{2+} at nanomolar levels with a 10 000-fold difference with the extracellular media, and this gradient is regulated by Ca^{2+} channels and Ca^{2+} pumps. Photoreceptors are compartmentalized cells with an outer segment deputed to phototransduction, an inner segment containing cytoplasm, nucleus and organelles and a synaptic region. The morphology of the rod

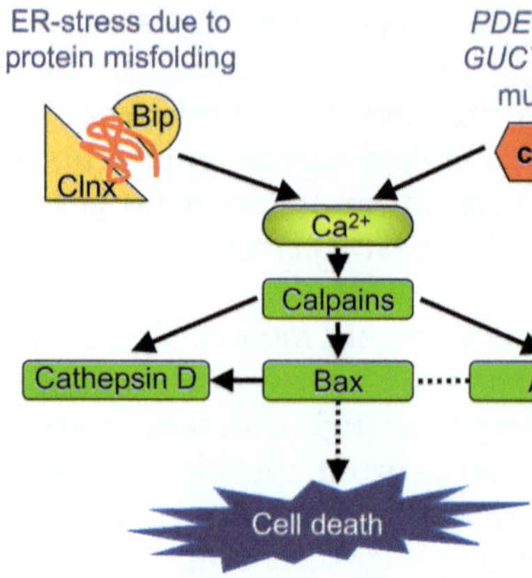

Figure 3.1 Cell death pathways downstream to calcium overload. Increases of intracellular calcium ions (Ca^{2+}) can be a consequence of inherited mutations in genes regulating intracellular levels of cGMP or of endoplasmic reticulum (ER) stress engaging chaperones such as Bip and Calnexin (Clnx). High Ca^{2+} ions can activate proteases, such as calpains, that cleave and activate cathepsin D and the apoptosis-inducing factor (Aif). Mitochondrial Aif then exits through pores formed by Bax and translocates into the nucleus together with EndoG or macrophage migration inhibitory factor inducing chromatin fragmentation and cell death.

photoreceptor mirrors the asymmetric distribution of Ca^{2+} regulators[20] with clear differences between the various parts (Figure 3.2). With respect to the outer segment, the CNG channels are kept open by cGMP, physiologically available in darkness, and allow the influx of Na^{+}, Mg^{2+} and Ca^{2+}. High levels of cGMP in photoreceptors have been first correlated to impaired function of the phosphodiesterase 6B (Pde6b) enzyme in the retinal degeneration 1 (*rd1*) mouse, a murine model of retinitis pigmentosa.[21,22] A consequence of high cGMP in pathological conditions is the abnormal Ca^{2+} influx in the photoreceptor outer segment.[23] The significance of this is underlined by the fact that concentrations of cytosolic Ca^{2+} >500 nM can trigger cell death mechanisms, as found in the *rd1*-mutant retina and other retinal degeneration models.[24–26] In another type of regulation, Ca^{2+} can be extruded from the photoreceptor outer segment by the Na^{+}/Ca^{2+}, K^{+} exchanger (NCKX).[27,28] The functions of these two Ca^{2+} regulators are very important for photoreceptor functionality and physiology, which is emphasized by the fact that mutations in the different subunits composing the CNG channel cause retinitis pigmentosa[29,30] and that mutations in the NCKX type 1 exchanger are linked to stationary night blindness.[31]

The inner segment expresses specific Ca^{2+} channels at the plasma membrane, but also contains mitochondria and endoplasmic reticulum, two important stores of intracellular Ca^{2+}. There is little transport and diffusion of Ca^{2+} between the outer and inner segment due to anatomical constraints and, in addition, the inner segment uses different systems for regulating Ca^{2+} homeostasis.[20] The Cav1.4 L-type calcium channels (VGCC), also present at

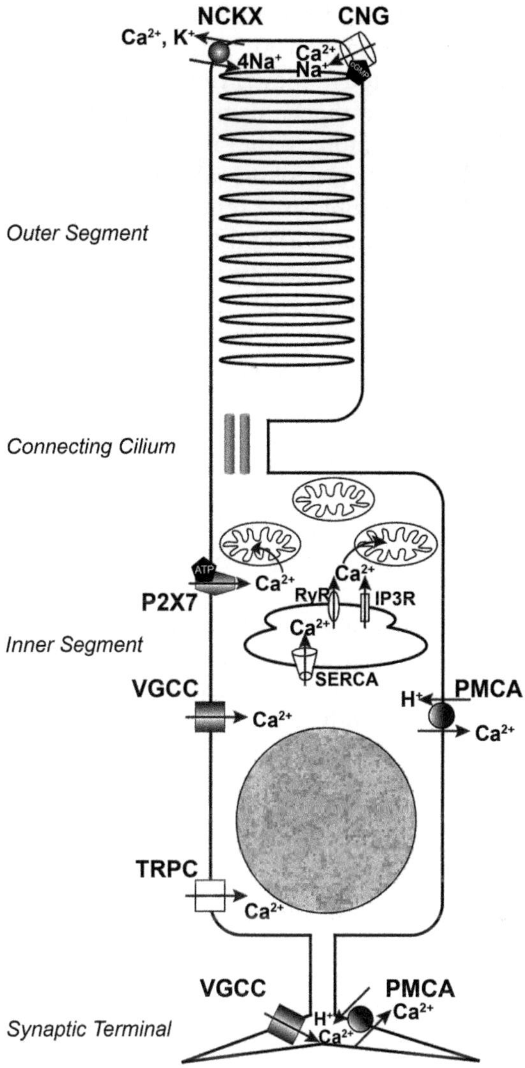

Figure 3.2 Calcium pumps and exchangers in rod photoreceptors. Schematic illustration of a rod photoreceptor with the outer segment separated from the inner segment by the connecting cilium. Calcium channels and exchangers are represented: cyclic nucleotide-gated channels (CNG) regulated by cGMP, Na^+/Ca^{2+}, K^+ exchanger (NCKX), Cav1.4 L-type calcium channels (VGCC), P2X7-type purinergic receptor regulated by ATP, transient receptor potential canonical channels (TRPC), plasma membrane Ca^{2+} ATPase (PMCA), sarco-endoplasmic reticulum calcium ATPase (SERCA), ryanodine receptors (RyR) and IP3 receptors (IP3R).

the synaptic terminals, and the P2X7-type purinergic receptors regulate Ca^{2+} influx into the inner segment. Over-activation of the P2X7-type purinergic receptor by ATP has been associated with retinal degeneration.[32] Ca^{2+} can also enter the cell through transient receptor potential canonical (TRPC) channels and extrusion of Ca^{2+} from the inner segment is regulated by the plasma membrane Ca^{2+} ATPase (PMCA).[33] Changes in the expression of TRPC and PMCA1 have been reported in the *rd1* retina, suggesting that photoreceptors activate some adaptation mechanisms to balance Ca^{2+} overload during degeneration.[34] The uptake of ions in the endoplasmic reticulum is regulated by SERCA (sarco-endoplasmic reticulum calcium ATPase) pumps isoform 2, whereas ions move from endoplasmic reticulum to cytoplasm through the ryanodine receptors (RyR isoform 2) and IP3 receptors (IP3R).[35] Ca^{2+} is also buffered in the cytoplasm by high-affinity Ca^{2+} binding proteins and sensors, such as the photoreceptor-specific guanylate cyclase activating protein operating in the modulation of the enzyme that synthesizes cGMP.[36] SERCA inhibition and endoplasmic reticulum stress are characterized by depletion of Ca^{2+} stores in the endoplasmic reticulum and activation of mechanisms that may lead to photoreceptor cell death.[35,37] Elevated Ca^{2+} in rods also affects another intracellular Ca^{2+} store, the mitochondrion, inducing depolarization and swelling.[24,38]

The contribution of unbalanced intracellular Ca^{2+} to photoreceptor degeneration has been demonstrated in several elegant studies. We generated double-mutant mice bearing the *rd1* mutation in the *Pde6b* gene and lacking expression of the Cngb1 subunit of the CNG channel. In this study, we depleted the outer segment of a functional CNG channel, thus interfering with Ca^{2+} influx through the cGMP-activated channel. Cell death and activation of calpains were significantly reduced in the double-mutant retina while cGMP levels remained high, supporting the hypothesis that entrance of Ca^{2+} through the CNG channel is one of the causes of photoreceptor cell death.[39] The application of fluorescent biosensors tracking intracellular Ca^{2+} allowed the study of Ca^{2+} dynamics in cones *in vivo*.[40,41] These studies suggested that mutations in the Pde enzyme and increases of cGMP altered Ca^{2+} intracellular dynamics, although they did not cause a sustained increase of the ion. In addition, the involvement of VGCCs was assessed by the generation of double-mutant mice bearing the *rd1* mutation and lacking a functional VGCC.[42,43] While VGCC depletion significantly impaired calpain activation, degeneration was only delayed. These results are in line with previous reports on the transient effects of D-*cis* diltiazem and other VGCC blockers.[44] We further confirmed these data showing that D-*cis* diltiazem interferes with the cell death pathway activated by calpain.[25] The limited effects of targeting CNG channels and VGCCs might have been due to drug dosing or delivery, but we cannot exclude the contribution of other membrane and intracellular Ca^{2+} regulators.

The P2X7-type purinergic receptor was targeted by treatment with saffron. The authors of this study showed that neuroprotection of saffron was mediated by reduced ATP-induced intracellular Ca^{2+} increase through the P2X7-type purinergic receptor.[45]

Recently, we found that the pigment epithelium-derived factor (PEDF), when intravitreally injected in *rd1*-mutant eyes, can decrease rod photoreceptor intracellular Ca^{2+} and cell death. By specifically interfering with the activity of different Ca^{2+} pumps/exchangers we defined that PEDF acts on PMCA, suggesting that at least some Ca^{2+} pumps/exchangers are promising targets for neuroprotective therapeutic approaches.[46]

3.3 Targeting Calpains

Molecular mechanisms triggered by imbalance of intracellular Ca^{2+} in degenerating photoreceptors were shown to activate common pathways in animal models of the disease caused by mutations in different genes. We and others identified calpains as key players during retinal degeneration.[11,17,25,26,47–56] Calpains belong to a family of calcium-dependent cysteine proteases. The two major isoforms, calpain 1 (μ-calpain) and calpain 2 (m-calpain), are expressed in photoreceptors and are activated by micromolar and millimolar Ca^{2+} concentrations, respectively. These proteases are localized in the endoplasmic reticulum as heterodimers composed of an 80 kDa catalytic subunit and a 28 kDa regulatory subunit.[57] In the endoplasmic reticulum the catalytic and regulatory subunits are associated with an endogenous inhibitor called calpastatin. The two calpain subunits dissociate upon intracellular Ca^{2+} increase and this structural modification causes the detachment from calpastatin and activation of the protease. Thus, downregulation of calpastatin during photoreceptor degeneration was suggested to contribute to the activation of the proteases.[53] Activation of calpains in the degenerating photoreceptors has several effects. We demonstrated that calpain 1 is a key player in the activation of cathepsin D and apoptosis-inducing factor (Aif) in the *rd1*-mutant retina.[51] Cathepsin D and calpains activate Bax, which forms pores in the mitochondrial outer membrane and allows the efflux of calpain-cleaved Aif.[58] This cell death pathway contributes to photoreceptor degeneration in several models of retinal degeneration[11,26,51] and inhibition of calpains was thus evaluated in several studies.

Small cell-permeable synthetic molecules with calpain inhibiting activity are commercially available and were tested in retinal explants or injected *in vivo* in rodent models of the disease. All the tested molecules provided neuroprotection to the degenerating retina, but unfortunately, none of them were specific for one type of calpain. We tested calpain inhibitor I (ALLN) and calpain inhibitor II (ALLM), which inhibit calpain 1 and calpain 2, but also cathepsin B and L and other cell pathways. ALLN was demonstrated to be more effective than ALLM when intravitreally injected into the *rd1*-mutant eye, but neither of them could completely block calpain activation and cell death.[25] In addition, we tested more potent and specific calpain inhibitors, such as calpain inhibitor VI (SJA6017) which targets calpain 1 and calpain 2, but also cathepsin B and L. SJA6017 reduced calpain activity, but was less potent than calpain inhibitor XI (CX295).[53] CX295 is more specific than other inhibitors because it blocks both calpain 1 and 2, but blocks cathepsin B and

L only weakly. Unfortunately, neither SJA6017 nor CX295 could completely interfere with calpain activation and cell death. Furthermore, long-term treatment of *rd1*-mutant retinal explants with CX295 resulted in toxicity. This toxicity might be caused by off-target activity such as inhibition of cathepsin B and L or other effects, that are still not well characterized.[59]

Because we detected a downregulation at the mRNA and protein level of calpastatin in the *rd1*-mutant retina, we assessed the hypothesis of delivering the calpastatin inhibitor to photoreceptors. Calpastatin is a multi-domain protein constituted by domain A- and C-binding calpains and by domain B bearing the inhibitory activity.[60] The ratio between calpain and calpastatin is strictly important in controlling calpain activity, and upon calpain activation calpastatin is cleaved by calpains or caspase 3 to block its inhibitory function.[61] Structural analysis of calpastatin bound to calpain identified a 27 amino acid peptide, called calpastatin peptide or CP1B, with highly specific and potent inhibitory activity for calpain 1 and 2.[62] This reversible calpain inhibitor was modified to favor cell penetration. We compared the effects of CX295 with those of the CP1B peptide. Differently from the toxic effects of CX295, CP1B had neuroprotective activity after short-term as well as long-term exposure of *rd1*-mutant retinal explants and no harmful outcome was detected.[59] This positive result prompted us to test the peptide *in vivo*, where we confirmed a significant reduction of cell death in different murine models of retinal degeneration.[26,51,59] Interestingly, CP1B could completely block calpain activation and cell death only in mice bearing a homozygous mutation in the *Rhodopsin* gene.[26] In the other retinal degeneration mice tested so far, a partial, although significant, decrease of calpain activation and cell death was observed. The failure to completely protect photoreceptors from cell death may indicate either activation of other calpains apart from calpain 1 and 2 in these diseased retinas, or suboptimal drug dosing or activation of other cell death pathways, as suggested by our studies.[26,63–65]

Other more specific calpain inhibitors were assessed in other studies. Because a key downstream target of calpains is mitochondrial Aif (Figure 3.1), inhibition of mitochondrial μ-calpain by intravitreal injection or topical application of a specific inhibiting peptide was tested. Mitochondrial μ-calpain inhibition could interfere with Aif activation and restrain cell death in retinas bearing dominant mutations in the *Rhodopsin* gene.[66]

3.4 Conclusions

We have discussed the molecular processes activated by Ca^{2+} overload and the contributions of different Ca^{2+} channels and pumps as well as the Ca^{2+}-activated proteases calpains. Several attempts to modulate these processes have been tried in different animal models of retinal degeneration. We believe that modulation of Ca^{2+} and calpains has good potentiality for the development of novel therapies for these blinding diseases. At the moment, only short-term treatments have been attempted and appropriate delivery systems for these drugs still need further improvement.

Acknowledgements

The authors thank for funding: European Union (DRUGSFORD; HEALTH-F2-2012-304963), the Charlotte and Tistou Kerstan Foundation, Fondazione Telethon (GGP11210A, GGP14180), E-RARE 2009 RHORCOD, Fondazione Roma (call for proposal 2013 sulla Retinite Pigmentosa), the Italian Ministry (Grants 2006053302_003, 20094CZ3M2), AFM Telethon project 16663, the Medical Faculty at Lund University and Programma di ricerca Regione-Università 2010–2012 of Regione Emilia Romagna (RARER).

References

1. J. W. B. Bainbridge, A. J. Smith, S. S. Barker, S. Robbie, R. Henderson and K. Balaggan, *et al.*, Effect of gene therapy on visual function in Leber's congenital amaurosis, *N. Engl. J. Med.*, 2008, **358**, 2231–2239.
2. A. M. Maguire, F. Simonelli, E. A. Pierce, E. N. Pugh, F. Mingozzi and J. Bennicelli, *et al.*, Safety and efficacy of gene transfer for Leber's congenital amaurosis, *N. Engl. J. Med.*, 2008, **358**, 2240–2248.
3. A. V. Cideciyan, T. S. Aleman, S. L. Boye, S. B. Schwartz, S. Kaushal and A. J. Roman, *et al.*, Human gene therapy for RPE65 isomerase deficiency activates the retinoid cycle of vision but with slow rod kinetics, *Proc. Natl. Acad. Sci. U. S. A.*, 2008, **105**, 15112–15117.
4. S. G. Jacobson, A. V. Cideciyan, A. J. Roman, A. Sumaroka, S. B. Schwartz and E. Heon, *et al.*, Improvement and decline in vision with gene therapy in childhood blindness, *N. Engl. J. Med.*, 2015, **372**, 1920–1926.
5. F. Testa, A. M. Maguire, S. Rossi, E. A. Pierce, P. Melillo and K. Marshall, *et al.*, Three-year follow-up after unilateral subretinal delivery of adeno-associated virus in patients with Leber congenital Amaurosis type 2, *Ophthalmology*, 2013, **120**, 1283–1291.
6. A. G. Bassuk, A. Zheng, Y. Li, S. H. Tsang and V. B. Mahajan, Precision medicine: genetic repair of retinitis pigmentosa in patient-derived stem cells, *Sci. Rep.*, 2016, **6**, 19969.
7. S. Botta, E. Marrocco, N. de Prisco, F. Curion, M. Renda and M. Sofia, *et al.*, Rhodopsin targeted transcriptional silencing by DNA-binding, *elife*, 2016, **5**, e12242.
8. M. C. Latella, M. T. Di Salvo, F. Cocchiarella, D. Benati, G. Grisendi and A. Comitato, *et al.*, In vivo editing of the human mutant Rhodopsin gene by electroporation of plasmid-based CRISPR/Cas9 in the mouse retina, *Mol. Ther. Nucleic Acids*, 2016, **5**, e389.
9. C. Mussolino, D. Sanges, E. Marrocco, C. Bonetti, U. Di Vicino and V. Marigo, *et al.*, Zinc-finger-based transcriptional repression of rhodopsin in a model of dominant retinitis pigmentosa, *EMBO Mol. Med.*, 2011, **3**, 118–128.
10. N. Overlack, T. Goldmann, U. Wolfrum and K. Nagel-Wolfrum, Gene repair of an Usher syndrome causing mutation by zinc-finger nuclease mediated homologous recombination, *Invest. Ophthalmol. Visual Sci.*, 2012, **53**, 4140–4146.

11. B. Arango-Gonzalez, D. Trifunović, A. Sahaboglu, K. Kranz, S. Michalakis and P. Farinelli, *et al.*, Identification of a common non-apoptotic cell death mechanism in hereditary retinal degeneration, *PLoS One*, 2014, **9**, e112142.
12. D. B. Farber, J. G. Flannery, A. C. Bird, T. Shuster and D. Bok, Histopathological and biochemical studies on donor eyes affected with retinitis pigmentosa, *Prog. Clin. Biol. Res.*, 1987, **247**, 53–67.
13. B. Kommonen, T. Kylma, R. J. Cohen, J. S. Penn, L. Paulin and M. Hurwitz, *et al.*, Elevation of cGMP with normal expression and activity of rod cGMP-PDE in photoreceptor degenerate labrador retrievers, *Ophthalmic Res.*, 1996, **28**, 19–28.
14. K. Sakamoto, M. McCluskey, T. G. Wensel, J. K. Naggert and P. M. Nishina, New mouse models for recessive retinitis pigmentosa caused by mutations in the Pde6a gene, *Hum. Mol. Genet.*, 2009, **18**, 178–192.
15. M. L. Suber, S. J. Pittler, N. Qin, G. C. Wright, V. Holcombe and R. H. Lee, *et al.*, Irish setter dogs affected with rod/cone dysplasia contain a nonsense mutation in the rod cGMP phosphodiesterase beta-subunit gene, *Proc. Natl. Acad. Sci. U. S. A.*, 1993, **90**, 3968–3972.
16. S. L. Semple-Rowland, N. R. Lee, J. P. Van Hooser, K. Palczewski and W. Baehr, A null mutation in the photoreceptor guanylate cyclase gene causes the retinal degeneration chicken phenotype, *Proc. Natl. Acad. Sci. U. S. A.*, 1998, **95**, 1271–1276.
17. D. Trifunović, K. Dengler, S. Michalakis, E. Zrenner, B. Wissinger and F. Paquet-Durand, cGMP-dependent cone photoreceptor degeneration in the cpfl1 mouse retina, *J. Comp. Neurol.*, 2010, **518**, 3604–3617.
18. J. Xu, L. Morris, A. Thapa, H. Ma, S. Michalakis and M. Biel, *et al.*, cGMP accumulation causes photoreceptor degeneration in CNG channel deficiency: evidence of cGMP cytotoxicity independently of enhanced CNG channel function, *J. Neurosci.*, 2013, **33**, 14939–14948.
19. V. Ramamurthy, G. A. Niemi, T. A. Reh and J. B. Hurley, Leber congenital amaurosis linked to AIPL1: a mouse model reveals destabilization of cGMP phosphodiesterase, *Proc. Natl. Acad. Sci. U. S. A.*, 2004, **101**, 13897–13902.
20. D. Krizaj and D. R. Copenhagen, Calcium regulation in photoreceptors, *Front. Biosci.*, 2002, 7, d2023–d2044.
21. R. N. Lolley, D. B. Farber, M. E. Rayborn and J. G. Hollyfield, Cyclic GMP accumulation causes degeneration of photoreceptor cells: simulation of an inherited disease, *Science*, 1977, **196**, 664–666.
22. C. Bowes, T. Li, M. Danciger, L. C. Baxter, M. L. Applebury and D. B. Farber, Retinal degeneration in the rd mouse is caused by a defect in the beta subunit of rod cGMP-phosphodiesterase, *Nature*, 1990, **347**, 677–680.
23. R. J. Davis, J. Tosi, K. M. Janisch, J. M. Kasanuki, N.-K. Wang and J. Kong, *et al.*, Functional rescue of degenerating photoreceptors in mice homozygous for a hypomorphic cGMP phosphodiesterase 6 b allele (Pde6bH620Q), *Invest. Ophthalmol. Visual Sci.*, 2008, **49**, 5067–5076.

24. L. He, A. T. Poblenz, C. J. Medrano and D. A. Fox, Lead and calcium produce rod photoreceptor cell apoptosis by opening the mitochondrial permeability transition pore, *J. Biol. Chem.*, 2000, **275**, 12175–12184.
25. D. Sanges, A. Comitato, R. Tammaro and V. Marigo, Apoptosis in retinal degeneration involves cross-talk between apoptosis-inducing factor (AIF) and caspase-12 and is blocked by calpain inhibitors, *Proc. Natl. Acad. Sci. U. S. A.*, 2006, **103**, 17366–17371.
26. A. Comitato, M. T. Di Salvo, G. Turchiano, M. Montanari, S. Sakami and K. Palczewski, *et al.*, Dominant and recessive mutations in rhodopsin activate different cell death pathways, *Hum. Mol. Genet.*, 2016, **25**, 2801–2812.
27. T. S. Kim, D. M. Reid and R. S. Molday, Structure-function relationships and localization of the Na/Ca-K exchanger in rod photoreceptors, *J. Biol. Chem.*, 1998, **273**, 16561–16567.
28. W. Haase, W. Friese, R. D. Gordon, H. Müller and N. J. Cook, Immunological characterization and localization of the Na+/Ca2(+)-exchanger in bovine retina, *J. Neurosci.*, 1990, **10**, 1486–1494.
29. T. P. Dryja, J. T. Finn, Y. W. Peng, T. L. McGee, E. L. Berson and K. W. Yau, Mutations in the gene encoding the alpha subunit of the rod cGMP-gated channel in autosomal recessive retinitis pigmentosa, *Proc. Natl. Acad. Sci. U. S. A.*, 1995, **92**, 10177–10181.
30. H. Kondo, M. Qin, A. Mizota, M. Kondo, H. Hayashi and K. Hayashi, *et al.*, A homozygosity-based search for mutations in patients with autosomal recessive retinitis pigmentosa, using microsatellite markers, *Invest. Ophthalmol. Visual Sci.*, 2004, **45**, 4433–4439.
31. F. Vinberg, T. Wang, R. S. Molday, J. Chen and V. J. Kefalov, A new mouse model for stationary night blindness with mutant Slc24a1 explains the pathophysiology of the associated human disease, *Hum. Mol. Genet.*, 2015, **24**, 5915–5929.
32. S. Notomi, T. Hisatomi, Y. Murakami, H. Terasaki, S. Sonoda and R. Asato, *et al.*, Dynamic increase in extracellular ATP accelerates photoreceptor cell apoptosis via ligation of P2RX7 in subretinal hemorrhage, *PLoS One*, 2013, **8**, e53338.
33. D. Križaj, Calcium stores in vertebrate photoreceptors, ed. M. S. Islam, in *Advances in Experimental Medicine and Biology*, Springer Netherlands: Dordrecht, 2012, pp. 873–889.
34. P. Barabas, C. C. Peck and D. Krizaj, Do calcium channel blockers rescue dying photoreceptors in the Pde6b rd1 mouse? *Adv. Exp. Med. Biol.*, 2010, **664**, 491–499.
35. J. Krebs, L. B. Agellon and M. Michalak, Ca^{2+} homeostasis and endoplasmic reticulum (ER) stress: an integrated view of calcium signaling, *Biochem. Biophys. Res. Commun.*, 2015, **460**, 114–121.
36. E. V. Olshevskaya, P. D. Calvert, M. L. Woodruff, I. V. Peshenko, A. B. Savchenko and C. L. Makino, *et al.*, The Y99C mutation in guanylyl cyclase-activating protein 1 increases intracellular Ca^{2+} and causes photoreceptor degeneration in transgenic mice, *J. Neurophysiol.*, 2004, **24**, 6078–6085.

37. V. Shinde, P. Kotla, C. Strang and M. Gorbatyuk, Unfolded protein response-induced dysregulation of calcium homeostasis promotes retinal degeneration in rat models of autosomal dominant retinitis pigmentosa, *Cell Death Dis.*, 2016, **7**, e2085.
38. M. M. Giarmarco, W. M. Cleghorn, S. R. Sloat, J. B. Hurley and S. E. Brockerhoff, Mitochondria maintain distinct Ca(2+) pools in cone photoreceptors, *J. Neurosci.*, 2017, **37**, 2061–2072.
39. F. Paquet-Durand, S. Beck, S. Michalakis, T. Goldmann, G. Huber and R. Mühlfriedel, *et al.*, A key role for cyclic nucleotide gated (CNG) channels in cGMP-related retinitis pigmentosa, *Hum. Mol. Genet.*, 2011, **20**, 941–947.
40. M. Kulkarni, D. Trifunović, T. Schubert, T. Euler and F. Paquet-Durand, Calcium dynamics change in degenerating cone photoreceptors, *Hum. Mol. Genet.*, 2016, **25**, 3729–3740.
41. E. Y. Ma, A. Lewis, P. Barabas, G. Stearns, S. Suzuki and D. Krizaj, *et al.*, Loss of Pde6 reduces cell body Ca(2+) transients within photoreceptors, *Cell Death Dis.*, 2013, **4**, e797.
42. D. S. Read, M. A. McCall and R. G. Gregg, Absence of voltage-dependent calcium channels delays photoreceptor degeneration in rd mice, *Exp. Eye Res.*, 2002, **75**, 415–420.
43. C. Schön, F. Paquet-Durand and S. Michalakis, Cav1.4 L-type calcium channels contribute to calpain activation in degenerating photoreceptors of rd1 mice, *PLoS One*, 2016, **11**, e0156974.
44. C. Schön, S. Asteriti, S. Koch, V. Sothilingam, M. Garcia Garrido and N. Tanimoto, *et al.*, Loss of HCN1 enhances disease progression in mouse models of CNG channel-linked retinitis pigmentosa and achromatopsia, *Hum. Mol. Genet.*, 2016, **25**, 1165–1175.
45. L. Corso, A. Cavallero, D. Baroni, P. Garbati, G. Prestipino and S. Bisti, *et al.*, Saffron reduces ATP-induced retinal cytotoxicity by targeting P2X7 receptors, *Purinergic Signal.*, 2016, **12**, 161–174.
46. A. Comitato, P. Subramanian, G. Turchiano, M. Montanari, S. P. Becerra and V. Marigo, Pigment epithelium-derived factor hinders photoreceptor cell death by reducing intracellular calcium in the degenerating retina, *Cell Death Dis.*, 2018, **9**, 560.
47. S. Marsili, S. Genini, R. Sudharsan, J. Gingrich, G. D. Aguirre and W. A. Beltran, Exclusion of the unfolded protein response in light-induced retinal degeneration in the canine T4R RHO model of autosomal dominant retinitis pigmentosa, *PLoS One*, 2015, **10**, e0115723.
48. S. Mizukoshi, M. Nakazawa, K. Sato, T. Ozaki, T. Metoki and S. Ishiguro, Activation of mitochondrial calpain and release of apoptosis-inducing factor from mitochondria in RCS rat retinal degeneration, *Exp. Eye Res.*, 2010, **91**, 353–361.
49. V. Gomez-Vicente, M. Donovan and T. G. Cotter, Multiple death pathways in retina-derived 661W cells following growth factor deprivation: crosstalk between caspases and calpains, *Cell Death Differ.*, 2005, **12**, 796–804.

50. M. Azuma and T. R. Shearer, The role of calcium-activated protease calpain in experimental retinal pathology, *Surv. Ophthalmol.*, 2008, **53**, 150–163.
51. A. Comitato, D. Sanges, A. Rossi, M. M. Humphries and V. Marigo, Activation of Bax in three models of retinitis pigmentosa, *Invest. Ophthalmol. Visual Sci.*, 2014, **55**, 3555–3562.
52. E. Nakajima, L. L. David, C. Bystrom, T. R. Shearer and M. Azuma, Calpain-specific proteolysis in primate retina: contribution of calpains in cell death, *Invest. Ophthalmol. Visual Sci.*, 2006, **47**, 5469–5475.
53. F. Paquet-Durand, S. Azadi, S. M. Hauck, M. Ueffing, T. van Veen and P. Ekström, Calpain is activated in degenerating photoreceptors in the rd1 mouse, *J. Neurochem.*, 2006, **96**, 802–814.
54. J. Kaur, S. Mencl, A. Sahaboglu, P. Farinelli, T. van Veen and E. Zrenner, *et al.*, Calpain and PARP activation during photoreceptor cell death in P23H and S334ter rhodopsin mutant rats, *PLoS One*, 2011, **6**, e22181.
55. N. Sanvicens, V. Gomez-Vicente, I. Masip, A. Messeguer and T. G. Cotter, Oxidative stress-induced apoptosis in retinal photoreceptor cells is mediated by calpains and caspases and blocked by the oxygen radical scavenger CR-6, *J. Biol. Chem.*, 2004, **279**, 39268–39278.
56. A. K. Sharma and B. Rohrer, Sustained elevation of intracellular cGMP causes oxidative stress triggering calpain-mediated apoptosis in photoreceptor degeneration, *Curr. Eye Res.*, 2007, **32**, 259–269.
57. D. E. Goll, V. F. Thompson, H. Li, W. Wei and J. Cong, The calpain system, *Physiol. Rev.*, 2003, **1990**, 731–801.
58. E. C. Cheung, L. Melanson-Drapeau, S. P. Cregan, J. L. Vanderluit, K. L. Ferguson and W. C. McIntosh, *et al.*, Apoptosis-inducing factor is a key factor in neuronal cell death propagated by BAX-dependent and BAX in dependent mechanisms, *J. Neurosci.*, 2005, **25**, 1324–1334.
59. F. Paquet-Durand, D. Sanges, J. McCall, J. Silva, T. van Veen and V. Marigo, *et al.*, Photoreceptor rescue and toxicity induced by different calpain inhibitors, *J. Neurochem.*, 2010, **115**, 930–940.
60. A. Wendt, V. F. Thompson and D. E. Goll, Interaction of calpastatin with calpain: a review, *Biol. Chem.*, 2004, **385**, 465–472.
61. M. I. Pörn-Ares, A. Samali and S. Orrenius, Cleavage of the calpain inhibitor, calpastatin, during apoptosis, *Cell Death Differ.*, 1998, **5**, 1028–1033.
62. S. Gil-Parrado, I. Assfalg-Machleidt, F. Fiorino, D. Deluca, D. Pfeiler and N. Schaschke, *et al.*, Calpastatin exon 1B-derived peptide, a selective inhibitor of calpain: enhancing cell permeability by conjugation with penetratin, *Biol. Chem.*, 2003, **384**, 395–402.
63. F. Paquet-Durand, S. M. Hauck, T. van Veen, M. Ueffing and P. Ekström, PKG activity causes photoreceptor cell death in two retinitis pigmentosa models, *J. Neurochem.*, 2009, **108**, 796–810.
64. J. Sancho-Pelluz, M. V. Alavi, A. Sahaboglu, S. Kustermann, P. Farinelli and S. Azadi, *et al.*, Excessive HDAC activation is critical for neurodegeneration in the rd1 mouse, *Cell Death Dis.*, 2010, **1**, e24.

65. F. Paquet-Durand, J. Silva, T. Talukdar, L. E. Johnson, S. Azadi and T. van Veen, *et al.*, Excessive activation of poly (ADP-ribose) polymerase contributes to inherited photoreceptor degeneration in the retinal degeneration 1 mouse, *J. Neurosci.*, 2007, **27**, 10311–10319.
66. T. Ozaki, S. Ishiguro, S. Hirano, A. Baba, T. Yamashita and H. Tomita, *et al.*, Inhibitory peptide of mitochondrial μ-calpain protects against photoreceptor degeneration in rhodopsin transgenic S334ter and P23H rats, *PLoS One*, 2013, **8**, e71650.

CHAPTER 4

CNS Targets for the Treatment of Retinal Dystrophies: A Win–Win Strategy

ENRIQUE J. DE LA ROSA* AND CATALINA HERNÁNDEZ-SÁNCHEZ

3D Lab, Dept. of Molecular Biomedicine, Centro de Investigaciones Biológicas CSIC, C/ Ramiro de Maeztu 9, 28040, Madrid, Spain
*E-mail: ejdelarosa@cib.csic.es

4.1 Introduction

The embryonic retina is a classical model for the study of central nervous system (CNS) development. Both the neuroretina and the retinal pigment epithelium are derived from the anterior medial neural plate of the embryo.[1] During embryonic development, the retinal field develops in parallel with the rest of the CNS. This process involves a series of complex morphogenetic movements, during which the developing retina field passes through the stages of the neural tube, cephalic vesicle, optic vesicle, and optic cup. The isolation of an already distinctive retina from the rest of the brain first becomes evident in the optic cup stage (Figure 4.1). Subsequently, the retinal cytoarchitecture follows molecular and cellular patterns that closely resemble those seen in other parts of the CNS.[2] Analyses of these developmental and anatomical similarities between the retina and the CNS,

Drug Discovery Series No. 66
Therapies for Retinal Degeneration: Targeting Common Processes
Edited by Enrique J. de la Rosa and Thomas G. Cotter

Published by the Royal Society of Chemistry, www.rsc.org

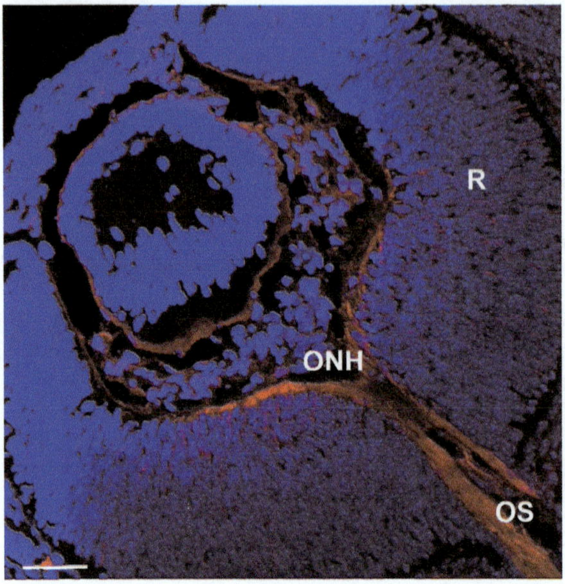

Figure 4.1 The developing retina. The relative isolation of the retina from the rest of the brain is first evident in an optic cup section from a mouse on embryonic day 12.5. The retina remains connected to the brain *via* the optic stalk and optic nerve. Early fibers of the optic nerve are immunostained for the detection of βIII-tubulin (red). Nuclei are labeled with DAPI (blue). ONH: optic nerve head; OS: optic stalk; R: neuroretina. Scale bar = 100 μm.

aided by the retina's accessibility to observation and experimental manipulation, have revealed numerous physiological and functional parallels between the retina and the brain.[3]

Surprisingly, the pathophysiological similarities between the retina and the rest of the CNS have been somewhat overlooked until recently, and drug development strategies for neurodegenerative conditions have primarily focused on those affecting the brain and spinal cord, and less on prevalent retinal diseases such as glaucoma and diabetic retinopathy. Recent findings have underscored the need for a new approach to the treatment of these two retinal pathologies. Because glaucoma was long considered a consequence of elevated intraocular pressure (IOP), therapeutic approaches traditionally focused on the management of IOP. However, the fact that the vision impairment persists even when IOP is controlled underscores the need for alternative neuroprotective therapies.[4] In the case of diabetic retinopathy, studies have shown that microvascular lesions, previously considered the primary lesion in this disease, are in fact preceded by morphological and functional changes.[5] Accordingly, the search for pharmacological therapies for these two diseases has shifted towards targets implicated in neurodegeneration.

Research attention has only recently turned to the presence of retinal alterations in patients with neurodegenerative disorders affecting the brain, such as Alzheimer's disease and Parkinson's disease, among others. Retinal screening using non-invasive imaging, electrophysiological, and behavioral techniques could potentially detect these putative early markers of disease in animal models and in patients with CNS neurodegeneration.[3,6] Retinal screening techniques provide a quantitative, objective output and could prove invaluable for the recruitment and follow-up of patients involved in clinical trials of drug candidates for the treatment of CNS neurodegenerative conditions, as well as providing a much less costly alternative to direct brain examination for population-wide screening.[7–10]

In this chapter, we review the evidence of pathophysiological parallels between neurodegenerative conditions of the retina and the brain. We also discuss the experimental approaches we use to assess the potential of glycogen synthase kinase (GSK)-3 inhibitors and the insulin precursor proinsulin as drug candidates for the treatment of these pathologies.

4.2 Retinal Alterations in Neurodegenerative Conditions of the Brain and Spinal Cord

We have created a list of several neurodegenerative diseases (Table 4.1) that collectively affect a considerable proportion of the population, and in which retinal alterations have been demonstrated. Preliminary studies suggest that several other conditions not included here, including Huntington's disease,

Table 4.1 Retinal alterations observed in neurodegenerative conditions of the brain and spinal cord.

Neurodegenerative conditions of the brain and spinal cord	Retinal alterations	References
Alzheimer's disease	Aβ deposition Imbalanced phosphorylation of Tau Decreased retinal perfusion Microglial activation and reactive gliosis Complement activation Neuroinflammation Thinning of the optic fiber layer and macula Reduced electroretinographic response in retinal ganglion cells	8 and 20–25
Parkinson's disease	α-synuclein deposition Dopamine deficiency and deterioration of the perifoveal dopaminergic plexus Microglial activation and reactive gliosis Thinning of the optic fiber layer Altered electroretinographic response	8,11 and 25
Multiple sclerosis	Optic neuritis and neuroinflammation Blood-retinal barrier dysfunction Retinal ganglion cell degeneration Thinning of the optic fiber layer and optic nerve	10

stroke, and psychiatric disorders, have manifestations in the retina.[3,11–14] The fact that these alterations are common to so many neurodegenerative conditions underscores the need for further studies to better understand the impact of CNS neurodegeneration on the retina.

Alzheimer's disease is a devastating condition characterized primarily by memory loss and a profound cognitive deficit. As demonstrated in both animal models and Alzheimer's disease patients, a key hallmark of this disease is neuronal loss accompanied by the deposition of amyloid beta (Aβ, which forms senile plaques) and hyperphosphorylated Tau (pTau, which forms neurofibrillary tangles) in the cerebral cortex. For many years, diagnosis of Alzheimer's disease has been primarily based on subjective evaluation of the patient's behavior. The search for objective and quantitative methods for diagnosis of Alzheimer's disease and the evaluation of disease progression has led to the characterization of putative biomarkers, such as Aβ and hyperphosphorylated Tau in cerebrospinal fluid, as well as the development of new brain imaging techniques. However, the effects of Alzheimer's disease are not confined to the brain, and also affect other CNS regions, including the retina (Table 4.1). Alterations in visual perception in Alzheimer's disease were initially attributed to defects in visual image processing by the brain, but studies subsequently revealed structural alterations in the retinas of Alzheimer's disease patients[15–19] and in animal models (Figure 4.2). The retinas of Alzheimer's disease patients not only exhibit the same molecular hallmarks seen in the brain (deposition of Aβ and pTau),[20–24] but also display neurodegenerative alterations in retinal cytoarchitecture, including loss of ganglion cells and thinning of the inner plexiform and retinal nerve fiber layers[8,25] (Table 4.1). Based on these findings, a growing number of studies have suggested that retinal examination could constitute a useful tool for the diagnosis and follow-up of Alzheimer's disease.[8]

Parkinson's disease is widely considered a motor dysfunction disease caused by degeneration of dopamine-producing neurons in the substantia nigra. However, in addition to tremor and slowness of movement, the most evident clinical signs, Parkinson's disease patients also exhibit biochemical, neurochemical, cellular, and functional changes in the retina (Table 4.1). These include deposition of α-synuclein, a Parkinson's disease-specific hallmark,[11] and changes in retinal structure, including a decrease in the thickness of the retinal nerve fiber layer and deterioration of the macula and fovea.[8,25]

In contrast to the primary neuronal damage and loss associated with Alzheimer's disease and Parkinson's disease, oligodendrocytes are the primary targets of the abnormal immune response that underlies multiple sclerosis (MS). The inflammation associated with MS leads to progressive demyelination of the central nerves.[3,10] Ocular inflammation and vision impairment are other recognized early signs of MS[3]. Visual impairment results from demyelination of the optic nerve, which is part of the CNS (Table 4.1), and consequent thinning of the retinal nerve fiber layer.

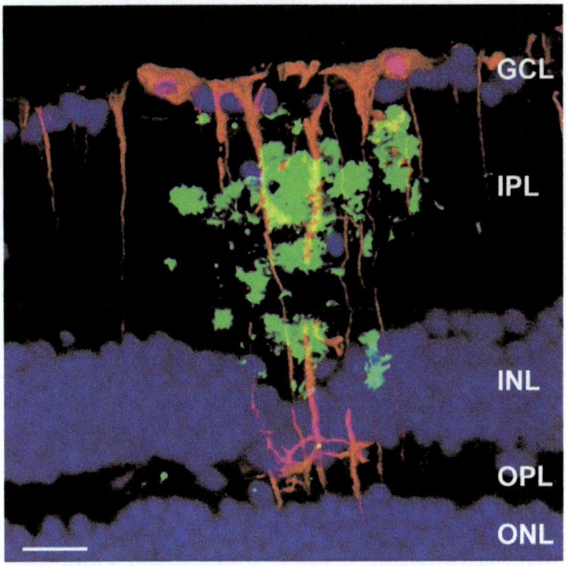

Figure 4.2 Retina from a mouse model of Alzheimer's disease. Immunostaining for abnormally processed amyloid (green) reveals the presence of Aβ plaques in a retinal section from a mouse carrying human familial Alzheimer's disease mutations in presenilin and amyloid precursor protein. The plaques are surrounded by reactive Müller glial cells and astrocytes, which are visualized by immunostaining for GFAP (red). Nuclei are labeled with DAPI (blue). GCL: ganglion cell layer; INL: inner nuclear layer; IPL: inner plexiform layer; ONL: outer nuclear layer; OPL: outer plexiform layer. Scale bar = 25 μm.

Despite the distinct etiologies of the neurodegenerative diseases described, retinal involvement appears to be common to all. As such, retinal analysis could constitute a useful strategy for the screening, diagnosis, and follow-up of a variety of neurodegenerative conditions, and for evaluation of potential therapies in clinical trials.

4.3 Neurodegenerative Conditions of the Brain and Retina Share Common Pathological Mechanisms

In contrast to the well documented retinal manifestations of CNS neurodegeneration (outlined in Section 4.2), retinal diseases do not appear to affect the brain or other regions of the CNS to the same extent. However, in line with the common origin and physiological similarities of both structures, neurodegenerative diseases of the brain and retina share common pathophysiological mechanisms, particularly at the molecular and cellular levels (Table 4.2). To illustrate these shared characteristics, we have

Table 4.2 Pathological signs common to neurodegenerative diseases of the retina and brain.

	Alzheimer's disease	Parkinson's disease
Glaucoma	Aβ deposition[6,28] pTau deposition[27] Neuroinflammation and microglial activation[8,25] Axonal atrophy and deficits in axonal transport[8,25] Neuronal cell death[8,25]	Synuclein deposition[32] Neuroinflammation and microglial activation[8,25] Axonal atrophy and deficits in axonal transport[8,25] Neuronal cell death[8,25]
Age-related macular degeneration	Aβ deposition[29–31] Complement activation[6] Neuroinflammation and microglial activation[6] Neuronal cell death[6]	Neuroinflammation and microglial activation[6] Neuronal cell death[6]

selected two retinal dystrophies, glaucoma and age-related macular degeneration (AMD), which differ etiologically and in terms of the retinal cells affected.

As mentioned earlier, elevated IOP is common feature and the greatest risk factor for glaucoma. However, a link between glaucoma and Alzheimer's disease has been proposed,[26] and the presence of protein deposits characteristic of Alzheimer's disease (Aβ and pTau) has been described in the retinas of glaucoma patients.[6,27,28] Moreover, in both diseases comparable Aβ and pTau profiles are observed in the compartments closest to the primary affected tissue. Thus, compared with unaffected individuals, lower levels of Aβ and higher levels of pTau are found both in the cerebrospinal fluid of Alzheimer's disease patients and in the vitreous humor of glaucoma patients.[6] Similarly, the aggregation of pathological proteins is observed in both Alzheimer's disease and AMD. The primary risk factor for AMD is aging, followed by other behavioral and genetic risk factors. Genetic risk factors include the presence of the *APOE* allele, which is associated with an increased risk of both AMD and Alzheimer's disease.[7] Aβ aggregates have also been detected inside drusen deposits in the retinas of AMD patients.[29–31] Protein aggregates are also an important feature of Parkinson's disease, and aberrant deposits of synuclein are found in both Parkinson's disease and glaucoma patients.[11,32]

Protein aggregation, a key pathological feature of Alzheimer's disease, is a likely trigger of neuroinflammation,[33,34] which is perhaps the most common trait of degenerative pathologies of the CNS (Table 4.2). Neuroinflammation has been well documented in CNS neurodegeneration, including retinal dystrophies[35,36] (Table 4.2), although whether this type of immune response contributes to the initiation and progression of neurodegeneration remains unclear.[33,34,37] Studies of the retina and the brain have shown that the inflammatory response to local neurodegeneration involves common cellular and molecular players. This response can be endogenous or peripheral in origin. Here we focus on the endogenous inflammatory response, which is common to neurodegenerative pathologies of the retina and the brain. Macroglia and microglia/perivascular monocytes are the two main endogenous cell types that drive CNS inflammation. In the brain, the reactive macroglial cells are astrocytes, while in the retina this role is shared by Müller glial cells and astrocytes.[35] Homeostatic macroglial cells play essential roles in the development, normal function, and wellbeing of neural tissue. Astrocytes and Müller cells provide trophic support for neurons, regulate the formation and activity of synapses, and are responsible for the extracellular clearance of ions and neurotransmitters.[38,39] Macroglia are also responsible for the maintenance of the blood–brain and blood–retinal barriers. Many of these "housekeeping" functions are also carried out by microglial cell populations consisting of resident macrophages of the CNS.[40] Retinal and brain glial cell populations react to neural tissue damage or dysfunction by altering their morphology and molecular signature. In these conditions, reactive glial cells produce proinflammatory cytokines, chemokines, and complement proteins that can either exacerbate or attenuate the pathology depending on the intensity and duration of the response.[34,35,37] However, interventions that impinge upon the inflammatory response may result in contrasting outcomes depending on the intervention period, most likely due to the complex interrelated and interdependent pathways that mediate inflammation.[41] Moreover, the efficacy of anti-inflammatory therapies depends on the degree to which neurodegeneration has progressed, since different mechanisms predominate depending on the disease stage.[37] Further research is therefore required to elucidate the specific roles of the molecules and cells involved and to identify the most suitable therapeutic targets and windows.

The next stage of the pathological process is characterized by the damage of neurons (or other cell types) at the axonal and/or synaptic levels, potentially resulting in neuronal death. This is a common feature of neurodegeneration of both the brain and retina, and is most likely the result of protein aggregation and neuroinflammation. In summary, despite the distinct etiological bases and functional alterations of neurodegenerative conditions of the brain and retina, these diseases are underpinned by common pathological mechanisms. Further studies of these common mechanisms are essential to better understand the underlying pathology and to aid the development of effective therapies.

4.4 GSK-3 as an Example of a Common Therapeutic Target for Neurodegenerative Conditions of the Brain and Retina

GSK-3 is a highly evolutionary conserved intracellular serine–threonine kinase that is expressed by almost all cells in the body. Initially identified as a glycogen synthesis enzyme (for which it is named), GSK-3 is currently considered a multitasking enzyme, for which more than 100 confirmed and 500 predicted substrates have been identified. Owing to its near ubiquitous expression and broad array of substrates, GSK-3 is well positioned as a key regulator of multiple and diverse biological processes. Perhaps unsurprisingly, aberrant GSK-3 activity has been implicated in several human diseases, including neurodegenerative and psychiatric disorders.[42,43] Despite their etiological differences, inflammation is a common feature of the onset and/or progression of these pathologies. GSK-3 plays a pivotal role in regulating the balance between pro-inflammatory and anti-inflammatory cellular responses. Because GSK-3 inhibitors shift the cellular response from pro-inflammatory to anti-inflammatory, GSK-3 is considered a potential therapeutic target for diseases with an inflammatory component.[44,45] Lithium, a weak inhibitor of GSK-3, was the first GSK-3 inhibitor identified, and has been used for the treatment of mood disorders.[46,47] Since then, researchers have sought to identify and design selective GSK-3 inhibitors, given their therapeutic potential in a variety of neurodegenerative diseases. Tideglusib, a highly specific GSK-3 inhibitor, has shown a broad therapeutic safety window for the treatment of both Alzheimer's disease and progressive supranuclear palsy in clinical trials,[48] and its efficacy for the treatment of autism spectrum disorders is currently being tested (https://clinicaltrials.gov/ct2/show/NCT02586935?term=tideglusib&rank=2).

Recent findings suggest that the therapeutic potential of GSK-3 inhibitors may extend to retinal dystrophies. Because neuroinflammation is common to neurodegenerative pathologies of both the brain and retina, attenuation of inflammatory signaling may be therapeutically beneficial in both cases. Recent studies by our group, in collaboration with Dr Ana Martinez's group (https://www.cib.csic.es/research/structural-and-chemical-biology/translational-medicinal-and-biological-chemistry), have shown that GSK-3 inhibition has neuroprotective effects in two different *ex vivo* models of retinal pathologies (glaucoma and retinitis pigmentosa). We found that treatment with structurally diverse GSK-3 inhibitors reduced ganglion cell death caused by *N*-methyl-D-aspartate excitotoxicity in retinal explants, a model of glaucoma-related damage of retinal ganglion cells (Figure 4.3).[49] Moreover, we showed that photoreceptor cells were preserved in retinal explants from the rd10 mouse model of retinitis pigmentosa which were treated with different GSK-3 inhibitors (Figure 4.3). Remarkably, these *in vitro* findings have been corroborated *in vivo* in rd10 mice, in which loss of visual function is attenuated by inhibiting GSK-3.[50]

Taken together with other published data, these findings support the clinical therapeutic potential of GSK-3 inhibitors for the treatment of retinal dystrophies. Intraperitoneal lithium has been used to reduce intraocular pressure in a rat model of glaucoma,[51] and the histone deacetylase valproic acid, a putative indirect inhibitor of GSK-3,[52] provides short-term visual function benefits in retinitis pigmentosa patients.[53,54] Repositioning

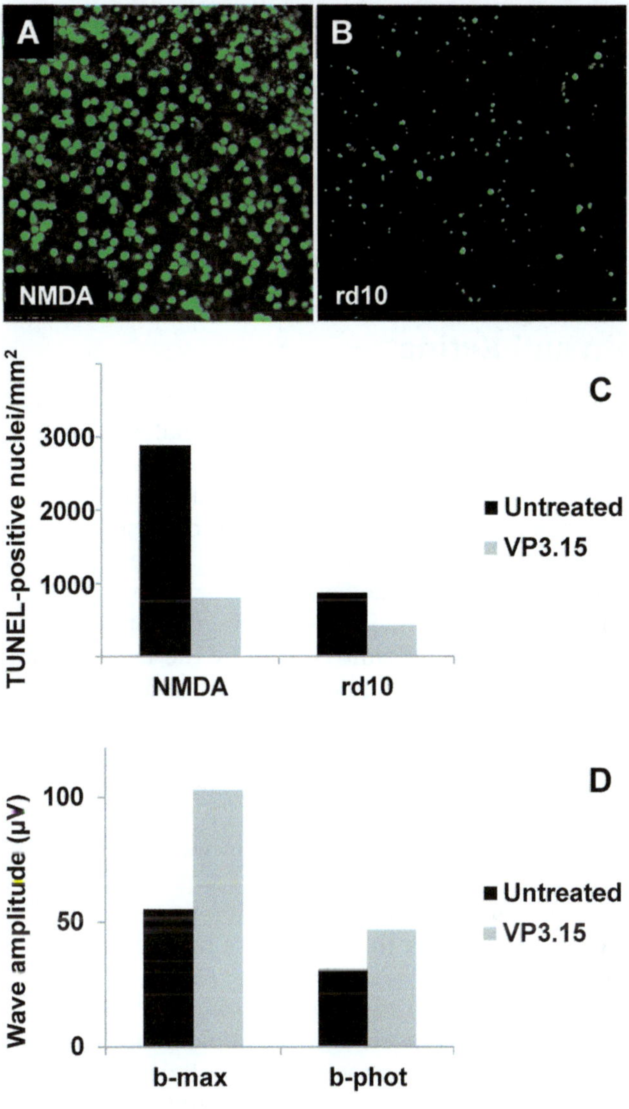

Figure 4.3 Effects of *in vitro* and *in vivo* glycogen synthase kinase (GSK)-3 inhibition on retinal neurodegeneration. (A) Retinal explant from a wild-type mouse treated with *N*-methyl-D-aspartate (NMDA) (glaucoma model). (B) Retinal explant from a rd10 mouse (retinitis pigmentosa model). TUNEL staining reveals ganglion and photoreceptor cell death (green nuclei in A and B, respectively). (C) Treatment of retinal explants for 24 h with the GSK-3 inhibitor VP3.15 attenuates cell death in both models. Experimental data from Marchena *et al.*[49] (D) *In vivo* treatment with VP3.15 partially preserves visual function, as determined by electroretinography (amplitudes of two relevant waves at P32 are shown). Experimental data from Sánchez-Cruz *et al.*[50]

of drugs approved for the treatment of CNS neurodegeneration could provide therapeutic alternatives for the neurodegenerative component of retinal dystrophies. Moreover, non-invasive, quantitative analysis of retinal structure and function could facilitate the development of novel treatments with no clearly sustained effects in clinical trials directed to brain pathologies.

4.5 Proinsulin, a Candidate Drug for Neurodegenerative Conditions of the Brain and Retina

The CNS is a well-recognized site of action of insulin and insulin-like growth factors-I and -II.[55] The insulin gene is transcriptionally active in tissues other than the pancreas, although at a much lower level, and local insulin gene expression has been demonstrated in the brain and retina.[56–58] Furthermore, potential autocrine/paracrine effects of insulin/proinsulin in the CNS have been suspected for some time, based on the local presence of moderate insulin/proinsulin levels that are independent of peripheral levels.[56,59] Specific functions of insulin in the brain include the regulation of food intake, body weight, reproduction, and glycemic control by the hypothalamus, and the promotion of hippocampus-dependent learning and memory.[60,61] The neuroprotective effects of insulin signaling are essential for healthy aging of the brain,[62] and defective insulin signaling has been demonstrated in the Alzheimer's disease brain.[63] These observations suggest that insulin may be a candidate therapeutic agent for Alzheimer's disease.[64] Proinsulin was long considered a low-activity precursor of insulin owing to its poor metabolic potential (5–10% that of insulin). However, recent studies in mammals and previous studies in chickens have revealed non-metabolic actions of proinsulin, and identified a role for this factor as a bioactive signaling molecule during development.[56,65,66]

In addition to its developmental role, proinsulin exhibits neuroprotective effects in mammalian models of retinitis pigmentosa, a form of retinal neurodegeneration that causes photoreceptor cell death and vision loss.[67–69] The therapeutic potential of human proinsulin (hPI) has been demonstrated in studies in which hPI levels in rodents were systemically increased. In the rd10 mouse, low-level constitutive transgenic expression of hPI in muscle delayed photoreceptor cell death and attenuated vision loss (Figure 4.4).[67] The potential neuroprotective effect of proinsulin has also been reported in two more clinically valid settings. In the rd10 mouse model, local administration of hPI microbeads attenuated photoreceptor cell death,[69] while in the P23H rat model of autosomal dominant retinitis pigmentosa, hPI administered by intramuscular injection of adeno-associated viral vector preserved the structure and function of photoreceptors and their contacts with postsynaptic neurons.[68] Given the aforementioned similarities between

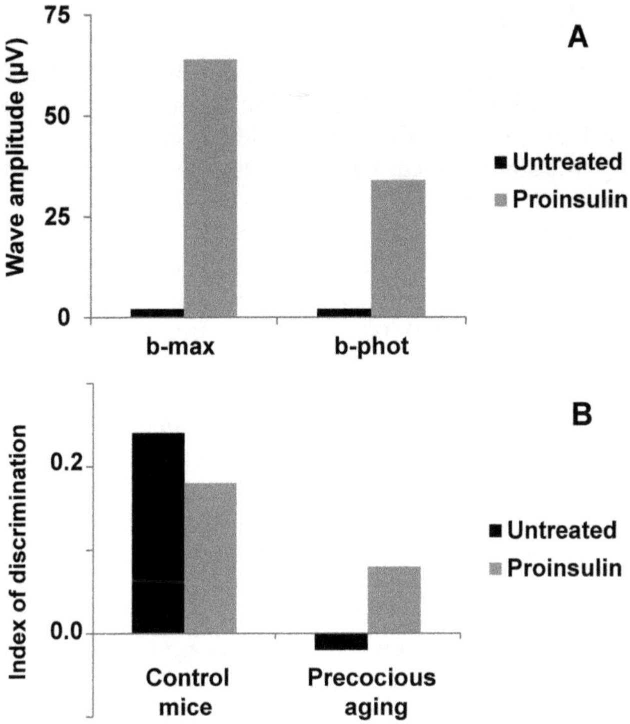

Figure 4.4 Neuroprotective effect of proinsulin in neurodegenerative conditions of the retina and brain. (A) Transgenic proinsulin expression in the rd10 mouse partially attenuates vision loss, as determined by electroretinography (amplitudes of two relevant waves at P35 are shown). Experimental data from Corrochano *et al.*[67] (B) AAV-mediated proinsulin expression partially preserves cognitive function, evaluated using the discrimination of the novel object in the recognition test (0 = no discrimination), in a mouse model of precocious aging. Experimental data from Corpas *et al.*[70]

neurodegenerative conditions of the retina and brain, we extended our studies to a mouse model of Alzheimer's disease-like cognitive impairment. In line its attenuation of vision loss in retinitis pigmentosa models, systemic proinsulin treatment attenuated cognitive deficits in the SamP8 mouse model of premature senescence (Figure 4.4), an effect that correlated with decreased brain inflammaging.[70]

Although the physiological relevance of proinsulin in adult mammals is yet to be determined, it constitutes a promising neuroprotective agent that warrants testing in diverse neurodegenerative conditions. Importantly, systemic proinsulin administration does not alter body weight or blood glucose levels,[67,70] two of the main adverse effects associated with systemic insulin treatment.

4.6 Conclusions

The molecular and cellular mechanisms underpinning degenerative conditions of the retina have received significantly less research attention than those of the brain, despite the striking developmental, anatomical, functional, and pathophysiological parallels described in this chapter between the retina and other parts of the CNS. Further investigation of the mechanisms underlying retinal diseases will generate knowledge that could facilitate the development of new therapies and further our understanding of the shared pathological bases of brain neurodegeneration and retinal dystrophies. Unfortunately, the development and testing of a plethora of putative pharmacological treatments for Alzheimer's disease and other neurodegenerative conditions of the brain has not yet rendered the expected results. These experimental treatments could be now applied to the treatment of retinal degeneration and, thus, the accessibility of the retina and the availability of non-invasive, quantitative techniques for retinal analysis may provide an alternative, excellent starting point for the development of putative treatment for neurodegenerative disorders that affect the CNS.

In summary, a unified view of CNS neurodegeneration that encompasses both the brain and the retina could further our understanding of neurodegeneration in general, and facilitate the development of new therapies for affected patients.

Acknowledgements

We thank Drs Noemí Álvarez-Lindo and Alberto M. Hernández-Pinto for providing Figures 4.1 and 4.2, respectively, and Dr Owen Howard for critical reading of the manuscript. Research in our lab is supported by grants CDS2010-00045 and SAF2016-75681-R from the Spanish Ministry of Economy, Industry and Competitiveness.

References

1. R. Sinn and J. Wittbrodt, *Mech. Dev.*, 2013, **130**, 347–358.
2. W. Heavner and L. Pevny, *Cold Spring Harbor Perspect. Biol.*, 2012, **4**, a008391.
3. A. London, I. Benhar and M. Schwartz, *Nat. Rev. Neurol.*, 2013, **9**, 44–53.
4. M. Almasieh and L. A. Levin, *Annu. Rev. Vis. Sci.*, 2017, **3**, 91–120.
5. A. J. Barber and B. Baccouche, *Vis. Res.*, 2017, **139**, 82–92.
6. V. Jindal, *Mol. Neurobiol.*, 2015, **51**, 885–892.
7. L. De Groef and M. F. Cordeiro, *J. Ocul. Pharmacol. Ther.*, 2018, **34**, 129–133.
8. J. Doustar, T. Torbati, K. L. Black, Y. Koronyo and M. Koronyo-Hamaoui, *Front. Neurol.*, 2017, **8**, 701.
9. E. Gordon-Lipkin and P. A. Calabresi, *J. Neuroimmunol.*, 2017, **304**, 93–96.
10. T. J. MacGillivray, E. Trucco, J. R. Cameron, B. Dhillon, J. G. Houston and E. J. van Beek, *Br. J. Radiol.*, 2014, **87**, 20130832.

11. I. Bodis-Wollner, P. B. Kozlowski, S. Glazman and S. Miri, *Ann. Neurol.*, 2014, **75**, 964–966.
12. C. Y. Cheung, M. K. Ikram, C. Chen and T. Y. Wong, *Prog. Retinal Eye Res.*, 2017, **57**, 89–107.
13. J. R. Pearl, L. M. Heath, D. E. Bergey, J. P. Kelly, C. Smith, M. Y. Laurino, A. Weiss, N. D. Price, A. LaSpada, T. D. Bird and S. Jayadev, *J. Huntington's Dis.*, 2017, **6**, 237–247.
14. D. Wang, Y. Li, C. Wang, L. Xu, Q. S. You, Y. X. Wang, L. Zhao, W. B. Wei, X. Zhao and J. B. Jonas, *Stroke*, 2014, **45**, 1651–1656.
15. J. C. Blanks, D. R. Hinton, A. A. Sadun and C. A. Miller, *Brain Res.*, 1989, **501**, 364–372.
16. J. C. Blanks, S. Y. Schmidt, Y. Torigoe, K. V. Porrello, D. R. Hinton and R. H. Blanks, *Neurobiol. Aging*, 1996, **17**, 385–395.
17. J. C. Blanks, Y. Torigoe, D. R. Hinton and R. H. Blanks, *Neurobiol. Aging*, 1996, **17**, 377–384.
18. D. R. Hinton, A. A. Sadun, J. C. Blanks and C. A. Miller, *N. Engl. J. Med.*, 1986, **315**, 485–487.
19. A. A. Sadun and C. J. Bassi, *Ophthalmology*, 1990, **97**, 9–17.
20. P. N. Alexandrov, A. Pogue, S. Bhattacharjee and W. J. Lukiw, *NeuroReport*, 2011, **22**, 623–627.
21. Y. Koronyo, D. Biggs, E. Barron, D. S. Boyer, J. A. Pearlman, W. J. Au, S. J. Kile, A. Blanco, D. T. Fuchs, A. Ashfaq, S. Frautschy, G. M. Cole, C. A. Miller, D. R. Hinton, S. R. Verdooner, K. L. Black and M. Koronyo-Hamaoui, *JCI Insight*, 2017, **2**, e93621.
22. M. Koronyo-Hamaoui, Y. Koronyo, A. V. Ljubimov, C. A. Miller, M. K. Ko, K. L. Black, M. Schwartz and D. L. Farkas, *NeuroImage*, 2011, **54**(Suppl. 1), S204–S217.
23. C. La Morgia, F. N. Ross-Cisneros, Y. Koronyo, J. Hannibal, R. Gallassi, G. Cantalupo, L. Sambati, B. X. Pan, K. R. Tozer, P. Barboni, F. Provini, P. Avanzini, M. Carbonelli, A. Pelosi, H. Chui, R. Liguori, A. Baruzzi, M. Koronyo-Hamaoui, A. A. Sadun and V. Carelli, *Ann. Neurol.*, 2016, **79**, 90–109.
24. C. Schon, N. A. Hoffmann, S. M. Ochs, S. Burgold, S. Filser, S. Steinbach, M. W. Seeliger, T. Arzberger, M. Goedert, H. A. Kretzschmar, B. Schmidt and J. Herms, *PLoS One*, 2012, **7**, e53547.
25. A. I. Ramirez, R. de Hoz, E. Salobrar-Garcia, J. J. Salazar, B. Rojas, D. Ajoy, I. Lopez-Cuenca, P. Rojas, A. Trivino and J. M. Ramirez, *Front. Aging Neurosci.*, 2017, **9**, 214.
26. Z. Yan, H. Liao, H. Chen, S. Deng, Y. Jia, C. Deng, J. Lin, J. Ge and Y. Zhuo, *Invest. Ophthalmol. Visual Sci.*, 2017, **58**, 5434–5443.
27. N. Gupta, J. Fong, L. C. Ang and Y. H. Yucel, *Can. J. Ophthalmol.*, 2008, **43**, 53–60.
28. S. J. McKinnon, D. M. Lehman, L. A. Kerrigan-Baumrind, C. A. Merges, M. E. Pease, D. F. Kerrigan, N. L. Ransom, N. G. Tahzib, H. A. Reitsamer, H. Levkovitch-Verbin, H. A. Quigley and D. J. Zack, *Invest. Ophthalmol. Visual Sci.*, 2002, **43**, 1077–1087.

29. D. H. Anderson, K. C. Talaga, A. J. Rivest, E. Barron, G. S. Hageman and L. V. Johnson, *Exp. Eye Res.*, 2004, **78**, 243–256.
30. T. Dentchev, A. H. Milam, V. M. Lee, J. Q. Trojanowski and J. L. Dunaief, *Mol. Vision*, 2003, **9**, 184–190.
31. L. V. Johnson, W. P. Leitner, A. J. Rivest, M. K. Staples, M. J. Radeke and D. H. Anderson, *Proc. Natl. Acad. Sci. U. S. A.*, 2002, **99**, 11830–11835.
32. I. Surgucheva, B. McMahan, F. Ahmed, S. Tomarev, M. B. Wax and A. Surguchov, *J. Neurosci. Res.*, 2002, **68**, 97–106.
33. A. Masuzzo, V. Dinet, C. Cavanagh, F. Mascarelli and S. Krantic, *Front. Neurol.*, 2016, **7**, 127.
34. H. Sarlus and M. T. Heneka, *J. Clin. Invest.*, 2017, **127**, 3240–3249.
35. N. Cuenca, L. Fernandez-Sanchez, L. Campello, V. Maneu, P. De la Villa, P. Lax and I. Pinilla, *Prog. Retinal Eye Res.*, 2014, **43**, 17–75.
36. R. M. Ransohoff, *Science*, 2016, **353**, 777–783.
37. T. Chitnis and H. L. Weiner, *J. Clin. Invest.*, 2017, **127**, 3577–3587.
38. S. Liddelow and B. Barres, *Cell*, 2015, **162**, 1170–1170.e1.
39. E. Vecino, F. D. Rodriguez, N. Ruzafa, X. Pereiro and S. C. Sharma, *Prog. Retinal Eye Res.*, 2016, **51**, 1–40.
40. A. Crotti and R. M. Ransohoff, *Immunity*, 2016, **44**, 505–515.
41. A. F. Keller, M. Gravel and J. Kriz, *Exp. Neurol.*, 2011, **228**, 69–79.
42. E. Beurel, S. F. Grieco and R. S. Jope, *Pharmacol. Ther.*, 2015, **148**, 114–131.
43. R. S. Jope, Y. Cheng, J. A. Lowell, R. J. Worthen, Y. H. Sitbon and E. Beurel, *Trends Biochem. Sci.*, 2017, **42**, 180–192.
44. E. Beurel, S. M. Michalek and R. S. Jope, *Trends Immunol.*, 2010, **31**, 24–31.
45. M. Martin, K. Rehani, R. S. Jope and S. M. Michalek, *Nat. Immunol.*, 2005, **6**, 777–784.
46. P. S. Klein and D. A. Melton, *Proc. Natl. Acad. Sci. U. S. A.*, 1996, **93**, 8455–8459.
47. V. Stambolic, L. Ruel and J. R. Woodgett, *Curr. Biol.*, 1996, **6**, 1664–1668.
48. H. Eldar-Finkelman and A. Martinez, *Front. Mol. Neurosci.*, 2011, **4**, 32.
49. M. Marchena, B. Villarejo-Zori, J. Zaldivar-Diez, V. Palomo, C. Gil, C. Hernandez-Sanchez, A. Martinez and E. J. de la Rosa, *J. Enzyme Inhib. Med. Chem.*, 2017, **32**, 522–526.
50. A. Sánchez-Cruz, B. Villarejo-Zori, M. Marchena, J. Zaldivar-Díez, V. Palomo, C. Gil, I. Lizasoain, P. de la Villa, A. Martínez, E. J. de la Rosa and C. Hernández-Sánchez, *Mol. Neurodegener.*, 2018, **13**, 19.
51. X. B. Sun, H. E. Lu, Y. Chen, X. H. Fan and B. Tong, *Pharmazie*, 2014, **69**, 889–893.
52. P. De Sarno, X. Li and R. S. Jope, *Neuropharmacology*, 2002, **43**, 1158–1164.
53. S. Iraha, Y. Hirami, S. Ota, G. A. Sunagawa, M. Mandai, H. Tanihara, M. Takahashi and Y. Kurimoto, *Clin. Ophthalmol.*, 2016, **10**, 1375–1384.
54. A. Kumar, N. Midha, V. Gogia, S. Gupta, S. Sehra and A. Chohan, *J. Ocul. Pharmacol. Ther.*, 2014, **30**, 580–586.
55. F. de Pablo and E. J. de la Rosa, *Trends Neurosci.*, 1995, **18**, 143–150.
56. C. Hernandez-Sanchez, A. Mansilla, E. J. de la Rosa and F. de Pablo, *Diabetologia*, 2006, **49**, 1142–1150.

57. C. Hernandez-Sanchez, E. Rubio, J. Serna, E. J. de la Rosa and F. de Pablo, *Diabetes*, 2002, **51**, 770–777.
58. T. Nemoto, F. Toyoshima-Aoyama, T. Yanagita, T. Maruta, H. Fujita, T. Koshida, T. Yonaha, A. Wada, A. Sawaguchi and M. Murakami, *Cell. Signalling*, 2014, **26**, 253–259.
59. J. Havrankova, D. Schmechel, J. Roth and M. Brownstein, *Proc. Natl. Acad. Sci. U. S. A.*, 1978, **75**, 5737–5741.
60. R. Ghasemi, A. Haeri, L. Dargahi, Z. Mohamed and A. Ahmadiani, *Mol. Neurobiol.*, 2013, **47**, 145–171.
61. A. Wada, H. Yokoo, T. Yanagita and H. Kobayashi, *J. Pharmacol. Sci.*, 2005, **99**, 128–143.
62. A. Taguchi and M. F. White, *Annu. Rev. Physiol.*, 2008, **70**, 191–212.
63. B. Chami, A. J. Steel, S. M. De La Monte and G. T. Sutherland, *Metab. Brain Dis.*, 2016, **31**, 497–515.
64. Y. Chen, J. Zhang, B. Zhang and C. X. Gong, *Curr. Top. Med. Chem.*, 2016, **16**, 485–492.
65. B. Diaz, J. Serna, F. De Pablo and E. J. de la Rosa, *Development*, 2000, **127**, 1641–1649.
66. A. I. Valenciano, S. Corrochano, F. de Pablo, P. de la Villa and E. J. de la Rosa, *J. Neurochem.*, 2006, **99**, 524–536.
67. S. Corrochano, R. Barhoum, P. Boya, A. I. Arroba, N. Rodriguez-Muela, V. Gomez-Vicente, F. Bosch, F. de Pablo, P. de la Villa and E. J. de la Rosa, *Invest. Ophthalmol. Visual Sci.*, 2008, **49**, 4188–4194.
68. L. Fernandez-Sanchez, P. Lax, C. Isiegas, E. Ayuso, J. M. Ruiz, P. de la Villa, F. Bosch, E. J. de la Rosa and N. Cuenca, *Hum. Gene Ther.*, 2012, **23**, 1290–1300.
69. C. Isiegas, J. A. Marinich-Madzarevich, M. Marchena, J. M. Ruiz, M. J. Cano, P. de la Villa, C. Hernandez-Sanchez, E. J. de la Rosa and F. de Pablo, *Invest. Ophthalmol. Visual Sci.*, 2016, **57**, 3610–3618.
70. R. Corpas, A. M. Hernandez-Pinto, D. Porquet, C. Hernandez-Sanchez, F. Bosch, A. Ortega-Aznar, F. Comellas, E. J. de la Rosa and C. Sanfeliu, *Neuropharmacology*, 2017, **123**, 221–232.

CHAPTER 5

Modulation of p75^{NTR}/Pro-NGF as a Therapeutic Approach for Degenerative Retinopathies

H. URI SARAGOVI*[a,b], ALBA GALÁN[a] AND PABLO F. BARCELONA[c]

[a]Lady Davis Institute – Jewish General Hospital, McGill University, Montreal, QC, Canada; [b]Departments of Pharmacology and Therapeutics, Ophthalmology, Integrated Program for Neuroscience, McGill University, Montreal, QC, Canada; [c]Centro de Investigaciones en Bioquímica Clínica e Inmunología, CIBICI-CONICET, Facultad de Ciencias Químicas, UNC, Córdoba, Argentina
*E-mail: Uri.saragovi@mcgill.ca

5.1 Mature NGF, Pro-NGF, and Their Receptors

Nerve growth factor (NGF) is a member of the neurotrophin family that includes brain-derived neurotrophic factor, neurotrophin (NT)-3, and NT4/5.[1] Mature NGF is derived by cleavage of the pro-peptide of the pro-NGF precursor.[2] NGF and pro-NGF, as well as other neurotrophins, bind to transmembrane receptors expressed predominantly in neurons and glia. There are two types of NGF receptors: TrkA and p75^{NTR}.

Drug Discovery Series No. 66
Therapies for Retinal Degeneration: Targeting Common Processes
Edited by Enrique J. de la Rosa and Thomas G. Cotter

Published by the Royal Society of Chemistry, www.rsc.org

TrkA is bound by mature NGF, to activate the intracellular tyrosine kinase domain of the receptor, resulting in signaling pathways (such as phosphoinositide 3-kinase, extracellular signal-regulated kinases or phospholipase C-γ) that in neurons promote survival, phenotypic maintenance or differentiation, and growth.[3–5] TrkA is not bound by pro-NGF.

The $p75^{NTR}$ is bound by mature NGF to positively or negatively regulate NGF affinity and TrkA activity.[6] The intrinsic $p75^{NTR}$ signals activated by mature NGF have been rather difficult to study. The intrinsic $p75^{NTR}$ signals, when $p75^{NTR}$ is in complex with sortilin, are best described after pro-NGF binding as a ligand.[7–9]

In neurons, pro-NGF binding $p75^{NTR}$ activates signals implicated in neuronal death, by ADAM17-dependent proteolysis to an extracellular fragment (p75-ECD) and a regulated intramembrane proteolysis by γ-secretase[10] to an intracellular fragment (p75-ICD) that mediates apoptotic cell death *via* JNK and caspase activation.[10]

In glia, pro-NGF binding $p75^{NTR}$ activates JNK and nuclear facto-κB, promoting production of pro-inflammatory cytokines such as α_2-macroglobulin (α_2M)[11–14] and tumor necrosis factor (TNF)α,[13–17] that at increased levels are neurotoxic and cause neuronal death in a paracrine fashion.[13,14,18,19] The secreted factor α_2M plays multiple roles, but noteworthy is its capability to increase the half-lives of pro-NGF and TNFα, hence magnifying toxicity.[13,19] In addition, $p75^{NTR}$/pro-NGF binding in glia promotes production of pro-NGF itself, creating a vicious positive feedback for $p75^{NTR}$ activation of toxic signals.

5.2 Localization of Mature NGF, Pro-NGF, and Their Receptors in the Retina

In the normal retina, $p75^{NTR}$ and pro-NGF levels are very low. However, in disease states $p75^{NTR}$ and pro-NGF are increased, causing neuronal death as described above by (1) the direct pro-apoptotic action of activated $p75^{NTR}$, and (2) promotion of synaptic pruning and neurodegeneration by glial production of neurotoxic cytokines such as α_2M, TNFα, and pro-NGF, which cause paracrine neuronal death. The changes in $p75^{NTR}$ and pro-NGF phenotype and the altered cytokine profile have been documented in human disease[20–22] and in rodent models of disease[23,24] including glaucoma and diabetic retinopathy, making $p75^{NTR}$ and pro-NGF potentially valuable therapeutic targets.

The lingering question was whether inhibition of $p75^{NTR}$ or pro-NGF would be disease modifying. We have shown that pharmacological inhibition of pro-NGF or $p75^{NTR}$, even after disease onset, effectively slows disease progression, reduces the levels of neurotoxic factors, and prevents neuronal loss for prolonged periods that surpass drug washout.[13,14,25] This is the emerging concept of "reduced neurotrophin toxicity" that molds the original concept of "neurotrophin neuroprotection".

The pro-NGF and mature NGF protein are present in embryonic retina[26] and in adult retina.[27–32] Mature NGF is localized in the retinal pigment epithelium,[33] photoreceptor outer segments, the outer nuclear layer (ONL), inner nuclear layer (INL), inner plexiform layer (IPL), and ganglion cell layer (GCL). NGF is detected in the somata of retinal ganglion cells (RGCs),[26,30] and on some bipolar cells.[30,34] In photoreceptors, NGF was found in the ONL and in photoreceptor outer segments.[28,30,31] Pro-NGF is localized in the retinal pigment epithelium (RPE), photoreceptor outer segments,[35] the ONL, INL, IPL, GCL, Müller cells, RGCs, and microglia.[29,36–39]

On adult healthy retina, the NGF receptor TrkA is expressed in RPE cells, photoreceptor outer segments, the ONL[30,31,40,41] specifically on bipolar and amacrine cells, and the INL in astrocytes[42] and Müller glial cells.[31,32] The NGF receptor p75^{NTR} is present on the INL[13,25,43] in Müller cells,[29,32,44] in astrocytes,[13,34,42] on the outer limiting membrane, in inner and outer photoreceptor segments, the ONL,[35,45] and *in vitro* in the Müller glial cell line rMC.[19] RGCs express TrkA and p75^{NTR} during embryonic and postnatal stages.[46–48] In the healthy adult RGCs, TrkA continues to be present,[16,30,32,40,44,49–51] and p75^{NTR} expression is lost.[16,43,52–55]

In retinal sections of the adult rodent retina Müller glial cells positive to vimentin, glutamine synthetase (GS) and cellular retinaldehyde-binding protein (CRALBP) confirmed the TrkA and p75^{NTR} expression.[13,16,19,30,43,52–54,56,57] Both receptors are expressed in cultured and isolated Müller cells.[30,56,58,59] During development, expression of p75^{NTR} by Müller cells has been found in mice and rats.[13,55,60] Cultured Müller cells from developing rats express p75^{NTR}, but not TrkA.[42,58] In addition, p75^{NTR} was observed in astrocytes in healthy mouse retinas.[52,61]

The presence of NGF and p75^{NTR} in astrocytes during retinal development suggests that the cells may play a role in the regulation of cell death implicated in the development of retinal vasculature. In the developing retina, NGF is released by microglial cells and induces cell death on progenitors that express p75^{NTR} [62] (Table 5.1).

In many diseases, oxidative stress, or inflammatory conditions, the retina exhibits an early imbalance of NGF processing, with accumulation of pro-NGF (*e.g.* diabetes,[37] glaucoma,[17] or ROP[13]). In parallel, p75^{NTR} levels increase in glial cells and neurons during glaucoma,[17] optic nerve damage,[16] retininis pigmentosa,[14] and diabetes.[13] This leads to activation of p75^{NTR} and pro-neurotoxic or neurodegenerative signals. In addition, p75^{NTR} is found on vascular endothelial cells[63,64] and pericytes,[13] which can lead to vascular deficits, leaky vessels and edema, and neovascularization as a secondary consequence[13] (Table 5.1).

In contrast, TrkA levels do not change significantly in early retinal disease, but the TrkA-protective signals are dampened or reduced (*e.g.* glaucoma and optic nerve axotomy[11,12,65]) likely due to the reduced levels of mature NGF ligand, and the toxic action of p75^{NTR}.

This results in a combination of loss of Trk-mediated neuroprotection/maintenance, and gain of toxic/pro-inflammatory action *via* the p75^{NTR}

Table 5.1 Description of the localization of nerve growth factor (NGF), pro-NGF, and the receptors TrkA and $p75^{NTR}$ on cells from the healthy and diseased retina.[a]

	Healthy retina				Diseased retina			
	Pro-NGF	NGF	TrkA	$p75^{NTR}$	Pro-NGF	NGF	TrkA	$p75^{NTR}$
Pigmented epithelium	+	+	+	ND	ND	ND	ND	ND
Photoreceptors	+	?	+	+	ND	ND	ND	ND
Horizontal cell	ND	ND	ND	ND	ND	ND	ND	ND
Bipolar cell	ND	−	+	−	ND	ND	ND	+
Müller cell	+	+	+	+	+/−	++	+	+++
Amacrine cell	ND	+	+	ND	ND	ND	ND	ND
Microglia cell	+	−	ND	ND	+	−	ND	+
Ganglion cell	+	+	+	−	+	+	+	−
Astrocyte	ND	−	+	−	ND	ND	+	++
Endothelial cells	ND	ND	+	ND	ND	ND	+	ND
Pericyte	ND	ND	ND	−	ND	ND	ND	+

[a]+: mRNA or protein expression detectable; −: mRNA or protein expression not detectable; ND: not determined; ?: contradictory data.

receptors. Toxicity is self-sustained due to pro-NGF production (as well as other factors) being promoted by activated $p75^{NTR}$ receptors.

5.3 Role of NGF, Pro-NGF, and Their Receptors in Retinal Degeneration Disease

We discuss the role and experimental therapy targeting NGF/NGF receptors in four retinal disease models.

5.3.1 Retinitis Pigmentosa

Retinitis pigmentosa[33] is a group of inherited retinal dystrophies with clinical similarities, despite arising from a large set of different genetic mutations. RCS rats[36,45,66,67] and rd mice[60] are models of retinitis pigmentosa, characterized by a progressive apoptosis of rod, cone photoreceptors, and RPE cells.[68,69]

In the rat retinitis pigmentosa model, $p75^{NTR}$ mRNA and protein are increased in all nuclear layers of the retina (ONL, INL, and GCL).[45] In the rd mice, $p75^{NTR}$ mRNA and protein are elevated mainly in the ONL,[35,56] mostly in Müller cells[56] and in photoreceptors.[35,45,70] A similar pattern of elevation is seen in retinas subjected to photic injury.[35,36,45,56,60,70]

Pro-NGF is expressed and released by photoreceptors and by microglia and Müller glial cells, and has been suggested to induce photoreceptor cell death *via* autocrine mechanisms[35,36] or by paracrine mechanisms release of TNFα and α_2M.[14,37,71] Oxidative stress and pro-inflammatory conditions, microglial activation, and reactive gliosis are implicated in the pathogenesis of retinitis pigmentosa.[68,72,73] In addition, the imbalance between pro-NGF and NGF may contribute to the induction of photoreceptor degeneration.[66]

Early analysis of the gene expression of the pro-NGF/p75^{NTR} system components in rd10 mouse retinas, before any sign of photoreceptor degeneration showed no differences in the mRNA levels of NGF, p75^{NTR}, TrkA, and sortilin compared to wild-type retinas. However, close to the beginning of photoreceptor cell death, there was a significant increase in the levels of pro-NGF, but not NGF. Interestingly, the elevation of pro-NGF preceded the increase of the mRNA levels of the pro-inflammatory cytokines (interleukin (IL)-1β, TNFα, and α_2M). The relevance of α_2M is that of α_2M and pro-NGF synergize to stimulate p75^{NTR}-dependent production of TNFα[19].

These cytokines up-regulate when the degenerative process is already detectable, concomitant to Müller gliosis. There is subsequent microglial infiltration to the outer plexiform layer, the ONL, and the outer segments of photoreceptors, where α_2M immunolabeling has been localized,[14] which may exacerbate degeneration.

These observations suggest that elevated pro-NGF levels may have a role in the pathogenesis of retinitis pigmentosa, most likely by p75^{NTR}-mediated induction of pro-inflammatory cytokines in Müller cells.

In vitro, rd10 retinal explants treated with different p75^{NTR} small-molecule antagonists[14] reduced photoreceptor cell death. Among the p75^{NTR} small-molecule antagonists, THX-B had the highest relative potency in decreasing photoreceptor cell death.[14] *In vivo*, a single intravitreal injection of THX-B at postnatal day 17 in rd10 mice elicited a neuroprotective effect in photoreceptor death, decreased the number of microglial cells and curbed the levels of expression of glial fibrillary acidic protein (GFAP) and the inflammatory molecules TNFα, α_2M, and IL-1β. In RHOP retinas, one single subconjunctival injection at postnatal day 18 rendered a significant neuroprotection of photoreceptors.

5.3.2 Glaucoma

Glaucoma is an optic neuropathy often associated with a high intraocular pressure (IOP), with progressive degeneration of nerve fiber layers and RGCs, causing progressive visual field loss, and leading to irreversible visual impairment or blindness.[74,75] Chronic glaucoma models can be induced by causing high IOP by interfering with intraocular pressure homeostasis (cauterization[51] or injection of hypertonic saline into episcleral veins[49,50,76]). In these models the retinal balance of pro-NGF/mature NGF is altered, with a net increase of pro-NGF and decrease of mature NGF[17,37] over the 5–7 weeks studied in these models.[47,49–51]

TrkA protein is maintained or increased slightly between 1 and 5 weeks after induction of high IOP,[49,51] but concomitant with significant RGC death it is decreased after 7 weeks.[50,51] The p75^{NTR} mRNA is up-regulated rapidly, from 1 week and up to 5 weeks after elevation of IOP.[49,51,76,77] Administration of a p75^{NTR} agonist caused progressive RGC death in normal retinas and accelerated RGC death in injured retinas, suggesting that p75^{NTR} activation is deleterious.

Effective glaucoma therapy was achieved by activation of TrkA or by inhibition of p75^{NTR} or pro-NGF. Exogenous administration of NGF as a therapeutic in experimental glaucoma did not exert a significant protective effect on RGC degeneration, likely due to p75^{NTR} or the mature NGF-inhibitory action of α_2M.[51,65] However, we note that other investigators report successful experimental therapeutic use of NGF.[50,78,79]

A slightly different rationale, using specific TrkA activating agents (mutant NGF proteins that do not activate p75^{NTR}, or small molecule TrkA activators not inhibited by α_2M) were more significantly neuroprotective,[17] suggesting a deleterious role for up-regulated retinal p75^{NTR}, as suggested by others.[16,17,20,24,28]

Delivery of p75^{NTR} small-molecule antagonists such as THX-B, or a neutralizing monoclonal anti-pro-NGF antibody (by intravitreal injection), afforded neuroprotection and inhibited the p75-downstream events such as up-regulation of the pro-inflammatory molecules TNFα and α_2M[17].

5.4 Optic Nerve Axotomy and Retinal Ischemia

In models sometimes purported as "acute glaucoma" where extremely high IOPs (>2.5-fold normal) cause retinal ischemia, or after injury induced by optic nerve crush/axotomy[16,53,77,80,81] there is up-regulation of pro-NGF protein[16,38,60] and decrease of mature NGF, likely due to poor proteolytic processing (as NGF mRNA is unchanged[77]) or due to stabilization of pro-NGF (possibly by α_2M). Like in glaucoma, optic nerve axotomy treatments are effective when they activate TrkA or inhibit p75^{NTR} or inhibit pro-NGF,[17,43,64,65] although the disease progression is more acute.

Another experimental therapy strategy is the transplantation of bone marrow mesenchymal stem cells and olfactory ensheathing cells or stem cells into the subretinal space of RCS rats. These cells secrete NGF,[43,82,83] and provide long-term neuroprotective effects.[82]

5.4.1 Diabetic Retinopathy

Diabetic retinopathy is a complication affecting >70% of all diabetics. The pathology includes inflammation, retinal neuronal death, and retinal edema, sometimes with neovascularization, leading to irreversible vision loss. In humans, diabetic retinopathy is primarily a microvascular disease of the retina, but neurodegenerative and inflammatory events occur even before vascular changes.[84,85] The diabetes model induced by streptozotocin show an early imbalance (starting at ~10 days of diabetes[13]), resulting in retinal increase of pro-NGF. There is a detectable decrease of retinal NGF protein from 4 weeks up to 8 weeks of diabetes.[13,86] In addition, cultured Müller cells stimulated with high glucose have increased secretion of pro-NGF.[37]

The p75^{NTR} protein is increased in the retina of diabetic mice and rats from ~7 days of diabetes.[13,84] Upregulation of p75^{NTR} was localized in CRALBP-,

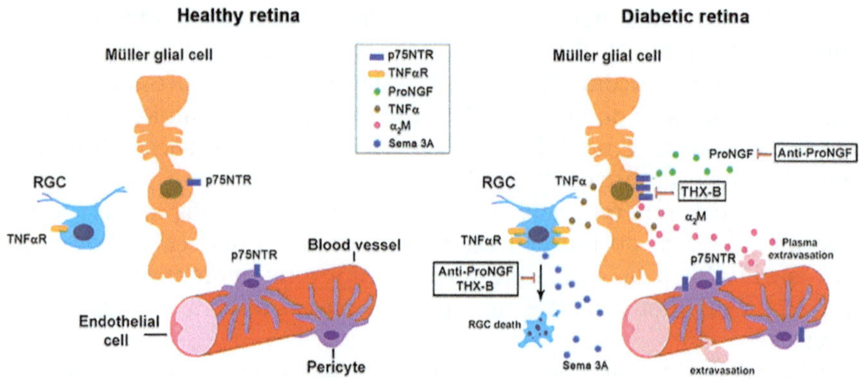

Figure 5.1 Model representing the main events in the etiology of p75^{NTR} in diabetic retinopathy. The mechanisms apply to other retinal diseases as well. In a healthy retina, the levels of p75^{NTR} are low, and mainly present in glia. In diabetic disease, pro-nerve growth factor (NGF) is elevated, and activates p75^{NTR} up-regulated in Müller glial cells and pericytes. Increased p75^{NTR} activity triggers the expression and secretion of α_2-macroglobulin (α_2M) and tumor necrosis factor (TNF)α by Müller cells. Up-regulated TNFα binds to receptors in retinal ganglion cells (RGCs) inducing the release of Sema3A as well as eventual RGC death. Sema3A released by RGCs binds to its receptor in the endothelial cells and provokes loosening of endothelial cell junctions in the blood–retina barrier (BRB), which ultimately leads to plasma extravasation. Intravitreal injections (or more recently a more benign subconjunctival delivery) of THX-B p75^{NTR} antagonist or an anti-pro-NGF-neutralizing monoclonal antibody in diabetic mice significantly reduced Müller glia activation, decreased TNFα and α_2M expression, decreased Sema3A production, maintained BRB integrity, prevented extravasation, and prevented RGC neuronal cell death.

GS- and GFAP-positive glial cells[13] (Figure 5.1). Likewise, Müller cells cultured in high-glucose media increase p75^{NTR} expression.

The consequence is that in the model of diabetes p75^{NTR}-dependent induction of pro-inflammatory agents TNFα[28,43,45,87] and α_2M[13] by Müller glial cells, disrupt the neuroglia vascular unit, cause blood–retina barrier (BRB) breakdown, edema, and ganglion cell neuronal death (Figure 5.1). An early and steady increase in the levels of a p75^{NTR} protein (from week 1 to week 6 after the induction of diabetes) was followed by the augmentation of the levels of pro-NGF at week 3. The pro-inflammatory cytokines α_2M and TNFα were robustly elevated at 6 weeks after the onset of diabetes. The combination of pro-NGF and p75^{NTR} up-regulation causes inflammatory events and neuronal death, both of which are pathologies of diabetic retinopathy (Figure 5.1).

Because neovascularization is sporadic in the rodent diabetic retina, an alternative model is the oxygen-induced retinopathy (OIR) where high oxygen exposure causes vaso-obliteration followed by a vaso-proliferation phase driven by Sema3A and HIF.[88] OIR was used to evaluate pro-NGF and p75^{NTR} neurotrophic pathways in the stressed vasculature. In OIR the pro-NGF and p75^{NTR} protein levels increase during the vaso-obliterative phase[13,26,78] and the vaso-proliferative phase.[13,78] The p75^{NTR} is expressed in perycites, causing perycite dysfunction. Without proper pericytes the vascular endothelium remains unstable, immature, and leaky.

Interestingly, the up-regulation of TNFα mRNA was mainly detected in the RGC layer and INL, the main retinal layers affected in diabetic retinopathy. In the RGC, TNFα mRNA partially co-localized with p75^{NTR} protein, most likely within the Müller end-feet. Furthermore, there was an increase in α_2M mRNA in the RGC layer and INL. Up-regulation of α_2M mRNA was also detected in the boundary between the INL and photoreceptor layers, likely associated with cells of the retinal vasculature.[13,25] Moreover, at week 8 after the induction of diabetes, there was an increase in the mRNA levels of Sema3A,[13] known to induce BRB breakdown, and an increase of plasma extravasation,[13,25] most likely induced by TNFα release. Given the known role of Sema3A in vascular remodeling, a model of OIR also evaluated the role of p75^{NTR}. Vaso-obliteration and neovascularization in OIR occurred in a p75^{NTR}-dependent and pro-NGF-dependent manner.[13,25]

These observations suggest that p75^{NTR} signaling promotes, in a pro-NGF-dependent manner, the production of the inflammatory cytokines TNFα and α_2M and Sema3A, leading to RGC death, BRB breakdown, and edema *via* a paracrine mechanism (Figure 5.1). These data identify a p75^{NTR} novel signaling mechanism implicating interplay between neurons and glia (and vasculature, in certain diseases). Inhibition of p75^{NTR} or inhibition of pro-NGF was therapeutic and reduced all these pathological features (inflammation, RGC death, vascular deficits) for prolonged periods.[13,14,17,25]

One single intravitreal of injection of THX-B or of an inhibitory pro-NGF antibody decreased Müller glial cell activation, diminished TNFα, α_2M, and Sem3A expression, maintained BRB integrity and prevented RGC death and fiber loss.[13] Interestingly, a similar beneficial outcome has been demonstrated by using one single subconjunctival injection of THX-B or anti-pro-NGF monoclonal antibody (mAb) (a less invasive method of administration) in the same model of diabetic retinopathy.[25] In an OIR model, both THX-B or anti-pro-NGF neutralizing mAb decreased vaso-obliteration, and curbed TNFα, and Sem3A expression.[13]

Altogether, these results demonstrate that p75^{NTR} and pro-NGF are key mediators of neuronal cell death by inducting the production of pro-inflammatory molecules by glial cells, in many different eye pathologies, and provide evidence for p75^{NTR} as a promising therapeutic target.

5.5 Clinical Application of Pro-NGF Antibody and p75^{NTR} Inhibitor

The pathological events described here are the result of p75^{NTR} up-regulation disrupting the homeostasis of the neuro–glia–vascular axis, triggering noxious self-sustained signals. The up-regulation of p75^{NTR}, TNFα, pro-NGF, and α_2M has been verified in human eye diseases[12–14,16,17,25] making the animal models and mechanisms relevant for medicinal translation.

Both, pro-NGF and $p75^{NTR}$ have been validated as therapeutic targets in a mouse model of diabetic retinopathy by intravitreal administration of a function-blocking anti-pro-NGF monoclonal antibody or of a $p75^{NTR}$ small-molecule antagonist. For medicinal translation, the administration of compounds using effective, but safer and less-invasive routes have been exhaustively investigated for the treatment of eye pathologies. For instance, in a model of diabetic retinopathy, a recent study has demonstrated that subconjunctival delivery of THX-B renders retinal THX-B exposure equal to intravitreal injection, without resulting in detectable drug in circulation; and provided therapeutic efficacy similar to intravitreal injections, with reduced inflammation and edema, decreased neuronal death, and a long-lasting protection of the retinal structure in diabetic retinopathy.[25] In retinitis pigmentosa, subconjunctival delivery of THX-B afforded photoreceptor cell death protection.[14]

Another recent investigation has explored alternative delivery methods for the administration of an anti-pro-NGF monoclonal antibody by using intravenous and subconjunctival routes. Both delivery routes afforded comparable retinal penetration of the antibody, without toxic side effects, reaching therapeutic doses. Systemic and subconjuctival delivery provided persistent antibody infiltration, and long-term therapeutic efficacy.

Hence, the subconjunctival and systemic routes to administer $p75^{NTR}$ antagonists offer an effective, advantageous, less invasive, and more compliant method to treat eye pathologies. The validation of pro-NGF and $p75^{NTR}$ as therapeutic targets, the validation of these targets as druggable (with agents that block toxicity), and availability of relatively benign routes of administration, appears to be promising for the development of therapeutics. These therapeutics can be applied to many diseases such as diabetes, glaucoma, retinitis pigmentosa, age-related macular degeneration, and other chronic retinal diseases that require long-lasting therapeutic efficacy and reduced frequency of intervention.

References

1. E. J. Huang and L. F. Reichardt, *Annu. Rev. Neurosci.*, 2001, **24**, 677.
2. M. A. Bruno and A. C. Cuello, *Proc. Natl. Acad. Sci. U. S. A.*, 2006, **103**, 6735.
3. P. P. Roux and P. A. Barker, *Prog. Neurobiol.*, 2002, **67**, 203.
4. O. Clewes, *et al.*, *J. Neurochem.*, 2008, **107**, 1124.
5. M. Bothwell, *Annu. Rev. Neurosci.*, 1995, **18**, 223.
6. S. Maliartchouk and H. U. Saragovi, *J. Neurosci.*, 1997, **17**, 6031.
7. L. F. Reichardt, *Philos. Trans. R. Soc. London, Ser. B*, 2006, **361**, 1545.
8. M. V. Chao, *Nat. Rev. Neurosci.*, 2003, **4**, 299.
9. L. C. Schecterson and M. Bothwell, *Dev. Neurobiol.*, 2010, **70**, 332.
10. S. Skeldal, D. Matusica, A. Nykjaer and E. J. Coulson, *BioEssays*, 2011, **33**, 614.
11. Z. Shi, *et al.*, *J. Biol. Chem.*, 2008, **283**, 29156.

12. Y. Bai, *et al.*, *Invest. Ophthalmol. Visual Sci.*, 2011, **52**, 5260.
13. P. F. Barcelona, *et al.*, *J. Neurosci.*, 2016, **36**, 8826.
14. M. Platon-Corchado, *et al.*, *Cell Death Dis*, 2017, **8**, e2922.
15. F. Lebrun-Julien, *et al.*, *J. Neurosci.*, 2009, **29**, 5536.
16. F. Lebrun-Julien, B. Morquette, A. Douillette, H. U. Saragovi and A. Di Polo, *Mol. Cell. Neurosci.*, 2009, **40**, 410.
17. Y. Bai, *et al.*, *J. Biol. Chem.*, 2010, **285**, 39392.
18. Y. Bai, *et al.*, *Invest. Ophthalmol. Visual Sci.*, 2010, **51**, 6639.
19. P. F. Barcelona and H. U. Saragovi, *Mol. Cell. Biol.*, 2015, **35**, 3396.
20. R. Schliebs and T. Arendt, *Behav. Brain Res.*, 2011, **221**, 555.
21. M. S. Tsai, *et al.*, *Exp. Mol. Med.*, 2018, **50**, e426.
22. L. W. Chen, K. K. Yung, Y. S. Chan, D. K. Shum and J. P. Bolam, *CNS Neurol. Disord.: Drug Targets*, 2008, **7**, 512.
23. S. Capsoni, C. Tiveron, D. Vignone, G. Amato and A. Cattaneo, *Proc. Natl. Acad. Sci. U. S. A.*, 2010, **107**, 12299.
24. M. W. VonDran, J. LaFrancois, V. A. Padow, W. J. Friedman, H. E. Scharfman, T. A. Milner, and B. L. Hempstead, *ASN Neuro*, 2014, **6**, 1759091414552185.
25. A. Galan et al, *Invest. Ophthalmol. Visual Sci.*, 2017, **58**, 2852.
26. X. Liu, *et al.*, *Invest. Ophthalmol. Visual Sci.*, 2010, **51**, 3749.
27. B. M. Braunger et al, *J. Neurosci.*, 2013, **33**, 14246.
28. E. Bronzetti, *et al.*, *Eur. J. Histochem.*, 2007, **51**, 251.
29. S. Chakrabarti, A. A. Sima, J. Lee, P. Brachet and E. Dicou, *Brain Res.*, 1990, **523**, 11.
30. T. B. Garcia, *et al.*, *J. Neurochem.*, 2014, **131**, 303.
31. X. Sun, *et al.*, *Ophthalmologica*, 2008, **222**, 58.
32. E. Vecino, E. Caminos, M. Ugarte, D. Martin-Zanca and N. N. Osborne, *Gen. Pharmacol.*, 1998, **30**, 305.
33. V. Gregorc, *et al.*, *Br. J. Cancer*, 2009, **101**, 219.
34. B. O. Balzamino, G. Esposito, R. Marino, F. Keller and A. Micera, *NeuroMol. Med.*, 2015, **17**, 314.
35. A. M. Santos, *et al.*, *PLoS One*, 2012, **7**, e36243.
36. B. Srinivasan, C. H. Roque, B. L. Hempstead, M. R. Al-Ubaidi and R. S. Roque, *J. Biol. Chem.*, 2004, **279**, 41839.
37. T. K. Ali, *et al.*, *Diabetologia*, 2011, **54**, 657.
38. Y. Wei, *et al.*, *China Med. J.*, 2012, **125**, 3875.
39. T. Harada, *et al.*, *J. Neurosci.*, 2002, **22**, 9228.
40. D. Kokona, I. Charalampopoulos, I. Pediaditakis, A. Gravanis and K. Thermos, *Neuropharmacology*, 2012, **62**, 2106.
41. M. L. Rocco, B. O. Balzamino, P. Petrocchi Passeri, A. Micera and L. Aloe, *PLoS One*, 2015, **10**, e0124810.
42. Y. S. Kim, D. H. Jo, H. Lee, J. H. Kim and K. W. Kim, *Biochem. Biophys. Res. Commun.*, 2013, **431**, 740.
43. F. Lebrun-Julien, *et al.*, *Proc. Natl. Acad. Sci. U. S. A.*, 2010, **107**, 3817.
44. G. Carmignoto, *et al.*, *Exp. Neurol.*, 1991, **111**, 302.
45. H. J. Sheedlo, *et al.*, *Brain. Res. Mol. Brain Res.*, 2002, **103**, 71.

46. C. Harada, *et al.*, *Dev. Biol.*, 2006, **290**, 57.
47. P. Jansen, *et al.*, *Nat. Neurosci.*, 2007, **10**, 1449.
48. A. Zanellato, M. C. Comelli, R. Dal Toso and G. Carmignoto, *Dev. Biol.*, 1993, **159**, 105.
49. M. Coassin, *et al.*, *Graefes Arch. Clin. Exp. Ophthalmol.*, 2008, **246**, 1743.
50. V. Colafrancesco, *et al.*, *J. Glaucoma*, 2011, **20**, 100.
51. M. Rudzinski, T. P. Wong and H. U. Saragovi, *J. Neurobiol.*, 2004, **58**, 341.
52. B. Hu, H. K. Yip and K. F. So, *Glia*, 1998, **24**, 187.
53. B. Hu, H. K. Yip and K. F. So, *NeuroReport*, 1999, **10**, 1293.
54. F. Xu, *et al.*, *Neurosci. Lett.*, 2009, **454**, 81.
55. J. Ding, B. Hu, L. S. Tang and H. K. Yip, *Dev. Neurosci.*, 2001, **23**, 390.
56. T. Harada, *et al.*, *Neuron*, 2000, **26**, 533.
57. W. Shen, L. Zhu, S. R. Lee, S. H. Chung and M. C. Gillies, *J. Neuroinflammation*, 2013, **10**, 137.
58. E. M. Wexler, O. Berkovich and S. Nawy, *Vis. Neurosci.*, 1998, **15**, 211.
59. J. Wang, *et al.*, *Mol. Vision*, 2016, **22**, 254.
60. K. Nakamura, *et al.*, *Mol. Vision*, 2005, **11**, 1229.
61. Y. Wei, *et al.*, *Neurosci. Lett.*, 2007, **429**, 169.
62. J. M. Frade and Y. A. Barde, *Neuron*, 1998, **20**, 35.
63. M. M. Al-Gayyar, *et al.*, *Diabetologia*, 2011, **54**, 669.
64. B. A. Mysona, *et al.*, *Diabetologia*, 2013, **56**, 2329.
65. Z. Shi, E. Birman and H. U. Saragovi, *Dev. Neurobiol.*, 2007, **67**, 884.
66. T. Amendola, M. Fiore and L. Aloe, *Neurosci. Lett.*, 2003, **345**, 37.
67. L. Lenzi, *et al.*, *Vis. Res.*, 2005, **45**, 1491.
68. J. Sancho-Pelluz, *et al.*, *Mol. Neurobiol.*, 2008, **38**, 253.
69. D. Athanasiou, *et al.*, *FEBS Lett.*, 2013, **587**, 2008.
70. B. Srinivasan, *et al.*, *Mol. Cell. Neurosci.*, 2007, **36**, 449.
71. B. L. Hempstead, *Neurotoxic. Res.*, 2009, **16**, 255.
72. D. T. Hartong, E. L. Berson and T. P. Dryja, *Lancet*, 2006, **368**, 1795.
73. N. Cuenca, *et al.*, *Prog. Retinal Eye Res.*, 2014, **43**, 17.
74. R. A. Bouhenni, J. Dunmire, A. Sewell and D. P. Edward, *J. Biomed. Biotechnol.*, 2012, **2012**, 692609.
75. Y. You, V. K. Gupta, J. C. Li, A. Klistorner and S. L. Graham, *Rev. Neurosci.*, 2013, **24**, 301.
76. Y. Guo, *et al.*, *Exp. Eye Res.*, 2009, **89**, 921.
77. C. Harada, *et al.*, *Am. J. Pathol.*, 2015, **185**, 3238.
78. N. Le Moan, D. M. Houslay, F. Christian, M. D. Houslay and K. Akassoglou, *Mol. Cell*, 2011, **44**, 476.
79. A. Lambiase, *et al.*, *Proc. Natl. Acad. Sci. U. S. A.*, 2009, **106**, 13469.
80. Q. Cui, L. S. Tang, B. Hu, K. F. So and H. K. Yip, *Invest. Ophthalmol. Visual Sci.*, 2002, **43**, 1954.
81. Z. Ahmed, *et al.*, *Neurobiol. Dis.*, 2010, **37**, 441.
82. Q. Jian, Z. Tao, Y. Li and Z. Q. Yin, *Vis. Res.*, 2015, **110**, 107.
83. B. Mead, A. Logan, M. Berry, W. Leadbeater and B. A. Scheven, *Invest. Ophthalmol. Visual Sci.*, 2013, **54**, 7544.

84. T. S. Kern and A. J. Barber, *J. Physiol.*, 2008, **586**, 4401.
85. M. S. Stem and T. W. Gardner, *Curr. Med. Chem.*, 2013, **20**, 3241.
86. M. S. Ola, *et al.*, *J. Mol. Neurosci.*, 2015, **56**, 440.
87. N. Agarwal, *et al.*, *Mol. Vision*, 2007, **13**, 1311.
88. A. Cerani, *et al.*, *Cell Metab.*, 2013, **18**, 505.

CHAPTER 6

Modulation of cGMP-signalling to Prevent Retinal Degeneration

VALERIA MARIGO[a], PER EKSTRÖM[b], FRANK SCHWEDE[c], ANDREAS RENTSCH[c] AND FRANÇOIS PAQUET-DURAND*[d]

[a]Dipartimento di Scienze della Vita, Università degli Studi di Modena e Reggio Emilia, Italy; [b]Ophthalmology, Department of Clinical Sciences Lund, Faculty of Medicine, Lund University, Sweden; [c]Biolog Life Science Institute, Bremen, Germany; [d]Institute for Ophthalmic Research, University of Tübingen, Germany
*E-mail: francois.paquet-durand@klinikum.uni-tuebingen.de

6.1 Photoreceptor Cell Death and Hereditary Retinal Degeneration

Retinal photoreceptors – rods and cones – are the primary neurosensory cells with which we perceive light. In the human retina, there are approximately 120 million rods and 6 million cones. While rods are more sensitive to low light, as in night vision, cones are more suited for colour discrimination and visual acuity in daylight, which is what we generally regard as our ability to see.[1]

Vision loss due to hereditary retinal degeneration is a devastating handicap with far-reaching effects on both the quality of life of the individuals affected and society as a whole. Retinal degeneration is caused by the death

Drug Discovery Series No. 66
Therapies for Retinal Degeneration: Targeting Common Processes
Edited by Enrique J. de la Rosa and Thomas G. Cotter

Published by the Royal Society of Chemistry, www.rsc.org

of photoreceptors[2] and since these are post-mitotic neurons they will not be replaced, which thus reduces the retina's ability to capture light forever and leads to severe visual impairment and often blindness. Retinal degeneration is in fact a collective term describing a group of diseases in which photoreceptors are lost, and among these diseases there are several important subgroups such as retinitis pigmentosa, Leber's congenital amaurosis (LCA), or achromatopsia.[3,4] Altogether, these diseases are considered the most common cause for visual loss in the working population of the industrialised world, and it is estimated that they affect 170 000 people in the European Union (EU) alone.[5] Being hereditary, retinal degeneration is caused by mutations in the genomic DNA and passed on from parent to child. While the mutations behind many of the disease types are known, there are yet no treatments available for these feared disabilities.

The primary retinal degeneration gene mutation could affect either the rods or the cones, where the loss of cones presents a greater threat to our visual ability than the loss of rods. However, when diseased rods degenerate and die, there will inevitably be a secondary loss of cones, even if these are not affected by the mutation as such, and the rod loss will hence compromise our general visual capacity. Conversely, this means that treatments aimed at protecting diseased rods will also protect the otherwise healthy cones and could therefore preserve functional human vision.[6]

6.2 Defining a Common Target for Genetically Heterogeneous Forms of Retinal Degeneration

The signalling molecule cyclic guanosine monophosphate (cGMP) occupies a key position in phototransduction and photoreceptor physiology. Interestingly, dysregulation of photoreceptor cGMP has emerged as a pathological hallmark of retinal degeneration, with a particular emphasis on the excessive accumulation found in many different animal models for retinitis pigmentosa, LCA, and achromatopsia.[7,8] In photoreceptors, cGMP is produced by retinal guanylyl cyclase, whose activity in turn is controlled by both the RD3 protein and high Ca^{2+} levels.[9,10] The corresponding hydrolysis of photoreceptor cGMP is mediated by phosphodiesterase-6 (PDE6), which is functionally assembled by aryl-hydrocarbon-interacting protein-like-1 (AIPL1).[11] Remarkably, all these genes involved in cGMP metabolism are also retinal degeneration disease genes. Certain genes that are likely to impact cGMP levels more strongly (*e.g. AIPL1*, *GUCY2D*, and *RD3*) cause the more severe, early onset, rapid progression phenotype of LCA,[4,12] while mutations that may reduce hydrolysis (*e.g. PDE6A*, *B*, *G*),[13,14] or that will affect cGMP synthesis only indirectly (*e.g.* cyclic nucleotide gated ion channels described later, *CNGB1*, and *CNGB3*)[15,16] result in the "milder" phenotype of retinitis pigmentosa, which progresses far more slowly (Figure 6.1).

Whatever the root cause, excessive levels of photoreceptor cGMP eventually result in photoreceptor loss. How exactly cGMP causes photoreceptor

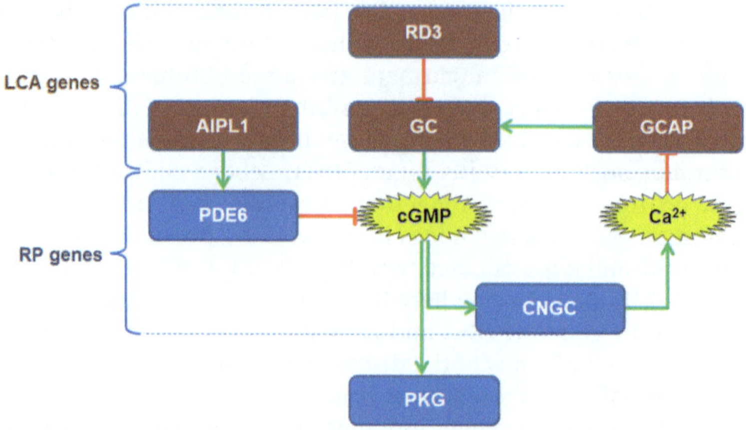

Figure 6.1 Photoreceptor cyclic guanosine monophosphate (cGMP) and Ca^{2+} signalling is controlled by Leber's congenital amaurosis (LCA) and retinitis pigmentosa (RP) genes/proteins. cGMP is produced by retinal guanylyl cyclase (GC), which is inhibited by the RD3 protein and high Ca^{2+} levels in a feedback fashion.[9,10] Expression of the cGMP hydrolysing enzyme phosphodiesterase-6 (PDE6) is regulated by the chaperone aryl-hydrocarbon-interacting protein-like-1 (AIPL1).[11] cGMP binding opens the cyclic nucleotide gated channel (CNGC), which allows for influx of cations, including Ca^{2+}, mediating negative feedback control of cGMP synthesis.[15,20] In addition, cGMP activates protein kinase G (PKG), which may be an important driver of cGMP-mediated cell death.[15,17] Mutations affecting the activity of PDE6 and CNGCs are known to cause the "milder" phenotype of RP (blue boxes), while mutations affecting AIPL1, RD3, and photoreceptor GC (brown boxes) cause the more severe LCA phenotype.

cell death remains unclear, but two main cGMP interactors – cyclic nucleotide gated ion channels (CNGCs) and cGMP-dependent protein kinase (PKG) – are known to be present in photoreceptors and may therefore be dysregulated by high cGMP. To this end, attenuation of PKG and CNGC activity can reduce photoreceptor cell death[17,18] and both types of proteins thus constitute potential targets to prevent retinal degeneration. In addition, blocking retinal guanylyl cyclase may also constitute a viable therapeutic approach.[19]

6.3 cGMP Analogues as Drugs to Prevent Retinal Degeneration

cGMP signalling is central in phototransduction, but cGMP levels elevated by excessive production or, conversely, by insufficient hydrolysis may trigger photoreceptor degeneration. The actions of cGMP and its targets can be altered by cGMP analogues including those that competitively inhibit cGMP-dependent target activation[21] and which may hence be used to rebalance dysregulated

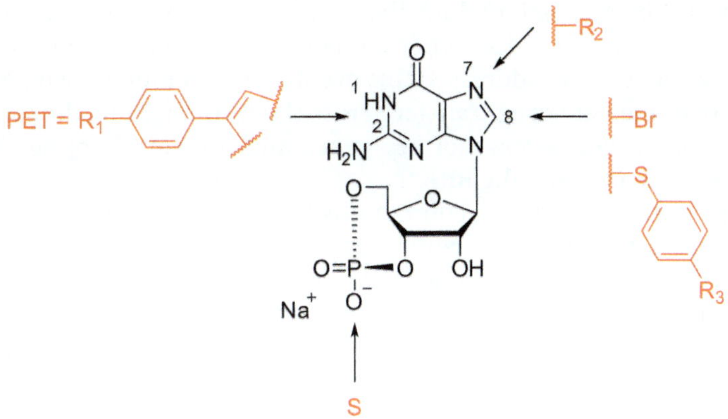

Figure 6.2 Structure and function relationships in cyclic guanosine monophosphate (cGMP) analogues. Structural modifications introduced at different positions in the cGMP molecule can alter the pharmacological properties in specific ways. R_P-configurated phosphorothioate-modified cGMP analogues (R_P-cGMPS, S substitution in figure) typically inhibit protein kinase G, regardless of other nucleobase modifications.[24,25] The addition of a β-phenyl-1,N^2-etheno-modification (PET) onto the R_P-cGMPS backbone inhibits photoreceptor cyclic nucleotide gated ion channels (CNGCs).[22] In contrast, R_P-cGMPS analogues with modifications in position 8 (C-8) or 7 (N-7) alone are CNGC activators.[23]

cGMP activities.[17] cGMP analogues can be tailored to specifically target either PKG or CNGC, or both cGMP effectors at once. For instance, β-phenyl-1,N^2-etheno- and Rp-configurated phosphoro-thioate-modified cGMP analogues provide a structural basis for inhibiting CNGCs[22,23] and PKG.[24,25] Interestingly, combinations of different substituents provide almost unlimited room for the development of novel substances,[26] which may show increased target binding or lipophilicity, for instance, and which are highly resistant towards hydrolysis by PDE6 or other enzymes (Figure 6.2).

In 2009 it was shown that inhibitory cGMP analogues that block or reduce activity of PKG and/or CNGCs can prevent photoreceptor degeneration in *rd1* and *rd2* mouse retinae.[17] Improved compound design and testing might then lead to the discovery of more potent cGMP analogues. Nevertheless, an important and entirely different challenge that remained to be addressed was the bioavailability, and in particular the access of the cGMP analogues to the diseased photoreceptors.

6.4 The Problem of Retinal Drug Delivery

To maintain its homeostasis and to shield from blood-borne toxins or infectious agents, for example, the retina is protected by the blood–retinal barrier (BRB).[27] The BRB not only engulfs all retinal blood vessels, but also forms the inner limiting membrane, which constitutes a barrier to the vitreous of the eye. The BRB is thus essential for the protection and viability of the retina. However, it also prevents access of therapeutic compounds,[28] and to date relatively few studies have addressed this problem in a comprehensive way. One possibility to mediate drug transfer across the BRB might be the temporary siRNA-mediated knock-down of key genes and proteins forming the tight junctions that constitute the BRB.[29]

Another option for helping compounds to cross the BRB may be to encapsulate these into liposomes conjugated with poly-ethylene-glycol chains, which in turn carry a glutathione (GSH) molecule at their external endings. In clinical studies, this nanoscale liposomal carrier was previously shown to improve delivery of the drug doxorubicin to the brain.[30] How exactly the liposomal cargo is transported across barriers is not entirely clear, but it appears to be an active process that is based on a GSH- and cytoskeletal rearrangement-dependent endocytosis of the liposome.[31] Since the blood–brain barrier and the BRB have essentially the same structural features,[32,33] it was reasonable to assume that this liposomal drug delivery system could also facilitate drug delivery to the retina and its photoreceptors.

6.5 Efficacy Testing of cGMP Analogues and Liposomal Drug Delivery

To test the potential therapeutic effects of novel cGMP analogues in combination with a suitable drug delivery system, we devised a screening approach in which the complexity of the test system increased with each consecutive step (Figure 6.3).

The experimental systems used were: (1) a cell-based *in vitro* system, in which photoreceptor-like cells were generated from retinal stem cells derived

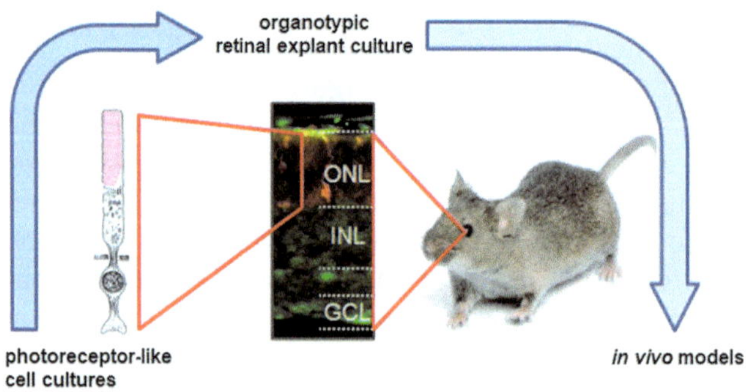

Figure 6.3 Photoreceptor neuroprotection tests in systems of increasing complexity: initial testing of drugs, drug delivery systems (DDS) and drug/DDS combinations was performed on photoreceptor-like cell cultures. After filtering for the most successful compounds, further testing was undertaken with explanted retina *in vitro*, and finally *in vivo*, in various animal models for retinal degeneration. ONL: outer nuclear layer; INL: inner retinal layer; GCL: ganglion cell layer.

from wild-type mice or disease models;[34,35] (2) an organotypic retinal explant culturing system, which, under serum-free conditions, conveniently monitors the effects of various treatments on disease models;[36,37] (3) *in vivo* systems using several different rodent animal models that fully exploit the unique diagnostic possibilities of the eye, *i.e.* direct, non-invasive visualisation and functional characterisation of neurodegenerative processes and treatment effects.[8,38] For instance, optic coherence tomography and scanning laser ophthalmoscopy were combined to assess retinal morphology and to trace the distribution of fluorescently labelled drugs *in vivo*. In parallel, electroretinography (ERG) allowed the assessment of the function of the retina.[18,26,39]

The initial screening in photoreceptor-like cell cultures filtered out the best candidates for further testing, either alone or encapsulated into liposomes, under more life-like *in vitro* conditions in retinal explant culture, to finally assess the therapeutic efficacy by studies *in vivo* (Figure 6.3). Separate experimentation assessed the compounds and drug delivery systems with respect to target inhibition and drug delivery efficacies, respectively, and addressed potential side-effects (*e.g.* off-target effects) and possible harmfulness. In combination, these tests and assays allowed to identify cGMP analogues with very promising photoreceptor protective properties.[26]

6.6 The "Drugs for Retinal Degeneration" (DRUGSFORD) Project

The workflow described above was exploited in the EU-funded DRUGSFORD project (www.drugsford.eu), which ran from 2012 to 2016. DRUGSFORD produced 221 cGMP analogues of which 187 were novel and initially tested in assays for PKG binding, lipophilicity, and hydrolytic stability. Then 20 analogues were chosen to be tested in photoreceptor-like cultures and from these the 14 compounds showing the strongest protective effects were forwarded to be tested in retinal explant cultures. Eventually, five cGMP analogues were tested *in vivo*, first in the rapidly degenerating *rd1* mouse.[40] The analogue showing the best photoreceptor protection in the *rd1* model – DF003 – and its liposomal formulation LP-DF003 (DF003 encapsulated in the drug delivery system) were then also tested in the more slowly degenerating *rd2*[41] and *rd10*[42] mouse models.

Administration of LP-DF003 clearly increased the survival of photoreceptors in all three retinal degeneration models highlighting LP-DF003's potential for mutation-independent treatment. For the sake of brevity, only *rd10* data is shown in the following (for a full report see Vighi *et al.*[26]). Intraperitoneal injection of LP-DF003, starting at post-natal day 14 (P14) and continued every day until P29, resulted in marked preservation of photoreceptors at P30, as shown by histologic preparations of the non-treated and LP-DF003-treated *rd10* retinae (Figure 6.4a). This morphological rescue of *rd10* photoreceptors

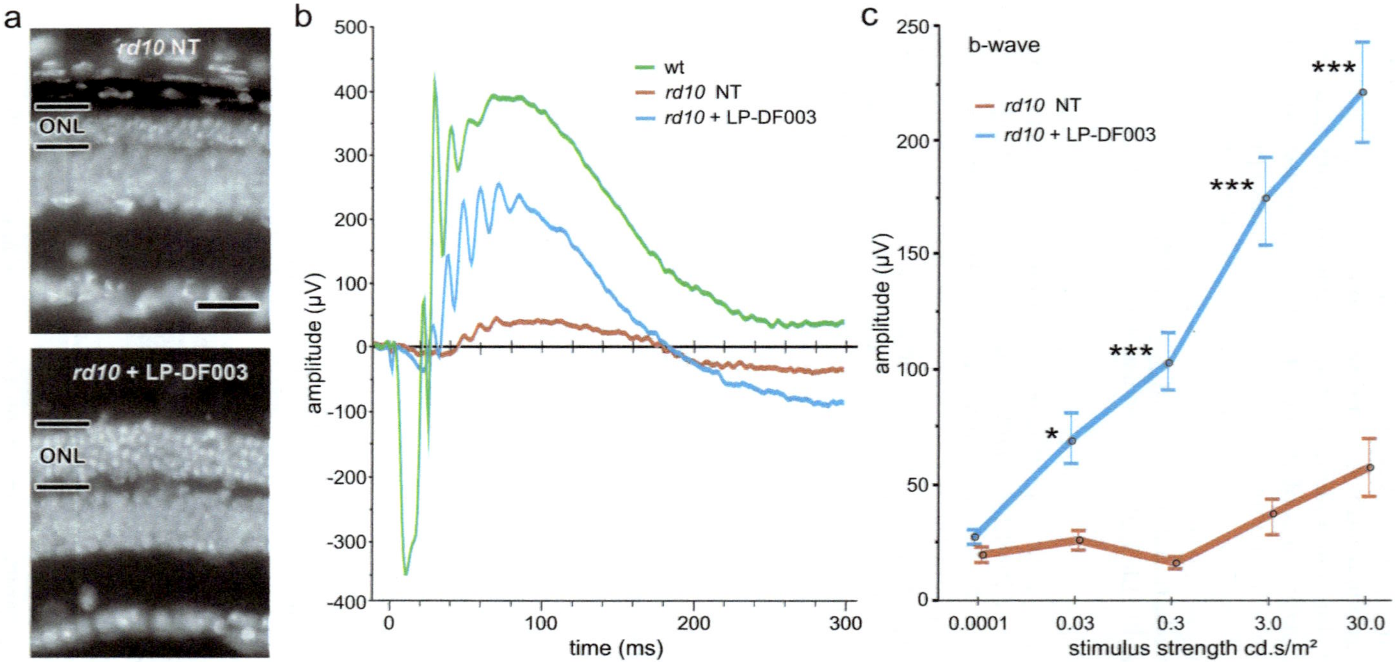

Figure 6.4 LP-DF003 preserves photoreceptor viability and retinal function in the *rd10* mouse *in vivo*. (a) Histological sections across non-treated (NT) and LP-DF003-treated *rd10* retinae at 30 days post-natal (P30) show how the treatment increased *rd10* photoreceptor survival. The thickness of the outer nuclear layer (ONL; *i.e.* the photoreceptor layer) is indicated. Scale bar = 50 μm. (b) Representative electroretinographic (ERG) responses (30.0 cd* s m^{-2}) in *rd10* animals; NT, and LP-DF003 treated. Traces from congenic, age-matched wild-type (wt) shown for comparison. (c) Quantification of b-wave ERG amplitudes *vs.* rising strength of light stimuli. Note the highly significant improvement of *rd10* retinal function with LP-DF003 treatment.

resulted in strongly improved *in vivo* functionality of the retina as assessed by ERG (Figure 6.4b). A quantification of b-wave ERG amplitudes over different light intensities revealed a highly significant LP-DF003 treatment effect, in particular in the photopic (*i.e.* daylight) range (Figure 6.4c).

Further ERG tests using scotopic and photopic flicker stimulation revealed that the functional response in treated *rd10* animals came almost exclusively from cone photoreceptors. Taken together, these results then indicated that the cGMP analogue DF003 had preserved *rd10* rod photoreceptor viability, while at the same time preserving cone photoreceptor viability and function. This is in line with the hypothesis that "sparing the rods saves the cones".[6]

6.7 Concluding Remarks

We propose cGMP-signalling as a key common denominator for the treatment of genetically very heterogenous retinal degeneration-type diseases. To target cGMP-signalling, we introduce cGMP analogues as a new class of pharmacologically active compounds. The combination with a nanoscale liposomal drug delivery vehicle efficiently transfers cGMP analogues to the retinal photoreceptors *in vivo*. In this way, we can preserve retinal function in relevant retinal degeneration animal models,[26] showcasing the therapeutic potential of cGMP analogues for currently untreatable blinding diseases.

Acknowledgements

This work was supported by the European Union (DRUGSFORD; HEALTH-F2-2012-304963), the Charlotte and Tistou Kerstan Foundation, and Programma di ricerca Regione-Università 2010–2012 of Regione Emilia Romagna (RARER), and the Medical Faculty at Lund University.

References

1. H. Kolb, How the retina works, *Am. Sci.*, 2003, **91**(1), 28–35.
2. A. Kennan, A. Aherne and P. Humphries, Light in retinitis pigmentosa, *Trends Genet.*, 2005, **21**(2), 103–110.
3. C. Hamel, Retinitis pigmentosa, *Orphanet J. Rare Dis.*, 2006, **1**, 40.
4. A. I. den Hollander, R. Roepman, R. K. Koenekoop and F. P. Cremers, Leber congenital amaurosis: genes, proteins and disease mechanisms, *Prog. Retinal Eye Res.*, 2008, **27**(4), 391–419.
5. M. Chizzolini, A. Galan, E. Milan, A. Sebastiani, C. Costagliola and F. Parmeggiani, Good epidemiologic practice in retinitis pigmentosa: from phenotyping to biobanking, *Curr. Genomics*, 2011, **12**(4), 260–266.
6. C. A. Curcio, C. Owsley and G. R. Jackson, Spare the rods, save the cones in aging and age-related maculopathy, *Invest. Ophthalmol. Visual Sci.*, 2000, **41**(8), 2015–2018.

7. B. Arango-Gonzalez, D. Trifunovic, A. Sahaboglu, K. Kranz, S. Michalakis and P. Farinelli, *et al.*, Identification of a common non-apoptotic cell death mechanism in hereditary retinal degeneration, *PLoS One*, 2014, **9**(11), e112142.
8. V. Sothilingam, M. Garcia-Garrido, K. Jiao, E. Buena-Atienza, A. Sahaboglu and D. Trifunovic, *et al.*, Retinitis pigmentosa: impact of different Pde6a point mutations on the disease phenotype, *Hum. Mol. Genet.*, 2015, **24**(19), 5486–5499.
9. E. V. Olshevskaya, A. N. Ermilov and A. M. Dizhoor, Factors that affect regulation of cGMP synthesis in vertebrate photoreceptors and their genetic link to human retinal degeneration, *Mol. Cell. Biochem.*, 2002, **230**(1–2), 139–147.
10. I. V. Peshenko, E. V. Olshevskaya and A. M. Dizhoor, Functional study and mapping sites for interaction with the target enzyme in retinal degeneration 3 (RD3) protein, *J. Biol. Chem.*, 2016, **291**(37), 19713–19723.
11. V. Ramamurthy, G. A. Niemi, T. A. Reh and J. B. Hurley, Leber congenital amaurosis linked to AIPL1: a mouse model reveals destabilization of cGMP phosphodiesterase, *Proc. Natl. Acad. Sci. U. S. A.*, 2004, **101**(38), 13897–13902.
12. S. Azadi, L. L. Molday and R. S. Molday, RD3, the protein associated with Leber congenital amaurosis type 12, is required for guanylate cyclase trafficking in photoreceptor cells, *Proc. Natl. Acad. Sci. U. S. A.*, 2010, **107**(49), 21158–21163.
13. C. Bowes, T. Li, M. Danciger, L. C. Baxter, M. L. Applebury and D. B. Farber, Retinal degeneration in the rd mouse is caused by a defect in the beta subunit of rod cGMP-phosphodiesterase, *Nature*, 1990, **347**(6294), 677–680.
14. S. H. Tsang, P. Gouras, C. K. Yamashita, H. Kjeldbye, J. Fisher and D. B. Farber, *et al.*, Retinal degeneration in mice lacking the gamma subunit of the rod cGMP phosphodiesterase, *Science*, 1996, **272**(5264), 1026–1029.
15. J. Xu, L. Morris, A. Thapa, H. Ma, S. Michalakis and M. Biel, *et al.*, cGMP accumulation causes photoreceptor degeneration in CNG channel deficiency: evidence of cGMP cytotoxicity independently of enhanced CNG channel function, *J. Neurosci.*, 2013, **33**(37), 14939–14948.
16. S. Huttl, S. Michalakis, M. Seeliger, D. G. Luo, N. Acar and H. Geiger, *et al.*, Impaired channel targeting and retinal degeneration in mice lacking the cyclic nucleotide-gated channel subunit CNGB1, *J. Neurosci.*, 2005, **25**(1), 130–138.
17. F. Paquet-Durand, S. M. Hauck, T. van Veen, M. Ueffing and P. Ekstrom, PKG activity causes photoreceptor cell death in two retinitis pigmentosa models, *J. Neurochem.*, 2009, **108**(3), 796–810.
18. F. Paquet-Durand, S. Beck, S. Michalakis, T. Goldmann, G. Huber and R. Muhlfriedel, *et al.*, A key role for cyclic nucleotide gated (CNG) channels in cGMP-related retinitis pigmentosa, *Hum. Mol. Genet.*, 2011, **20**(5), 941–947.

19. J. Tosi, R. J. Davis, N. K. Wang, M. Naumann, C. S. Lin and S. H. Tsang, shRNA knockdown of guanylate cyclase 2e or cyclic nucleotide gated channel alpha 1 increases photoreceptor survival in a cGMP phosphodiesterase mouse model of retinitis pigmentosa, *J. Cell Mol. Med.*, 2011, **15**(8), 9.
20. M. Kulkarni, D. Trifunovic, T. Schubert, T. Euler and F. Paquet-Durand, Calcium dynamics change in degenerating cone photoreceptors, *Hum. Mol. Genet.*, 2016, **25**, 3729–3740.
21. S. Wolfertstetter, J. P. Huettner and J. Schlossmann, cGMP-dependent protein kinase inhibitors in health and disease, *Pharmaceuticals*, 2013, **6**(2), 269–286.
22. J. Y. Wei, E. D. Cohen, Y. Y. Yan, H. G. Genieser and C. J. Barnstable, Identification of competitive antagonists of the rod photoreceptor cGMP-gated cation channel: beta-phenyl-1,N^2-etheno-substituted cGMP analogues as probes of the cGMP-binding site, *Biochemistry*, 1996, **35**(51), 16815–16823.
23. R. H. Kramer and G. R. Tibbs, Antagonists of cyclic nucleotide-gated channels and molecular mapping of their site of action, *J. Neurosci.*, 1996, **16**(4), 1285–1293.
24. E. Butt, M. Eigenthaler and H. G. Genieser, (Rp)-8-pCPT-cGMPS, a novel cGMP-dependent protein kinase inhibitor, *Eur. J. Pharmacol.*, 1994, **269**(2), 265–268.
25. E. Butt, D. Pohler, H. G. Genieser, J. P. Huggins and B. Bucher, Inhibition of cyclic GMP-dependent protein kinase-mediated effects by (Rp)-8-bromo-PET cyclic GMPS, *Br. J. Pharmacol.*, 1995, **116**(8), 3110–3116.
26. E. Vighi, D. Trifunovic, P. Veiga-Crespo, A. Rentsch, D. Hoffmann and A. Sahaboglu, *et al.*, Combination of cGMP analogue and drug delivery system provides functional protection in hereditary retinal degeneration, *Proc. Natl. Acad. Sci. U. S. A.*, 2018, **115**(13), E2997–E3006.
27. M. J. Hogan, J. A. Alvarado and J. E. Weddell, *Histology of the Human Eye: An Atlas and Textbook*, Saunders Company, Philadelphia, USA, 1971.
28. P. J. Gaillard, C. C. Visser, C. C. Appeldoorn and J. Rip, Targeted blood-to-brain drug delivery –10 key development criteria, *Curr. Pharm. Biotechnol.*, 2012, **13**(12), 2328–2339.
29. M. Campbell, M. M. Humphries and P. Humphries, Barrier modulation in drug delivery to the retina, *Methods Mol. Biol.*, 2013, **935**, 371–380.
30. T. Birngruber, R. Raml, W. Gladdines, C. Gatschelhofer, E. Gander and A. Ghosh, *et al.*, Enhanced doxorubicin delivery to the brain administered through glutathione PEGylated liposomal doxorubicin (2B3-101) as compared with generic Caelyx,((R))/Doxil((R))–a cerebral open flow microperfusion pilot study, *J. Pharm. Sci.*, 2014, **103**(7), 1945–1948.
31. D. Maussang, J. Rip, J. van Kregten, A. van den Heuvel, S. van der Pol and B. van der Boom, *et al.*, Glutathione conjugation dose-dependently increases brain-specific liposomal drug delivery in vitro and in vivo, *Drug Discovery Today Technol.*, 2016, **20**, 59–69.

32. M. Campbell, E. Ozaki and P. Humphries, Systemic delivery of therapeutics to neuronal tissues: a barrier modulation approach, *Expert Opin. Drug Delivery*, 2010, **7**(7), 859–869.
33. A. Trost, S. Lange, F. Schroedl, D. Bruckner, K. A. Motloch and B. Bogner, *et al.*, Brain and retinal pericytes: origin, function and role, *Front. Cell. Neurosci.*, 2016, **10**, 20.
34. D. Sanges, A. Comitato, R. Tammaro and V. Marigo, Apoptosis in retinal degeneration involves cross-talk between apoptosis-inducing factor (AIF) and caspase-12 and is blocked by calpain inhibitors, *Proc. Natl. Acad. Sci. U. S. A.*, 2006, **103**(46), 17366–17371.
35. G. C. Demontis, C. Aruta, A. Comitato, M. A. De and V. Marigo, Functional and molecular characterization of rod-like cells from retinal stem cells derived from the adult ciliary epithelium, *PLoS One*, 2012, **7**(3), e33338.
36. A. R. Caffe, P. Ahuja, B. Holmqvist, S. Azadi, J. Forsell and I. Holmqvist, *et al.*, Mouse retina explants after long-term culture in serum free medium, *J. Chem. Neuroanat.*, 2001, **22**(4), 263–273.
37. D. Trifunovic, B. Arango-Gonzalez, A. Comitato, M. Barth, E. M. Del Amo and M. Kulkarni, *et al.*, HDAC inhibition in the cpfl1 mouse protects degenerating cone photoreceptors in vivo, *Hum. Mol. Genet.*, 2016, **25**(20), 4462–4472.
38. A. Sahaboglu, N. Tanimoto, J. Kaur, J. Sancho-Pelluz, G. Huber and E. Fahl, *et al.*, PARP1 gene knock-out increases resistance to retinal degeneration without affecting retinal function, *PLoS One*, 2010, **5**(11), e15495.
39. N. Tanimoto, V. Sothilingam, M. Kondo, M. Biel, P. Humphries and M. W. Seeliger, Electroretinographic assessment of rod- and cone-mediated bipolar cell pathways using flicker stimuli in mice, *Sci. Rep.*, 2015, **5**, 10731.
40. C. E. Keeler, The inheritance of a retinal abnormality in white mice, *Proc. Natl. Acad. Sci. U. S. A.*, 1924, **10**(7), 329–333.
41. S. Sanyal and H. G. Jansen, Absence of receptor outer segments in the retina of rds mutant mice, *Neurosci. Lett.*, 1981, **21**(1), 23–26.
42. B. Chang, N. L. Hawes, R. E. Hurd, M. T. Davisson, S. Nusinowitz and J. R. Heckenlively, Retinal degeneration mutants in the mouse, *Vision Res.*, 2002, **42**(4), 517–525.

CHAPTER 7

PEDF Peptides in Retinal Degenerations

ALBERTO M. HERNÁNDEZ-PINTO[a] AND S. PATRICIA BECERRA*[b]

[a]3D Lab, Dept Molecular Biomedicine, Centro de Investigaciones Biológicas CSIC, C/ Ramiro de Maeztu 9, 28040 Madrid, Spain; [b]NIH-NEI, Bg. 6, Rm. 134, 6 Center Drive, Bethesda, MD 20892-0608, USA
*E-mail: becerrap@nei.nih.gov

7.1 Introduction

Protein multifunctionality unifies several individual functions on a single polypeptide chain. It seems to be a general characteristic of proteins in which each function is assigned to a different region or domain of the polypeptide chain. One example is in pigment epithelium-derived factor (PEDF), a natural protein found in the retina.[1] Its multifunctionality has combined independent neurotrophic and antiangiogenic activities and several binding functions on a single polypeptide.[2] It has been termed the "ocular guardian" due to its properties and localization, because it naturally protects the retina from degeneration by cell death and pathological neovascularization.[3] In addition, it has antitumorigenic, antimetastatic, antioxidant and anti-inflammatory properties.[4] Mapping of these functions reveals that they are localized in separate and distinct domains on

Drug Discovery Series No. 66
Therapies for Retinal Degeneration: Targeting Common Processes
Edited by Enrique J. de la Rosa and Thomas G. Cotter

Published by the Royal Society of Chemistry, www.rsc.org

the primary and tertiary structure of PEDF.[2,5] However, studies of the independent activities of PEDF are challenged by the presence of other properties – sometimes opposing – of the same molecule. An approach for such investigations has been to use peptide synthesis, protein chemistry and recombinant DNA technologies based on the three-dimensional structure of the protein to separate and/or alter them individually. This chapter summarizes the approaches to study this multifunctional protein within the context of the retina.

7.2 A Member of the Serpin Superfamily

Serpins (serine proteinase inhibitors) form a large and broadly distributed group of proteins, most of which are protease inhibitors. This superfamily currently includes more than 1500 members present in animals, fungi, bacteria, archaea and viruses.[6] They share a common overall protein conformation with a three-dimensional structure with three β-sheets and eight to ten α helices.[6] The majority of serpins exhibit protease inhibitory functions, and regulate processes such as inflammation, blood coagulation or extracellular matrix remodeling.[7] The region responsible for interaction with proteases is an exposed loop termed the "reactive center loop" (RCL), which is critical in inhibitory serpins because it acts as bait to attract target serine proteases.[8] Upon cleavage of the RCL by target proteases, inhibitory serpins undergo a conformational change that traps the enzyme inhibiting it in an irreversible fashion.[6]

There is a subgroup of serpins that lack protease inhibition activities, although they fold like their inhibitory counterparts. PEDF is a member of the serpin superfamily with no demonstrable inhibitory activity against serine proteases, and belongs to the subgroup of non-inhibitory serpins.[8–10] The 50 kDa glycoprotein is encoded by the human *SERPINF1* gene (Gene ID: 5176 and OMIM: 172860) located in chromosome 17, within the region 17p13.1.[10,11] The *SERPINF1* gene (~16 kb) contains eight exons separated by seven introns and codes for a 1.5 kb transcript widely detected during embryo development and adulthood in many tissues, including retinal pigment epithelium (RPE), retina, liver, testis, ovaries, stomach, prostate, heart, colon, brain, spinal cord, *etc.*[12] Figure 7.1 illustrates that PEDF sequences of mammalian species are highly conserved. It also shows a map of exon junctions on the amino acid sequence of human PEDF. Among ocular tissues, the RPE expresses the highest levels of the *SERPINF1* gene.[13] The polarized monolayer of RPE cells secretes PEDF protein preferentially toward its apical side into the interphotoreceptor matrix (IPM), where it has a role in the survival of photoreceptors and maintaining the avascularity of this region. PEDF is found surrounding the outer segments of photoreceptors.[13] In the inner retina, PEDF is detected at a lower level in the retinal ganglion cell layer, Müller glia and the inner nuclear layer.[14–16] Interestingly, *Serpinf1* expression was not detected in the outer nuclear layer of mouse retinas (personal observation). The ciliary body

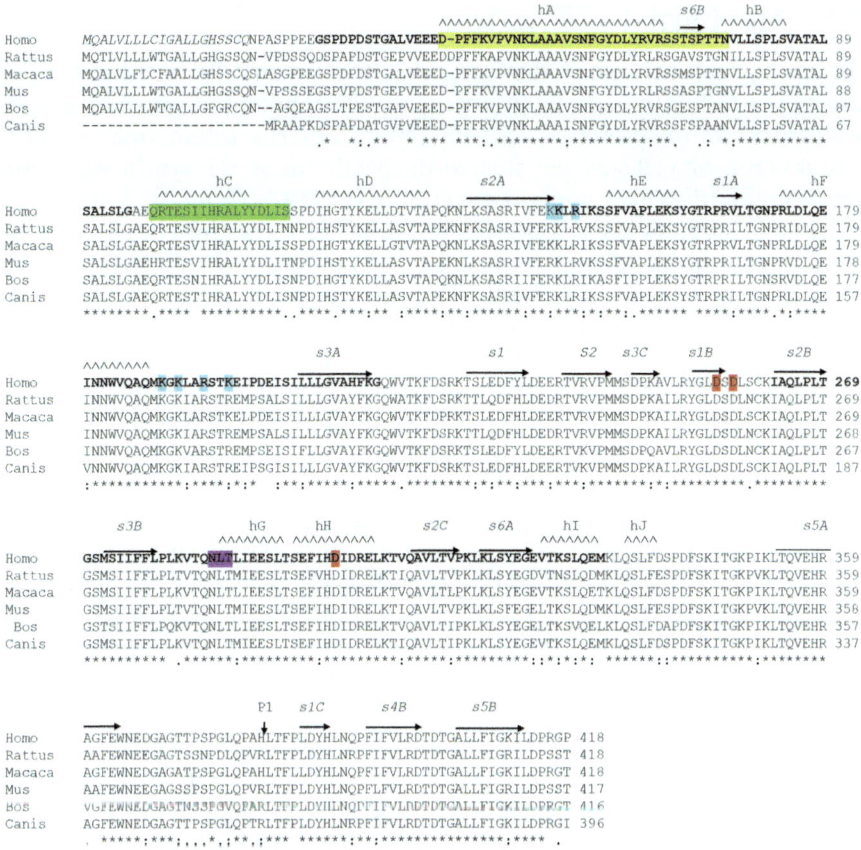

Figure 7.1 Sequence alignment of PEDF sequences among different mammalian species. The amino acid sequence of human PEDF was mapped to the exons of the SERPINF1 gene, and given in bold/non-bold code throughout the sequence. Alignment of the human, rat, monkey, mouse, bovine and dog PEDF amino acid sequences was done with CLUSTALW. The secondary structures of the human amino acid sequence from Protein Data Bank ID 1MV was used and mapped on the line above the sequence. Functional sites are color coded, as follows. Yellow: 34-mer – antiangiogenic domain; green: 17-mer – neurotrophic domain; cyan: glycosaminoglycan binding site; red: collagen binding site; purple: NLT *N*-glycosylation consensus site; in italics: the secretion signal peptide; P1: homologous cleavage site for serine proteases; ∧: α helix; →: β sheet; *: identical; :: conserved; .: semi-conserved.

also expresses the *SERPINF1* gene, which explains the presence of PEDF protein in the vitreous and aqueous humor.[17,18] PEDF binds effectively to components of the extracellular matrix. Its affinity for glycosaminoglycans (heparan sulfate, hyaluronan) and collagens serves for its localization and deposit in IPM, vitreous and aqueous humor.[2,5,19,20]

7.3 Mechanism of Action

The biological functions of PEDF depend on interactions with the surface of target cells rather than on its protease inhibition potential, given its lack of protease inhibitory properties. Several reports describe interactions of PEDF with proteins on cell surfaces such as the β subunit of ATP synthase,[21] laminin receptor,[22,23] plexin domain containing (PLXDC) 1 and 2,[24] low-density lipoprotein receptor-related protein 6 (LRP6)[25] and the protein encoded by *PNPLA2* (patatin-like phospholipase A2) gene termed PEDF-R.[26] While most of these receptors are associated to the pro-apoptotic or cell death inducing properties of PEDF, the *PNPLA2* gene is crucial for PEDF cell survival activity.[27,28] Altogether, the complexity of these receptors and subsequent downstream cascades activated by PEDF may explain its multifunctionality in living tissues.

The presence of the β subunit of ATP-synthase, laminin receptor and/or LRP6 in endothelial cells each contribute to PEDF antiangiogenic properties. There are lines of evidence for the interactions of the 34-mer region of PEDF with these receptors to trigger such activity. Upon PEDF binding there is a demonstrable inhibition of ATP synthesis and Wnt signaling, and regulation of the interaction of cell and matrix, which in turn promote apoptosis, cell death and inhibition of migration, tube formation and angiogenesis, *etc*. PEDF binding to PLXDC1 and PLXDC2 complexes disassembles the heterodimers and activates different cascades that vary depending on the cell lineage.[24] Whereas PLXDC2 initiates programmed cell death in endothelial cells, in neural cells PLXDC1 activates the antioxidant defense system and protect cells against oxidative insult, normally present in the retina due to its special physiology and oxygen consumption.[29] Mainly PLXDC1, and to a low degree, PLXDC2 can promote anti-inflammatory interleukin-10 expression in macrophages after PEDF binding, contributing to the maintenance of the immune privilege of the eye.

Upon binding PEDF-R, PEDF induces the antiapoptotic *Bcl2* gene, and in turn prevents the translocation of apoptosis inducing factor to the nucleus.[28,30] PEDF binding enhances the phospholipase A_2 activity of PEDF-R, releasing bioactive lipids from plasma membranes.[26] Docosahexaenoic acid (DHA)-containing phospholipids are found in high levels in the outer retina, in particular in photoreceptors.[31] PEDF-R can be found in cell surfaces of RPE cells and inner segments of photoreceptors, and consequently DHA-containing phospholipids become available substrates for the PEDF-R phospholipase activity.[26] Neuroprotectin (NPD1) is a neuroprotectant derived from DHA and PEDF enhances its release from RPE cells. Like PEDF, NPD1 has antiangiogenic, antioxidant and anti-inflammatory properties in the retina and central nervous system (CNS).[32,33] Moreover, it exerts neurotrophic activity by delaying the onset of the apoptotic pathway, including increases in antiapoptotic *Bcl-2* protein levels.[28,30,35] These observations imply a plausible upstream mechanism for transmitting PEDF signal in retina cells that involve DHA release by PEDF-R at the cell surface.

The region that confers neurotrophic activity to the PEDF polypeptide is located away from the homologous serpin active site.[34,35] This PEDF region interacts with the cell surface receptor PEDF-R that mediates the cytoprotective properties of PEDF in photoreceptors.[28,35] The interface that mediates the PEDF/PEDF-R protein–protein interactions has been mapped to a 17-mer region of PEDF and an ectodomain in PEDF-R,[28,35] as discussed later.

7.4 PEDF Isoforms

The mature extracellular PEDF protein lacks the first 20 amino acids encoded by the *SERPINF1* transcript, which correspond to a secretion signal peptide.[10,13,36] The polypeptide is *N*-glycosylated at asparagine position 285 within the *NLT* consensus motif for *N*-glycosylation that increases its molecular weight by ~5 kDa.[36,37] The three-dimensional structure of PEDF (Figure 7.2) has a globular protein conformation that confers protease resistance, except for a protease-sensitive exposed peptide loop located towards its *C*-end that contains the homologous RCL.[8] An amino acid leucine occupies the homologous P1 position within the RCL, which in the serpin antichymotrypsin gives specificity to inhibit chymotrypsin and cathepsin G. However, during evolution, PEDF lost its ability to inhibit proteases, and gained multiple biological activities.

Protein isoforms may explain the multifunctional modality of proteins. The distinct biochemical forms of PEDF vary in overall charge and have distinct biological efficacies on various cell types.[38,39] Upon maturation and secretion, PEDF versions may have adopted modifications post-translationally that confer charge differences, and in turn, grant diversity in biological activities to the protein. Some recombinant PEDF isoforms have differences in carbohydrate composition at its glycosylation site.[38] PEDF isoforms reported in the cerebrospinal fluid of patients with peripheral neuropathies with and without pain have differences in the presence of fucose residues on the glycosidic chain of PEDF, being lower in those without pain.[40] Additionally, differences in sialylation and generation of pGlu at the amino-terminus of the mature protein after the signal peptide is removed have been reported in PEDF isoforms from these patients. Another source of PEDF isoforms derives from phosphorylation of the polypeptide, as described by the conversion of PEDF from a neurotrophic to an antiangiogenic factor after extracellular phosphorylation.[41] In this regard, differential phosphorylation induces variable effects of PEDF, and therefore can contribute to the complexity of PEDF action.[42] Another example is the triple phosphomimetic PEDF, EEE-PEDF, with three negatively charged glutamic acids instead of serines, which possesses significantly increased antiangiogenic and anticancer activities in comparison to wild-type PEDF.[43–45] Naturally occurring PEDF isoforms have been identified in ocular, tumor and nontumor tissues, and, interestingly, their distribution varies between normal and tumor tissues.[36,46,47] These observations point to regulation of the biological activities

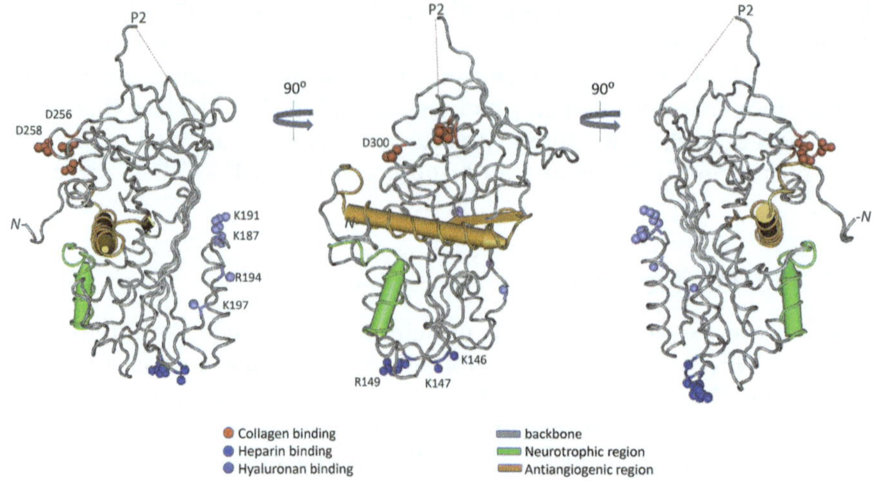

Figure 7.2 Three-dimensional structure of human PEDF protein (from Protein Data Bank ID 1 MV) illustrating structural features associated to function. The three structures are rotated ~90° from each other from left to right with highlighted domains for the neurotrophic and antiangiogenic regions and positions of amino acids critical for binding extracellular matrix components, as indicated at the bottom of the figure. Blue: basic amino acids K146, K147 and R149 residues located in a turn between β-strand s2A and α-helix E, and K189, K191, K194 and R197 located in another turn between α-helix F and β-strand s3A, both within B*X*7B HA-binding sites corresponding to heparin- and hyaluronan-binding sites, respectively; red: acidic amino acids D256, D258 and D300 corresponding to the collagen-binding site; orange: amino acid positions F44–N77 for the antiangiogenic domain; green: the 17-mer neurotrophic region E/Q98–S114; P2: the residue next to the homologous serpin reactive site, P1; *N*–: the position of the amino-end terminus of the polypeptide in the three-dimensional structure corresponding to position 26. Structures were visualized and reproduced using Cn3D (National Center for Biotechnology Information).

of PEDF by post-translational modifications, which can change the overall charge of the protein and/or expose a discrete active region of the folded PEDF protein to make it available for interactions with partners to trigger an individual function.

7.5 Mapping of Active Domains

Mapping the active domains of a multifunctional protein is useful for individualizing different activities. In cases in which the sites are comprised by nonconsecutive sequence, this becomes more complicated. The three-dimensional structure of PEDF has ~60% of its amino acid positions forming three major β-sheets and 10 α-helices[48] (Figures 7.1 and 7.2). A PEDF signature sequence VEEEDPF[FY] is found toward the amino-terminal region preceding the first α-helix A as a negatively charged peptide. A striking feature of the structure is an asymmetric charge distribution, with high density of basic residues concentrated on one side of the molecule and of acidic residues on the opposite side. Binding domains for extracellular matrix components map to specific sites on the three-dimensional structure. Protein chemical and mutational studies point out positively charged amino acids Lys^{166}, Lys^{167} and Arg^{149} necessary for heparin binding, and Lys^{189}, Lys^{191}, Arg^{194} and Lys^{197} for hyaluronan binding, while negatively charged amino acids Asp^{256}, Asp^{258} and Asp^{300} for collagen binding. Sequences of PEDF among mammalian species are highly conserved (see Figure 7.2).

Two regions confer distinct activities to the PEDF polypeptide, which are located away from the homologous RCL[12,34,35] and are responsible for binding or activating several of the above-described signaling cascades. The linear structure of human PEDF and the different domains and features are illustrated in Figure 7.3. The first region corresponds to the sequence of the human PEDF spanning between Asp^{44}–Asn^{77} (termed 34-mer) that has antiangiogenic properties. A peptide derived from this region can bind to the β-subunit of ATP synthase, laminin receptor and LRP6; can activate peroxisome proliferator-activated receptor-γ and impair vascular endothelial growth factor-mediated signaling pathways.[49–51] Two small peptide regions within the 34-mer region have been designed and characterized for antiangiogenic properties. Mirochnik *et al.* described P18 peptide (Asn^{60}–Asn^{77}), which impairs vascular endothelial growth factor (VEGF) signaling in endothelial cells[52] and Bernard *et al.* described P46 peptide (Phe^{46}–Ser^{70}) as the maximum binding domain to laminin receptor.[22] The second domain spanning the sequence between Val^{78}–Thr^{121} termed 44-mer has demonstrable neuroprotective and cytoprotective activities in a variety of cells.[8,53–56] It interacts with PEDF-R to mediate the cytoprotective properties of PEDF in photoreceptors.[21,28,35] A peptide derived from the 44-mer region (positions 82–121) can protect the retina from ischemic injury when delivered in PLGA nanospheres.[57] An even smaller region of 17 amino acids (positions 98–114) within the 44-mer contains critical residues for interacting with the PEDF neurotrophic receptor PEDF-R.[35]

7.6 Biomimetic Peptides as Drugs

The future potential of peptides in drug development is increasing. Peptides containing fewer than 50 amino acids are most interesting, as they are discriminated from proteins, referred to as "biologics". The advantages over biologics are several. Peptides are components of the very proteins that are in patients, and are amenable to site-specific modifications that might be used to tailor therapeutics to individual patients, as in personalized medicine. They have the specificity and potency of larger protein biologics, but are smaller in size and more accessible and cheaper to manufacture, combining the advantages of large biologics (>5000 kDa) and small drugs

Figure 7.3 Human PEDF amino acid sequence showing reported peptides. Scheme based on and modified from the study performed by Simonovic *et al.*[48]

(<500 kDa). They bind with exquisite specificity to their *in vivo* targets with relatively few off-target side effects. A small peptide is easily made in large quantities chemically, compared to the large, complex protein, which must be manufactured biologically. The peptide can be chemically synthesized for clinical use and scaled up very easily. With this idea in mind, peptides designed from retinoprotective factors have prospective high experimental value. Several PEDF peptides have been designed and tested on retina cells *in cellulo* and *in vivo*.

Fragmentation of the multifunctional PEDF polypeptide has been achieved by chemical peptide synthesis as well as by recombinant DNA technology (see Table 7.1). The peptide 34-mer exhibits antiangiogenic activity on an experimental model of choroidal neovascularization in rats.[49] Even when the peptide is injected after the onset of neovascularization, the neovascular lesions are significantly smaller than without the peptide. In this regard, biomimetic peptides derived from PEDF can provide sustained, multitargeted therapy that may provide advantages over intraocular injections of specific VEGF-neutralizing proteins.

Conversely, peptides derived from the neurotrophic region 44-mer of PEDF exhibit activity on their own like that of the full-length protein.[34,53] Alanine scanning of the small fragment from the neurotrophic region of PEDF of 17 amino acid residues reveals key interacting residues for binding PEDF-R and alternative retinoprotective peptide versions with higher efficacy.[35] Peptides

Table 7.1 Reported PEDF peptides, its properties and signaling pathways.[a]

PEDF domain	Effect	Signaling pathway/ receptor involved	Reference
34-mer	ATP decrease	β-subunit ATP synthase inhibition on EC	21 and 65
34-mer	Inhibition of EC migration and angiogenic gene expression	Wnt and PDGFR inhibition *via* LRP6	25, 66 and 70
34-mer	Inhibition of tubule network formation	FGF impairment *via* LR	22
34-mer	EC apoptosis	P53 and FasL expression *via* PPARγ activation	6
P46	EC apoptosis	Binding to LR	22
P18	EC apoptosis	Inhibition of VEGFR2 phosphorylation	52
Unknown	VEGFR impairment	Activation of γ-secretase	67
Unknown	EC apoptosis	PLXDC2	24
StVOrth-1	Increase cell adhesion and inhibit proliferation	Not fully described	68
StVOrth-2	Inhibit cell proliferation	Not fully described	68
StVOrth-3	Increase cell adhesion and inhibit VEGF expression	Not fully described	68
StVOrth-4	Inhibit cell invasion and VEGF expression	Not fully described	68
Unknown	Anti-inflammatory	IL-10 expression *via* PLXDC1 and PLXDC2	24
Unknown	Antioxidant	PLXDC1	24
44-mer	Neuroprotection	PEDF-R binding Neuroprotectin D1 expression	28 and 69
17-mer	Neuroprotection	Similar to 44-mer	35

[a]EC: endothelial cells; FGF: fibroblast growth factor; IL: interleukin; LR: laminin receptor; LRP6: low-density lipoprotein receptor-related protein 6; PDGFR: platelet-derived growth factor receptor; PLXDC: plexin domain containing protein; FasL: first apoptosis signal receptor Ligand or cluster of differentiation 95 (CD95); PPARγ: peroxisome proliferator-activated receptor gamma; PEDF-R: pigment epithelium-derived factor receptor; VEGF: vascular endothelial growth factor; VEGFR: vascular endothelial growth factor receptor.

with higher binding affinity for the receptor (*e.g.* 17-mer [H105A]) had higher efficacy for survival of retina-like cells R28 in culture and of photoreceptors in the rd1 mouse model of retinitis pigmentosa.

Given the structural homology that PEDF has with other serpins, we selected those with neurotrophic and antiangiogenic activities for comparison to identify novel peptides with relevant retinoprotective properties. Protease nexin-1 (PN-1), also known as glia-derived nexin, is a serpin with neurotrophic and antiangiogenic properties in the brain and retina.[58,59] It is a 43–50 kDa glycoprotein encoded by the human *SERPINE2* gene (Gene ID 5270 and OMIM 177010).[60] The adult brain, bone, seminal vesicle, cauda epididymis and ovary, as well as the retina and RPE express *SERPINE2*.[58,61,62] In contrast to PEDF, the secreted pericellular PN-1 inhibits thrombin, plasmin and plasminogen activators by forming complexes with these target proteases. Interactions with polysaccharides accelerate the PN-1-mediated thrombin inhibition by >100-fold.[63] Aside from a primary role of thrombin inhibition in the blood, PN-1 serves as a neurotrophic factor in the brain, CNS and peripheral nervous system and prevents cell death in cases of brain injury, disturbances of the blood–brain barrier and ischemia, all of which induce the upregulation of PN-1.[64] We have recently reported that PN-1 has retinoprotective activity and contains a neurotrophic active region that overlaps with that in the PEDF polypeptide.[58] The addition of recombinant human PN-1, a version of PN-1 referred to as PN-1[R346A], lacking serine protease inhibitory activity, and synthetic peptides designed from PN-1 sequences from the 17-mer region of PN-1 to serum-starved retina cells in culture decreased the number of TUNEL-positive nuclei relative to the untreated cells. PN-1, PN-1[R346A] and PN-1-17-mer treatments increased the antiapoptotic *Bcl2* transcript levels in serum-starved cells, as seen with PEDF. These observations point out that the mechanisms of action of the survival activity of PN-1 are not dependent on serine protease inhibition and that PN-1-derived peptides also have potential in retinoprotection.

7.7 Conclusions

The generation of retinoprotective peptides that activate specific and selective targets in the eye is of interest due to their high potential therapeutic value in retinal dystrophies. The well-established multiple actions of native PEDF, its involvement in controlling retina homeostasis, in addition to its downregulation in several retinal dystrophies make this serpin a prime candidate for future ocular therapeutic applications. The "ocular guardian" exerts neurotrophic, neuroprotective, antiangiogenic, gliastatic, antioxidant and antiangiogenic effects in the retina. It protects the retina from degeneration processes induced by cell death, pathological neovascularization, tumorigenesis and inflammation. Studies on protein structure and function have demonstrated that the multiple actions of PEDF rely on specific epitopes distributed throughout the protein and interactions with several targets, including specific surface receptors, orphan receptors or other proteins. Mapping of the biological active regions has made possible the isolation of individual and specific activities of this multifunctional protein. Protein fragmentation and chemical peptide synthesis have been employed in the design of small peptides that have retained independent activities of PEDF *in cellulo* and *in vivo*. Further studies on peptide chemistry, stability, pharmacokinetics, diffusibility and delivery systems of these peptides will prove useful in generating novel therapeutic agents for retinal dystrophies.

Acknowledgements

This work was supported by National Eye Institute, National Institutes of Health Intramural Research Program (to S.P.B.). This work was also supported by the Spanish Ministry of Economy, Industry and Competitiveness (CDS2010-00045 and SAF2016-75681-R projects), and by an ASBMB, PABMB and IUBMB PROLAB award (to A.M.H-P.).

References

1. S. P. Becerra, *Exp. Eye Res.*, 2006, **82**, 739.
2. S. P. Becerra, L. A. Perez-Mediavilla, J. E. Weldon, S. Locatelli-Hoops, P. Senanayake, L. Notari, V. Notario and J. G. Hollyfield, *J. Biol. Chem.*, 2008, **283**, 33310.
3. N. Bouck, *Trends Mol. Med.*, 2002, **8**, 330.
4. S. P. Becerra and V. Notario, *Nat. Rev. Cancer*, 2013, **13**, 258.
5. N. Yasui, T. Mori, D. Morito, O. Matsushita, H. Kourai, K. Nagata and T. Koide, *Biochemistry*, 2003, **42**, 3160.
6. R. H. Law, Q. Zhang, S. McGowan, A. M. Buckle, G. A. Silverman, W. Wong, C. J. Rosado, C. G. Langendorf, R. N. Pike, P. I. Bird and J. C. Whisstock, *Genome Biol.*, 2006, **7**, 216.
7. D. van Gent, P. Sharp, K. Morgan and N. Kalsheker, *Int. J. Biochem. Cell Biol.*, 2003, **35**, 1536.
8. S. P. Becerra, A. Sagasti, P. Spinella and V. Notario, *J. Biol. Chem.*, 1995, **270**, 25992.
9. P. G. Gettins, M. Simonovic and K. Volz, *Biol. Chem.*, 2002, **383**, 1677.
10. F. R. Steele, G. J. Chader, L. V. Johnson and J. Tombran-Tink, *Proc. Natl. Acad. Sci. U. S. A.*, 1993, **90**, 1526.
11. J. Tombran-Tink, H. Pawar, A. Swaroop, I. Rodriguez and G. J. Chader, *Genomics*, 1994, **19**, 266.
12. S. Filleur, T. Nelius, W. de Riese and R. C. Kennedy, *J. Cell. Biochem.*, 2009, **106**, 769.
13. S. P. Becerra, R. N. Fariss, Y. Q. Wu, L. M. Montuenga, P. Wong and B. A. Pfeffer, *Exp. Eye Res.*, 2004, **78**, 223.
14. K. C. Behling, E. M. Surace and J. Bennett, *Mol. Vision*, 2002, **8**, 449.
15. P. C. Karakousis, S. K. John, K. C. Behling, E. M. Surace, J. E. Smith, A. Hendrickson, W. X. Tang, J. Bennett and A. H. Milam, *Mol. Vision*, 2001, **7**, 154.

16. J. D. Unterlauft, W. Eichler, K. Kuhne, X. M. Yang, Y. Yafai, P. Wiedemann, A. Reichenbach and T. Claudepierre, *Neurochem. Res.*, 2012, **37**, 1524.
17. J. Ortego, J. Escribano, S. P. Becerra and M. Coca-Prados, *Invest. Ophthalmol. Visual Sci.*, 1996, **37**, 2759.
18. Y. Q. Wu and S. P. Becerra, *Invest. Ophthalmol. Visual Sci.*, 1996, **37**, 1984.
19. E. Alberdi, C. C. Hyde and S. P. Becerra, *Biochemistry*, 1998, **37**, 10643.
20. C. Meyer, L. Notari and S. P. Becerra, *J. Biol. Chem.*, 2002, **277**, 45400.
21. L. Notari, N. Arakaki, D. Mueller, S. Meier, J. Amaral and S. P. Becerra, *FEBS J.*, 2010, **277**, 2192.
22. A. Bernard, J. Gao-Li, C. A. Franco, T. Bouceba, A. Huet and Z. Li, *J. Biol. Chem.*, 2009, **284**, 10480.
23. T. Matsui, Y. Higashimoto and S. Yamagishi, *Biochem. Biophys. Res. Commun.*, 2014, **443**, 847.
24. G. Cheng, M. Zhong, R. Kawaguchi, M. Kassai, M. Al-Ubaidi, J. Deng, M. Ter-Stepanian and H. Sun, *elife*, 2014, **3**, e05401.
25. K. Park, K. Lee, B. Zhang, T. Zhou, X. He, G. Gao, A. R. Murray and J. X. Ma, *Mol. Cell. Biol.*, 2011, **31**, 3038.
26. L. Notari, V. Baladron, J. D. Aroca-Aguilar, N. Balko, R. Heredia, C. Meyer, P. M. Notario, S. Saravanamuthu, M. L. Nueda, F. Sanchez-Sanchez, J. Escribano, J. Laborda and S. P. Becerra, *J. Biol. Chem.*, 2006, **281**, 38022.
27. R. Gonzalez, L. L. Jennings, M. Knuth, A. P. Orth, H. E. Klock, W. Ou, J. Feuerhelm, M. V. Hull, E. Koesema, Y. Wang, J. Zhang, C. Wu, C. Y. Cho, A. I. Su, S. Batalov, H. Chen, K. Johnson, B. Laffitte, D. G. Nguyen, E. Y. Snyder, P. G. Schultz, J. L. Harris and S. A. Lesley, *Proc. Natl. Acad. Sci. U. S. A.*, 2010, **107**, 3552.
28. P. Subramanian, S. Locatelli-Hoops, J. Kenealey, J. DesJardin, L. Notari and S. P. Becerra, *J. Biol. Chem.*, 2013, **288**, 23928.
29. D. Athanasiou, M. Aguila, D. Bevilacqua, S. S. Novoselov, D. A. Parfitt and M. E. Cheetham, *FEBS Lett.*, 2013, **587**, 2008.
30. Y. Murakami, Y. Ikeda, Y. Yonemitsu, M. Onimaru, K. Nakagawa, R. Kohno, M. Miyazaki, T. Hisatomi, M. Nakamura, T. Yabe, M. Hasegawa, T. Ishibashi and K. Sueishi, *Am. J. Pathol.*, 2008, **173**, 1326.
31. S. J. Fliesler and R. E. Anderson, *Prog. Lipid Res.*, 1983, **22**, 79–131.
32. N. G. Bazan, *Brain Pathol.*, 2005, **15**, 159.
33. W. J. Lukiw, J. G. Cui, V. L. Marcheselli, M. Bodker, A. Botkjaer, K. Gotlinger, C. N. Serhan and N. G. Bazan, *J. Clin. Invest.*, 2005, **115**, 2774.
34. E. Alberdi, M. S. Aymerich and S. P. Becerra, *J. Biol. Chem.*, 1999, **274**, 31605.
35. J. Kenealey, P. Subramanian, A. Comitato, J. Bullock, L. Keehan, F. Polato, D. Hoover, V. Marigo and S. P. Becerra, *J. Biol. Chem.*, 2015, **290**, 25241.
36. Y. Q. Wu, V. Notario, G. J. Chader and S. P. Becerra, *Protein Expression Purif.*, 1995, **6**, 447.
37. E. Stratikos, E. Alberdi, P. G. Gettins and S. P. Becerra, *Protein Sci.*, 1996, **5**, 2575.

38. E. J. Duh, H. S. Yang, I. Suzuma, M. Miyagi, E. Youngman, K. Mori, M. Katai, L. Yan, K. Suzuma, K. West, S. Davarya, P. Tong, P. Gehlbach, J. Pearlman, J. W. Crabb, L. P. Aiello, P. A. Campochiaro and D. J. Zack, *Invest. Ophthalmol. Visual Sci.*, 2002, **43**, 821.
39. P. Subramanian, M. Deshpande, S. Locatelli-Hoops, S. Moghaddam-Taaheri, D. Gutierrez, D. P. Fitzgerald, S. Guerrier, M. Rapp, V. Notario and S. P. Becerra, *J. Biomed. Biotechnol.*, 2012, **2012**, 425907.
40. A. Conti, P. Ricchiuto, S. Iannaccone, B. Sferrazza, A. Cattaneo, A. Bachi, A. Reggiani, M. Beltramo and M. Alessio, *Proteomics*, 2005, **5**, 4558.
41. G. Maik-Rachline, S. Shaltiel and R. Seger, *Blood*, 2005, **105**, 670.
42. G. Maik-Rachline and R. Seger, *Blood*, 2006, **107**, 2745.
43. A. Konson, S. Pradeep, C. W. D'Acunto and R. Seger, *J. Biol. Chem.*, 2011, **286**, 3540.
44. Y. Feng, W. Bao, Y. Luo, L. Tian, X. Chen, M. Yi, H. Xiong and Q. Huang, *Invest. Ophthalmol. Visual Sci.*, 2012, **53**, 6793.
45. A. Konson, S. Pradeep and R. Seger, *Cancer Res.*, 2010, **70**, 6247.
46. A. M. Rodriguez-Pineiro, S. Blanco-Prieto, N. Sanchez-Otero, F. J. Rodriguez-Berrocal and M. P. de la Cadena, *J. Proteomics*, 2010, **73**, 1511.
47. J. Tombran-Tink, S. M. Shivaram, G. J. Chader, L. V. Johnson and D. Bok, *J. Neurosci.*, 1995, **15**, 4992.
48. M. Simonovic, P. G. Gettins and K. Volz, *Proc. Natl. Acad. Sci. U. S. A.*, 2001, **98**, 11131.
49. J. Amaral and S. P. Becerra, *Invest. Ophthalmol. Visual Sci.*, 2010, **51**, 1318.
50. S. Filleur, K. Volz, T. Nelius, Y. Mirochnik, H. Huang, T. A. Zaichuk, M. S. Aymerich, S. P. Becerra, R. Yap, D. Veliceasa, E. H. Shroff and O. V. Volpert, *Cancer Res.*, 2005, **65**, 5144.
51. R. Longeras, K. Farjo, M. Ihnat and J. X. Ma, *Exp. Diabetes Res.*, 2012, **2012**, 518426.
52. Y. Mirochnik, A. Aurora, F. T. Schulze-Hoepfner, A. Deabes, V. Shifrin, R. Beckmann, C. Polsky and O. V. Volpert, *Clin. Cancer Res.*, 2009, **15**, 1655.
53. M. M. Bilak, S. P. Becerra, A. M. Vincent, B. H. Moss, M. S. Aymerich and R. W. Kuncl, *J. Neurosci.*, 2002, **22**, 9378.
54. X. Gao, H. Zhang, W. Zhuang, G. Yuan, T. Sun, X. Jiang, Z. Zhou, H. Yuan, Z. Zhang and H. Dong, *Sci. Rep.*, 2014, **4**, 5637.
55. T. Sanagi, T. Yabe and H. Yamada, *J. Neuropathol. Exp. Neurol.*, 2010, **69**, 224.
56. S. C. Shih, T. C. Ho, S. L. Chen and Y. P. Tsao, *PLoS One*, 2016, **11**, e0157647.
57. H. Li, V. V. Tran, Y. Hu, W. Mark Saltzman, C. J. Barnstable and J. Tombran-Tink, *Exp. Eye Res.*, 2006, **83**, 824.
58. P. N. Winokur, P. Subramanian, J. L. Bullock, V. Arocas and S. P. Becerra, *Mol. Vision*, 2017, **23**, 372.
59. S. Selbonne, F. Azibani, S. Iatmanen, Y. Boulaftali, B. Richard, M. Jandrot-Perrus, M. C. Bouton and V. Arocas, *Mol. Cell. Biol.*, 2012, **32**, 1496.
60. R. E. Carter, K. M. Cerosaletti, D. J. Burkin, R. E. Fournier, C. Jones, B. D. Greenberg, B. A. Citron and B. W. Festoff, *Genomics*, 1995, **27**, 196.

61. M. Z. Djie, S. R. Stone and B. F. Le Bonniec, *J. Biol. Chem.*, 1997, **272**, 16268.
62. C. S. Simpson, H. M. Johnston and B. J. Morris, *Neurosci. Lett.*, 1994, **170**, 286.
63. B. Richard, M. C. Bouton, S. Loyau, D. Lavigne, D. Letourneur, M. Jandrot-Perrus and V. Arocas, *Thromb. Haemostasis*, 2006, **95**, 229.
64. C. Nitsch, A. L. Scotti, D. Monard, C. Heim and K. H. Sontag, *Eur. J. Neurosci.*, 1993, **5**, 292.
65. M. Deshpande, L. Notari, P. Subramanian, V. Notario and S. P. Becerra, *Int. J. Oncol.*, 2012, **41**, 219.
66. P. Protiva, J. Gong, B. Sreekumar, R. Torres, X. Zhang, G. S. Belinsky, M. Cornwell, S. E. Crawford, Y. Iwakiri and C. Chung, *Cell. Mol. Gastroenterol. Hepatol.*, 2015, **1**, 535e514.
67. J. Cai, L. Wu, X. Qi, S. Li Calzi, S. Caballero, L. Shaw, Q. Ruan, M. B. Grant and M. E. Boulton, *PLoS One*, 2011, **6**, e21164.
68. E. T. Ek, C. R. Dass, K. G. Contreras and P. F. Choong, *J. Orthop. Res.*, 2007, **25**, 1671.
69. J. He, M. S. Cortina, A. Kakazu and H. E. Bazan, *Invest. Ophthalmol. Visual Sci.*, 2015, **56**, 3505.
70. T. H. Tsai, S. C. Shih, T. C. Ho, H. I. Ma, M. Y. Liu, S. L. Chen, and Y. P. Tsao, *PLoS One.*, 2014, **9**, e95443.

CHAPTER 8

Beyond Anti-inflammation: Steroid-induced Neuroprotection in the Retina

ALICE C. WYSE-JACKSON, GILLIAN GROEGER AND THOMAS G. COTTER*

Cell Death and Development Laboratory, Department of Biochemistry, Biosciences Institute, University College Cork, Cork, Ireland
*E-mail: t.cotter@ucc.ie

8.1 Introduction

Many retinal diseases are caused by a loss of photoreceptor cells. Triggered by a number of different genetic aetiologies,[1] common disease features include night blindness, visual field loss and subsequent deterioration of visual acuity. Although recent advances in molecular tools have facilitated the successful implementation of gene therapy for some forms of retinal degenerations,[2–5] retinal diseases can be highly heterogeneous. For example, more than 3100 different genetic mutations have been reported to cause retinitis pigmentosa alone.[1] Therefore, a neuroprotective, mutation-independent treatment strategy is highly desirable in the fight against degenerative blindness. As such, the concept of "neuroprotection" is one therapy currently under investigation for the treatment of retinal disease.[6,7] Quite simply,

Drug Discovery Series No. 66
Therapies for Retinal Degeneration: Targeting Common Processes
Edited by Enrique J. de la Rosa and Thomas G. Cotter

Published by the Royal Society of Chemistry, www.rsc.org

neuroprotective strategies in the retina aim to conserve the photoreceptor layer, leading to preservation of photoreceptor cell function. These strategies involve the administration of differing molecules ranging from growth factors[8] to antioxidants[9] and steroids,[10] the focus of this review.

Steroids are a family of organic compounds containing a four cycloalkane ring core (Figure 8.1). There are hundreds of different types of steroids present in nature and all have varied function. However, in the human body, the majority of steroids are derived from the membrane lipid "cholesterol", which can form a number of different steroid derivatives through a series of enzymatic modifications. This review focuses on two such steroid groups associated with retinal cell neuroprotection: glucocorticoids and sex steroids.

Glucocorticoids are a class of corticosteroid. Binding to the glucocorticoid receptor,[11] the name glucocorticoid comes from the role that they play in glucose metabolism regulation, their synthesis in the adrenal cortex and the steroid structure: *Glu*cose + *cor*tex + ster*oid*. Three glucocorticoids used

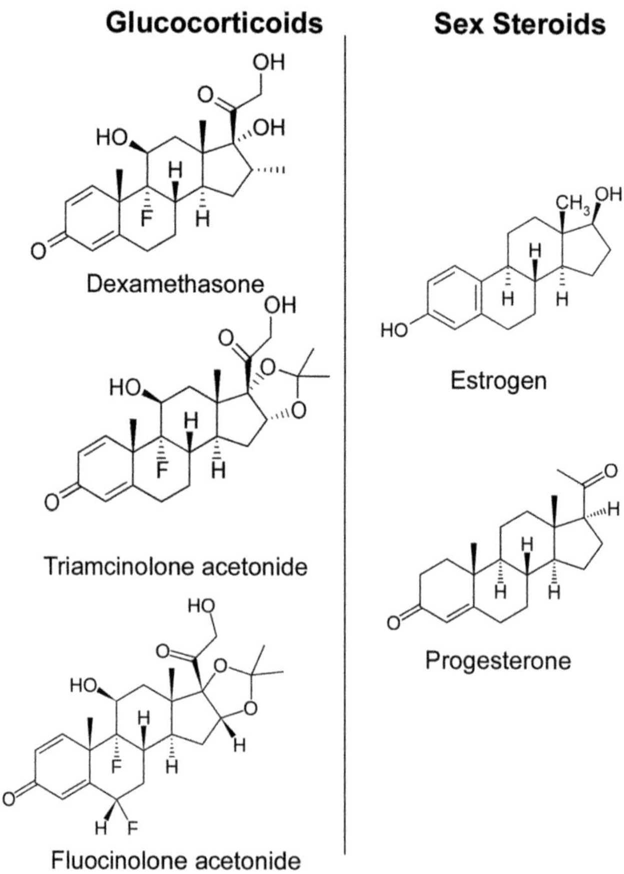

Figure 8.1 Chemical structures of steroids with recognized neuroprotective potential in the retina.

therapeutically in the retina are dexamethasone (DEX), triamcinolone acetonide (TA) and fluocinolone acetonide (FA) (Figure 8.1). Similar to other corticosteroids, these glucocorticoids are renowned for their anti-inflammatory properties, and it was this characteristic that first led to their use in the retina several decades ago.[11] More recently, increased clinical use has shown these glucocorticoids to have neuroprotective properties in retinal cells, though far less is known of the molecular mechanism leading to this protection.

The sex steroids can be divided into three functional groups: oestrogens, progestogens and androgens. The late 1970s saw the emergence of this area of research[12–14] and oestrogen and progestins have since been shown to promote neuronal survival and growth.[15,16] While androgens were investigated in relation to eye diseases in the 1950s,[17] these studies were not continued and so are not be discussed in this review. Instead, we focus upon a comparison of the two sex steroid groups, oestrogen and progestogen.

The glucocorticoids and sex steroids share a very similar structure (Figure 8.1). As such, it is likely that there are observations and subsequent advances to be made from studying the two steroid groups in parallel. Through emphasizing similarities and differences between the glucocorticoids DEX, TA and FA, and the sex steroids oestrogen and progesterone, insights gained from one group will be highlighted with the aim of aiding future studies of the other. Through this, we hope to facilitate the development of these steroids as neuroprotective therapies for retinal diseases.

8.2 Glucocorticoids

8.2.1 Evidence for Protection

In certain retinal degeneration disease models, glucocorticoids have been shown to be neuroprotective to photoreceptors. For example, in the rodent light damage model, DEX prevents photoreceptor cell death.[18–22] TA is protective to photoreceptors in a rabbit model of subretinal haemorrhage[23] and preserves the inner segment/outer segment photoreceptor interface in patients with diabetic macular oedema (DMO).[24] Glybina *et al.* also demonstrated specific protection of photoreceptors using retinal implants of FA in two inherited retinal degeneration models, the Royal College of Surgeons (RCS) rat model[25] and the s-334ter-4 rat model.[26] These studies (Table 8.1) were completed over the same time period as several different human clinical trials of varying degrees of success. These studies all looked at the potential for glucocorticoids to act as anti-inflammatories in the eye, and were nicely reviewed in 2015 by Cebeci and Kir.[27]

Clinical trials used various formulations of these steroids, demonstrating protection of vision in humans. DEX is five times and 20 times more potent an anti-inflammatory agent than TA and FA, respectively,[11,28] however, its short half-life and high solubility means that it is not retained in the eye.[29] To combat this, Ozurdex (Allergan, Irvine, CA, USA), a slow-release DEX

Table 8.1 Summary of studies which used the steroids under review as therapeutics in models of retinal degeneration.[a]

Steroid	Model used	Dose + route of administration	Degree of protection; variable[b]	Reference
DEX	Mouse light damage	52 mg kg^{-1} in single IP injection	Complete; ONL Complete; ERG	22
	Mouse light damage	52 mg kg^{-1} in single IP injection	Qualitative; retinal neuronal layer integrity	19
	Dog light damage	3 × SC injections (2 @ 0.18 mg kg^{-1}, 1 @ 1.8 mg kg^{-1})	None found in retinal sections	47
TA	Rabbit subretinal haemorrhage	2 mg in single intravitreal injection	Quantitative; ONL	23
	Human-diabetic macular oedema	4 mg in single intravitreal injection	Quantitative; IS/OS as measured by OCT	24
FA	Royal College of Surgeons rat	0.2 µg day^{-1} equivalent release from intravitreal implant	Quantitative; ONL Quantitative; ERG	25
	S334ter-4 rat	0.2 µg day^{-1} equivalent release from intravitreal implant	Quantitative; ONL	26
Oestrogen and derivatives	Rat light damage	2.5 mg	Qualitative; retina	115
	S334ter-3 rats	2 µL of 10 mM in single intravitreal injection	Quantitative; ONL	84
	Rat light damage	500 µg kg^{-1} in single IP injection	Qualitative; ONL Qualitative; ERG	85
Progestins	RD10 mouse	100 mg kg^{-1} IP injection on alternate days	Quantitative; ONL Qualitative; ERG	45
	Mouse light damage	100 mg kg^{-1} IP injection every 3 days	Qualitative; ONL	45
	Rat retinal ischaemia reperfusion	4 mg kg^{-1} in 2 × IP injections	Qualitative; INL	120

[a]ERG: electroretinogram; DEX: dexamethasone; FA: fluocinolone acetonide; INL: inner nuclear layer; IP: intraperitoneal; IS/OS: inner segment/outer segment of photoreceptors; OCT: optical coherence tomography; ONL: outer nuclear layer; SC: subcutaneous; TA: triamcinolone acetonide.

[b]Qualitative: study reports images only; quantitative: study undertook statistical analysis, which proved significance.

intravitreal implant, was developed. At 0.46 mm in diameter and 6 mm in length, this rod-like implant is used to treat patients with uveitis, and to protect vision in patients with DMO and macular oedema related to branch or central retinal vein occlusion.[30,31] Approved by the US Food and Drug Administration in 2009 and the European Commission in 2010, long-term evaluation of patients receiving multiple implants are now available and seem promising.[32]

TA, due its more lipophilic nature, is better retained within the eye post-injection.[33] Abraldes *et al.*[34] reviewed all the clinical trials of TA in the treatment of DMO and found that TA can increase visual acuity in certain cases. For example, I-vation (SurModics, Eden Prairie, MN, USA), a helical sustained-release implant containing 0.925 μg TA, demonstrated the effectiveness of TA in treating DMO. However, major complications involving increased intraocular pressure (IOP) and cataract development[35] meant that phase 2b trials were terminated.[36] Finally, similar to Ozurdex and I-vation, a slow-release implant of FA has been developed: Retisert (Bausch & Lomb, Rochester, NY, USA). This implant had some success in clinical trials for uveitis; however as with I-vation, patients showed increased cataract formation and raised IOP, which has consequently restricted its use.[37]

In general, advanced research in this area is directed towards the development of slow-release systems.[25,26,37] These enable steroids to be retained better within the eye, thereby reducing the number of intravitreal injections required, and consequently decreases associated complications. In addition, intravitreal implant systems negate the need for a vehicle, *i.e.* the solution in which the steroid suspended before injection. These vehicles can often cause complications: a case well illustrated by the first studies of TA. When TA was first injected intravitreously, it was suspended in benzyl-alcohol using a product originally designed for intramuscular injection to treat dermatological conditions. These studies not only found the vehicle to be toxic to the retina,[38] but also found TA itself to be toxic at high doses (4–20 mg of TA in 0.1 mL injection). TA is only safe and neuroprotective in humans at the lower dose of 0.5–1 mg.[39] Difficulties associated with administration of individual systems serve to highlight the need for specific testing of various drug delivery routes, rather than simply trialling the compound itself.

In all of the aforementioned trials, visual acuity was improved through corticosteroid administration. Accordingly, there must have been some degree of photoreceptor protection. However, the trouble with this deduction is that many of these trials were completed in patients suffering from macular oedema.[30,33] Macular oedema occurs when fluid and protein deposits collect on or under the macula of the eye, resulting in poor communication between the photoreceptors and other retinal neurons. This means that photoreceptors lose out on vital signalling and thus loss of vision occurs thereafter.[40] Since macular oedema is not associated with photoreceptor death in its early stages, it is difficult to determine if glucocorticoids affect photoreceptor cells directly. Glucocorticoids promote a healthy retina through excess fluid reduction. Therefore, the photoreceptors are not as stressed and so are less likely to die, as normally occurs in the later stages of this disease. Human studies into macular oedema and animal trials into other retinal diseases all agree that glucocorticoids can protect vision. However, the mechanism of action by which this occurs, be it directly or indirectly on the photoreceptors, has not yet been fully elucidated.

8.2.2 Mechanism of Action

Retinal cells have a bi-modal response to glucocorticoids, due to signalling through differing glucocorticoid receptor (GR) isoforms.[41,42] Alternative splicing of the GR gene transcript generates two *C*-terminal glucocorticoid receptor isoforms: GRα and the classic GR. In addition, the GR gene has several initiation sites, which can give rise to isoforms of differing lengths: GRβ, GRγ, GR-A and GR-P.[42] The classical GR and GRα initiate a response through genomic signalling.[11] This mechanism allows for ligand–receptor complex binding and sequential internalization to the nucleus. Here, they bind to glucocorticoid response elements, directly regulating the transcription of various genes. For example, GRα may inhibit key transcription factors, nuclear factor-κB (NF-κB) and activator protein-1 (AP-1).[43] However, as with many steroid hormones, glucocorticoids may also activate a non-genomic signalling system. This means that a membrane-bound form of GR may directly activate an intracellular signalling cascade, thus transmitting signalling information at a much faster speed than the slower classical GR signalling.[44] Precisely how these two signalling mechanisms result in increased cell survival is still ill-defined, though what is known will be discussed below.

With regard to the direct action of glucocorticoids on photoreceptors, Wenzel *et al.* attempted to explicate their mechanism through use of the light damaged mouse model of retinal degeneration. This is a method whereby adult, albino balb/c mice are exposed to a bright a damaging light over a length of time and their photoreceptors degenerate as a result.[45] Wenzel *et al.* found that intraperitoneal injection of DEX activated the GR, which in turn reduced the levels of the transcription factor AP-1 within the retina[22] (Figure 8.2A). Conversely, AP-1 has been shown under certain conditions to induce photoreceptor death.[46] Hao *et al.* further investigated this mechanism in mice. They found that DEX only protected against bright light death, where AP-1 was strongly implicated. However, in dim light induction where AP-1 did not play a role, DEX had no effect.[19]

In contrast to these mouse investigations, a study completed in dogs using bright light showed no protection of photoreceptors with dexamethasone, although the concentration of DEX used was far less than in the aforementioned mouse trials (Figure 8.1).[47] This dog trial found again that despite a DEX-induced decrease in AP-1 activation, and an inhibited phosphorylation of c-Fos and extracellular signal-regulated kinases 1/2 (Figure 8.2B), neither intraocular nor systemic injection of DEX could protect against light-induced photoreceptor degeneration. Gu *et al.* hypothesised that the discrepancy in

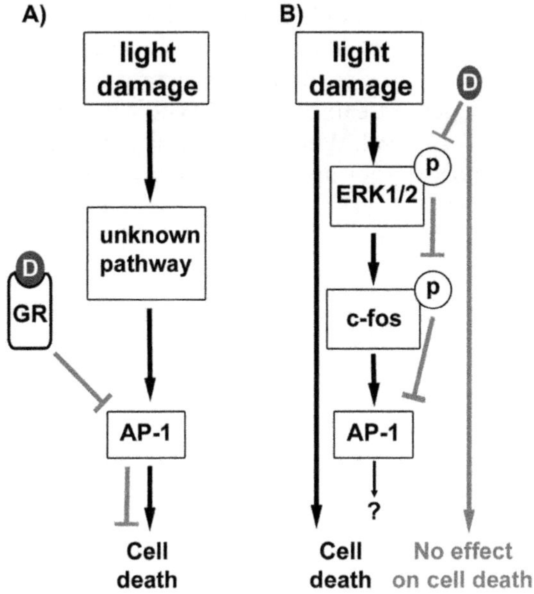

Figure 8.2 Known signalling pathways of dexamethasone in the retina. (A) Dexamethasone (D) activates the glucocorticoid receptor (GR) and inhibits activator protein-1 (AP-1), which prevents cell death.[22] This was determined in a mouse light damage model. (B) Light damage causes cell death in a dog model and this is concurrent with an increase in the phosphorylation (p) of extracellular signal-regulated kinase (ERK)1/2 and c-fos and an increase in AP-1 activity. While dexamethasone inhibits these three events, it has no effect on cell death.[47]

DEX efficacy between the mouse and dog studies had to do with the differing role that AP-1 may play in photoreceptor cell death.[47] Certainly, since these studies were completed, AP-1 has been recognized to promote cell death or survival, depending on its dimer composition.[48] Therefore, it is too simplistic to say that AP-1 activation will always lead to cell death. Thus, DEX-induced reduction of AP-1 may not always be neuroprotective to photoreceptor cells (Figure 8.2).

In a recent study, it was shown that DEX administration to light-damaged balb/c mice significantly increased levels of protective retinal BCL-X_L in photoreceptor inner segments,[49] as is the case in other tissues whereby DEX has anti-apoptotic actions.[50] In comparison, high levels of cleaved caspase-3 and Bax, both drivers of apoptosis,[51] and phosphorylated histone H2AX, a marker of DNA breakage and repair[52] were found in mifepristone GR antagonist-treated controls.[49] This study is one of the first to elucidate the signalling pathways associated with DEX-induced photoreceptor cell survival *in vivo*.

While it is evident that some progress is being made in understanding the actions of DEX in promoting neuronal survival, very little is known about the intracellular signalling mechanisms of TA and FA in neurons.

Using microarrays[53] and through analysis of selected genes,[54] both have been shown to regulate genes in a characteristic anti-inflammatory manner. However, these studies did not manage to determine the pathway leading to transcription changes. FA can bind to the GR,[55] but it is not known if this promotes retinal cell survival. What is known is that when TA has cytotoxic effects (*i.e.* at high doses), they are independent of classical GR signalling,[56] suggesting a toxic non-GR mechanism. Unfortunately, due to the lack of published studies in this area, it is impossible to conclude how TA and FA might signal to protect photoreceptor cells directly.

It is believed that many of the effects that glucocorticoids have on photoreceptors may occur indirectly. Working on other retinal cell types, the photoreceptors are consequently affected downstream of these actions. Glucocorticoids have been widely used to inhibit inflammation, a finding recently validated in the retina.[11] For example, Glybina *et al.* found that a retinal implant releasing the equivalent of 0.2 $\mu g\ day^{-1}$ FA was effective in protecting photoreceptors in RCS and transgenic S334ter-4 rat models.[25,26] They demonstrated that this protection results from a decrease in activation and recruitment of microglia to damaged photoreceptors – a process now commonly accepted to potentiate photoreceptor cell degeneration.[57,58] This diminished microglial response leads to reduced production of pro-inflammatory cytokines and ultimately prolongs photoreceptor cell survival.[51,59–61] In addition, TA has been found to reduce the number of microglial cells within the retina in response to a variety of different injurious stimuli.[62–64] Taken together, these results suggest that the glucocorticoids are highly successful mediators of damaging microglial cell activation in the retina.

Microglia are not the only glial cell type in the retina. The retina also has glial cells known as Müller glial cells, which, like all glial cells serve as support cells for the neurons of the eye.[57] Spanning the width of the retina, these cells respond in a corticosteroid-specific manner. For example, TA has no influence on Müller cell gliosis,[62] a process occurring after neuronal damage. Reactive gliosis is characterised through Müller glial cell proliferation and changes in specific gene expression. Consequently, this may lead to glial scarring and inhibition of axonal regeneration.[65] Alternatively, DEX has been shown not only to supress reactive gliosis by Müller glial cells,[47] but through the GR may also alter their mitochondrial metabolism.[66] Two studies have demonstrated that DEX and TA can modify the number of ion and water channels expressed by Müller glial cells, changing their osmotic swelling capacity, thereby aiding the resolution of oedema.[53,67] The effects of glucocorticoids on retinal cell types other than photoreceptors reflect their use in clinical trials. Through modification of glial cell signalling, the glucocorticoids contribute to an anti-inflammatory-response in the retina, a process which has been proven conclusively to delay photoreceptor cell death.[51,59–61]

In addition to retinal cells, intravitreally administered steroids may impact other ocular cell types. In a comprehensive microarray study of the human trabecular meshwork, Fan *et al.* found that 1 mg mL^{-1} of TA upregulated 36 genes and downregulated 21 others.[68] This was dose dependent, as 0.1 mg mL^{-1} resulted in fewer gene alterations. While gene alterations within trabecular cells may not impact on photoreceptors directly, they contribute to the overall health of the eye, as indicated by the association between their malfunction and glaucoma.[69] This study by Fan *et al.* is important for another reason. It directly compared genes altered by TA and dexamethasone in the same cell line. Even though the structural differences between these two glucocorticoid molecules are relatively small (Figure 8.1), they differentially regulate several genes. TA changed the expression of 57 genes in total, seven of which it shared with DEX, while DEX affected a separate 19 genes.[68]

The evidence presented in this section serves to illustrate what little is known about the pro-survival signalling mechanisms employed by glucocorticoids within the retina. Nevertheless, glucocorticoid-induced increase in visual acuity is evident in almost all the discussed studies, and thus highlights the potential of these compounds to provide a successful therapy for certain inflammatory retinal diseases.

8.3 Sex Steroids

8.3.1 Evidence for Protection

Steroid hormones and their metabolites in the central nervous system have been shown to have neuroprotective properties.[70,71] Commonly referred to as "neurosteroids", the sex steroids oestrogen and progesterone are now well documented to provide protection in several different systems. Progesterone, for instance, has been trialled for a treatment of spinal cord injury, brain ischaemia, stroke, Alzheimer's disease and perhaps most prominently, traumatic brain injury (TBI).[70,72–77] Indeed, studies highlighting its viability in a number of different models of TBI have ensured progesterone's entry into human clinical trials.[78–80] Oestrogen has been trialled for protection against spinal cord injuries, cognitive decline following cerebral ischaemia and several neurodegenerative disorders.[81–83] These studies seem promising, making it likely that oestrogen will soon be assessed clinically.

Clearly, both progesterone and oestrogen are emerging as potential therapeutics in the treatment of neurodegenerative disorders. As such, research in to the protective properties of these compounds has moved in to the retina. Oestrogen and its derivatives in particular have shown promising results in a number of systems and models of retinal disease. These include the s334ter-3 transgenic model of retinitis pigmentosa,[84] light-damaged rats,[85] 661W cone photoreceptor cells,[86] human Müller glial cells[87] and rat retinal neuronal primary cultures[85,88] (Table 8.1). Interestingly, despite the success of progesterone treatment in disorders of the central nervous system, conflicting data exist as to the protective potential of progesterone in the eye. As such, retinal studies have been slower to start. Norgestrel, a synthetic progesterone analogue, has been shown to exert neuroprotective effects both *in vitro* and *ex vivo*;[89–92] and in two distinct models of retinitis pigmentosa: the rd10 mouse model, in a preventative situation;[93,94] and the light-damage model, in a rescue situation.[45,95] In addition, progesterone was shown to prevent the rapid photoreceptor cell death seen in the rd10 mouse and to inhibit osmotic swelling following ocular ischaemia.[96,97]

While there have been no human clinical trials wherein oestrogen or progesterone has been specifically administered to preserve vision, some trials have been designed to take into consideration hormone replacement therapy. This is where women are supplemented with exogenous hormones. In 2008 for example, a large-scale study was undertaken in 74 996 postmenopausal women. This study looked at the incidence of age-related macular degeneration (AMD) in those taking postmenopausal hormones (PMH) consisting of a combined treatment of progesterone and oestrogen; and those who had previously taken oral contraceptives. Most interestingly, PMH users had a notable 48% lower risk of neovascular AMD compared with those were not using PMH. Moreover, the risk of developing neovascular AMD was lowest amongst PMH users who had used oral contraceptives in the past.[98]

In 2001, a study of 58 women looked at the effect of the PMH tibolone, a steroid drug with progesterone, oestrogen and weak androgen actions. This research showed that the functional status of neurovisual transmissions was significantly improved after a 3-month daily dose of 2.5 mg tibolone.[99] Interestingly, no difference in effects on visual acuity, IOP, tear functions and visual evoked potential were seen in a more prolonged experiment totalling 6 months, just a few years later.[100] More recently, the effects of PMH were investigated with respect to blood flow and protection of the optic nerve. Researchers found that PMH users (of both oestrogen-only and oestrogen plus progesterone PMH preparations) had a significantly increased retinal perfusion and greater preservation of the optic nerve fibre layer compared to controls.[101] These studies suggest a neuroprotective role of PMH in the retina. Taken together with a number of neuronal based studies showing that PMH users have a decreased risk of developing neurodegenerative disease, *e.g.* Alzheimer's disease,[102] Parkinson's disease[103] or stroke,[104] these data lead us to conclude that the area of PMH supplementation for postmenopausal retinal protection deserves further attention.

8.3.1.1 Oestrogen-specific Neuroprotection

With regard to oestrogen-specific retinal neuroprotection, one of the earliest studies was carried out using the phytoestrogen genistein. This naturally occurring, oestrogen-like compound was found to prevent neuronal degeneration in the inner retina of rats occurring as a result of ischaemic injury.[105] The most potent endogenous oestrogen, 17β-oestradiol, was soon found to prevent retinal cell death induced by the excitotoxin pregnenolone sulphate in male Wistar rats, through prevention of a pregnenolone sulphate-induced reduction in pro-forms of caspase-3 and caspase-2.[106] 17β-estradiol was subsequently and sequentially found to protect against hydrogen peroxide-induced cell death in primary cell cultured retinal neurons[85] and in ARPE-19 human retinal pigment epithelial cells;[107,108] against retinal ganglion cell loss in the DBA/2J mouse model of inherited glaucoma;[109] and against glutamate-induced neurotoxicity (5 mM) in 661W cells, a mouse cone photoreceptor-like cell line.

In 2004, Yu *et al.* tested both 17α-oestradiol and 17β-oestradiol *in vivo*, using light-damaged, ovariectomized (an animal model for menopause), female Sprague-Dawley albino rats. Systemic administration of 17β-oestradiol was found to significantly protect the structure and function of light-damaged rat retinas, inhibiting photoreceptor apoptosis. Most recently, Deschênes *et al.* continued this 17β-oestradiol study of light-damaged, ovariectomized rats, alongside their study of postmenopausal hormone therapy on blood flow and optic nerve protection (discussed earlier). Through this paralleled research, they found that not only did 17β-oestradiol treatment significantly increase retinal perfusion in the range 22–45%, but also that perfusion flow and diameter was greater in the oestrogen-only therapy group than in the oestrogen plus progestogen therapy group. However, the differences were not statistically significant.[101]

The positive results obtained both *in vitro* and *in vivo* for cellular protection following oestrogen administration paved the way for human clinical studies of postmenopausal women taking oestrogen-only PMH. Fraser-Bell *et al.* found that exogenous oestrogen was protective against soft drusen formation, associated with dry AMD.[110] This finding was similar to a study performed around the same time by Haan *et al.*, whereby oestrogen was administered concomitantly with progestin to postmenopausal women. Combined, these steroids effectively conferred against soft drusen formation and neovascular AMD.[111] In 2012, clinical trials using oestrogen-derivative phytoestrogen for the treatment of postmenopausal dry eye verified the safety and efficacy of direct oestrogen-to-eye administration in humans.[112] Altogether, this litany of studies suggest that exogenous oestrogen should be further explored as a potential treatment for retinal cell survival.

8.3.1.2 Progesterone-specific Neuroprotection

Despite clear protective effects in the brain, conflicting studies exist of the efficacy of progesterone in retinal cell neuroprotection. Early research into the actions of progesterone in the eye began in the 1970s.[113,114] In 1977, one of the first major experiments using the light-damage model of retinal degeneration was carried out in ovariectomized albino rats. Rats were treated with 2.5 mg progesterone before exposure to damaging light; however, no difference was seen in photoreceptor cell death in progesterone-treated rats compared to control.[115] The negative results of this study ensured that interest in the neuroprotective efficacy of progesterone in the retina waned in following years, until its sudden reappearance in retinal literature in the mid-2000s.[85,116,117] Much of this research demonstrated that progesterone could not protect against cell death in photoreceptors[85,116] and indeed may actually promote vascular endothelial growth factor-induced neovascularization within the retina.[117] And yet, despite these results, promising studies have recently been released detailing the potential of progesterone to suppress retinal cell death.

Research from our group has extrapolated the capabilities of the oral contraceptive levonorgestrel (commonly referred to as "norgestrel") to protect damaged photoreceptors.[45,89–95,118,119] These studies verify previous literature looking at progesterone-induced protection in models of diabetic and ischaemic retinopathies. In 2008, two 4 mg kg^{-1} intraperitoneal injections of progesterone were shown to preserve the inner nuclear and nerve fibre layers (Table 8.1) in a rat model of retinal ischaemia-reperfusion.[120] Following on from this, progesterone was shown to reduce swelling of retinal glial cells in models of diabetic and ischaemic retinopathies, implying a consequential overall reduction of cytotoxic retinal oedema.[96]

These retinal studies, along with research undertaken by the broader neuroscience research field,[70,72–80] provide ample evidence for progesterone's protective role in neurodegeneration. Indeed, in a recent paper by Allen *et al.*, the authors detail how progesterone is protective following middle cerebral artery occlusion, causing transient ischaemia in the retina and brain. However, Allen *et al.* also showed that progesterone cannot protect against rodent anterior ischaemic optic neuropathy, which induces permanent monocular optic nerve stroke.[97] Simply, they postulate that the disparity in these data may arise from mechanistic differences with injury type and the therapeutic action of progesterone; a hypothesis that may fit the conflicting results seen in other studies.

8.3.2 Mechanism of Action

Knowledge of the mechanisms underlying the protection provided by oestrogen, progesterone and their derivatives are arguably more advanced than that of glucocorticoids. However, the temporal sequencing of the precise intracellular signalling events remain to be resolved.[121] In the following sections, we attempt to divulge the potential mechanisms through which these steroids exert their neuroprotective effects.

8.3.2.1 Oestrogen-specific Mechanism of Action

In brief, there are four possible mechanisms through which oestrogen may have its effects, although they are not mutually exclusive:

1. Genomic signalling *via* its cognate nuclear receptors, oestrogen receptors α and β (ERα/ERβ)
2. Non-genomic signalling through membrane receptors, *i.e.* its cognate receptor, GPR30, and atypical oestrogen receptors

3. Through its free-radical scavenging properties
4. Where the molecule itself integrates into and stabilizes membranes (Figure 8.3)

The evidence for each of these four mechanisms, highlighting the studies completed in the retina, is described in the following paragraphs. For additional information about the protective effects of oestrogen in brain diseases, the reader is referred to recent and comprehensive reviews.[121–123]

Generally, it is accepted that oestrogen acts through its two nuclear receptors ERα/ERβ to upregulate growth factors.[122] In three studies involving retinal pigment epithelial cells, ranging from the human ARPE-19 cell line to animal knockout mouse models, investigators found that ERβ was essential

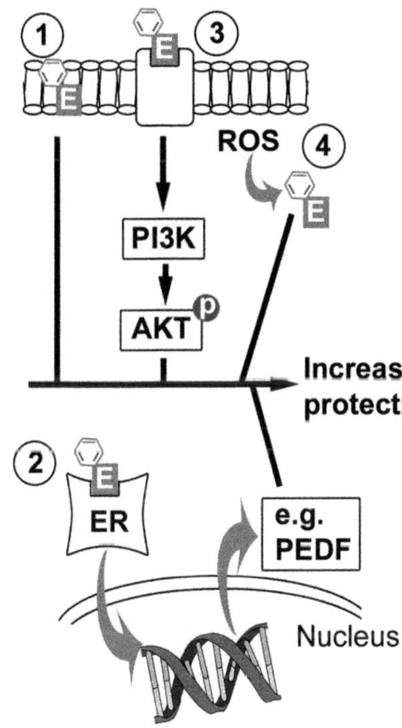

Figure 8.3 Recognized neuroprotective signalling pathways of estrogen (E). (1) Intercalation of estrogen into cellular membranes, *via* its phenolic ring, stabilizes these membranes.[84,133,134] (2) Interaction with its cognate cytoplasmic receptors (oestrogen receptors: ER), which act as ligand-dependent transcription factors, upregulating pro-survival factors, *e.g.* pigment epithelium derived factor (PEDF), and downregulating pro-death factors.[87,88] (3) Interaction with membrane receptors resulting in changes in phosphorylation (p) of pro-survival signalling pathways, such as PI3K/AKT.[85,86,121] (4) Scavenging of reactive oxygen species (ROS), thereby limiting oxidative stress.[122,132] (1–3 have been confirmed in the retina and it is highly probable that 4 occurs there also.).

to conferring maximal protection from oestrogen.[108,124,125] However, although in these three cases ERβ was the critical receptor, the ultimate response to oestrogen varied, from changing matrix metalloproteinase-2 activity[124,125] to the upregulation of antioxidant genes.[108]

In retinal ganglion cells, the situation is less clear. One study showed that 17β-oestradiol (βE2) ligand binding to the ER was critical preventing glaucoma development in the DBA/2J mouse, a mouse model of an inherited (pigmentary) glaucoma. In this research, mice pre-treated with tamoxifen, an oestrogen receptor (ERα/β) antagonist, did not see any improvement in ganglion cell survival following βE2-administration.[109] Similarly, two separate studies showed the capability of oestrogen receptor (ERα/β) inhibitor ICI 182-780 to reduce the protection conferred by oestrogen in two rat models of glaucoma.[126,127]

Unfortunately, Kumar *et al.* found that the same inhibitor had no effect on the protection elicited by oestrogen in a retinal ganglion cell line,[128] although the results of this study and all other studies employing the RGC-5 retinal ganglion cell must now be regarded carefully. The RGC-5 cell line has since been found to be nearly identical to the 661W cone photoreceptor cell line and thus is likely not to be a representative of retinal ganglion cells *in vivo*.[129,130] However, on a similar note, a large-scale human clinical study (3842 participants) showed that in men, polymorphisms in ERβ were associated with a higher risk of open angle glaucoma.[131] While these studies implicate the oestrogen family of receptors in ganglion cell survival, they do not differentiate between ERα- *versus* ERβ-mediated signalling. Nor do they conclusively show the direct role that these receptors play in anti-apoptotic signalling. Indeed, in the following paragraphs, we dissect the possibility that the actions of oestrogen in pro-survival signalling may instead be primarily independent of ERα/β.

Oestrogen is a female hormone. Therefore, to improve the suitability of using a feminizing[†] hormone as a therapy in neuronal disease for both sexes, efforts have ensued to generate non-feminizing derivative. These derivatives have been structurally altered to prevent their binding to ERα/β, for it is this interaction that will propagate the "feminizing" effects. Consequently, any such altered oestrogen derivative must now confer its protection in an ER-independent manner.

There is evidence that membrane oestrogen receptors perform a neuroprotective role within photoreceptors. Nixon and Simpkins used the 661W photoreceptor cell line to show that an agonist of GPR30 protected against a glutamate insult.[86] However, they also demonstrated that oestrogen and two of its analogues did not activate GPR30 or ERα/β. Besides GPR30, oestrogen is known to bind to other types of membrane receptors.[121] Yu *et al.* demonstrated that the insulin receptor β (IRβ) was partly responsible for the pro-survival effects of oestrogen in primary cultures of retinal cells and in the rat light damage model.[85] While they did not elucidate

[†]Feminizing effects of oestrogen include breast growth, decreases in sexual libido and sperm formation, redistribution of body fat, skin softening and decreased hair growth.[168]

specifically how oestrogen activates IRβ, they hypothesised that Src may play a role. Downstream from this receptor, they demonstrated the phosphoinositide 3-kinase–AKT signalling pathway is important in protecting photoreceptors.

The studies discussed all stand to highlight that the receptor with which oestrogen interacts is important in determining how protection is achieved. Nonetheless, this interaction does not form the focus of every study in this area. For example, there are two studies which did not determine the receptor used by oestrogen in the retina, but still demonstrated a positive, neuroprotective effect. In the first, it was found that exogenous oestrogen resulted in the differential regulation of 69 genes in human Müller cell cultures.[87] These included some genes with known links to survival signalling, such as mitogen-activated protein 3 and pigment epithelium derived factor. A separate study demonstrated different downstream effectors of this sex steroid.[88] Pre-treatment of primary retinal cultures with oestrogen resulted in a down-regulation of apoptotic BAX and caspase 3/9, which ultimately resulted in a reduction of mitochondrial-regulated cell death. Certainly, Zhou *et al.* found the signalling following 17β-oestradiol (βE2) administration to the DBA/2J mouse model of glaucoma to be diverse and far reaching. They showed that βE2 treatment reduced retinal ganglion cell apoptosis, preventing the subsequent loss of retinal neurofibres. They demonstrated the ability of βE2 to activate pro-survival AKT and cAMP-responsive-element-binding-protein, increase the expression of anti-oxidant molecule thioredoxin-1 and to decrease the activation of mitogen-activated protein kinases and NF-κB. Finally, they found that βE2 administration successfully inhibited pro-inflammatory interleukin-18 release.[109] Altogether, these data suggest the critical role that oestrogen may play in the regulation of multiple biochemical events.

The final ways in which oestrogen can confer protection to retinal neurons is, interestingly, not through interacting with any specific receptor, but simply due to its molecular structure. This occurs *via* two distinct mechanisms. Oestrogen can act as a free-radical scavenger,[122,132] thereby protecting against oxidative stress. While this has yet to be proven in retinal tissues, the effect should not be tissue specific. Alongside this, oestrogen is also known to protect cells through integrating into cellular membranes.[133,134] This mechanism was investigated in retinal cells by Dykens *et al.*, who found that MITO-4565, an oestrogen derivative, has protective properties in a model of retinitis pigmentosa.[84] It accomplishes this by integrating into cellular membranes and preventing the spread of lipid peroxidation, thus preserving membrane integrity. Key to the protection of cells was the ability of MITO-4565 to stabilize mitochondrial membranes, thereby reducing the likelihood of cells undergoing mitochondrial collapse and associated apoptosis. In order for an oestrogen derivative to possess this membrane intercalation capability, it must possess a steroid A-ring phenol[133,134] (Figure 8.1). This was also found to hold true in the retina, as when Dykens *et al.* substituted the phenol ring of MITO-4565 with a methoxy group, its neuroprotective properties were abrogated.[84]

8.3.2.2 Progesterone-specific Mechanism of Action

In among the multifaceted capabilities of progesterone is its capacity to inhibit apoptosis and inflammation:[135–137] critical in the prevention of most retinal diseases. As with most neurosteroids, progesterone acts through specific binding to cellular receptors.[138] Classical progesterone signalling occurs throughout the central nervous system through the classic progesterone receptors A and B (PR A/B). These receptors are widely expressed and can modulate a variety of different physiological functions.[139,140] Regulating gene transcription,[141,142] PRs A/B are found at the plasma membrane, but will localise to the nucleus upon ligand-binding.[139] However, just like oestrogen, progesterone is a complex hormone and can also regulate its pro-survival activities through a variety of non-classical receptors.[77,143–145] The progesterone receptor membrane component (PGRMC) family, featuring PGRMC1 and PGRMC2[146,147] and the progestin and adipoQ (PAQR) family of receptors, featuring membrane progesterone receptors α, β and γ (mPRα, mPRβ and mPRγ),[148] have both been shown to facilitate survival signalling in neuronal cells (reviewed by Pang and Thomas,[143] Cahill[146] and Brinton *et al.*[149]). The expression of all of these receptors has been demonstrated in the C57 and rd10 mouse retina, and all but classical PR A/B were also found to be present in the 661W cone photoreceptor-like cell line.[90]

In addition to these "typical" progesterone receptors, Neumann *et al.* demonstrated that in the retina, progesterone may also act *via* atypical membrane receptors. They identified these to be metabotropic glutamate receptors: the purinergic (P2Y1) receptor and the adenosine A1 receptor. They postulated that this is because exogenous progesterone results in the release of glutamate and ATP, thereby activating these receptors and facilitating the reduction in glial cell swelling.[96] Therefore, these receptors are unlikely to form the apex of signalling in this model. In fact, so far in the literature, it is only PGRMC1 that has been definitively linked to progesterone-induced retinal cell survival. In a recent paper, siRNA knockdown of PGRMC1 decreased cell viability of healthy 661W photoreceptor cells by approximately 20%.[90] Not only this, but both knockdown and specific inhibition by PGRMC1 inhibitor AG-205[77] prevented the synthetic progesterone analogue norgestrel from protecting stressed photoreceptors both *in vitro* and *ex vivo*.[90,92] These examples stand to highlight the fact that it is probable that one receptor type does not act in isolation in response to progestins. Instead, it is likely that both the nuclear and membrane receptors are responsible for triggering different cell-survival signalling cascades, over different temporal scales. This would stand to reason, for this has already been shown to be the case with oestrogen.

To expand upon this, we examine some of the signalling events associated with these receptors. Studies have shown that progesterone will bind to PR A/B found at the plasma membrane, subsequently protecting against apoptosis.[138] Here, PR A/B can interact with signal transduction kinases such as proto-oncogene tyrosine-protein kinase Src and can activate the mitogen-activated protein kinase pathway.[150] In support of this, PR A/B mRNA is present in the human and rodent retina.[151] However, seeing as both the mRNA and protein expression of PR A/B was found to be relatively low in the C57 and rd10 mouse retinas and is completely absent in the 661W cone photoreceptor cell line,[90] we suggest that it unlikely that these are the primary progesterone receptors associated with the neuroprotective properties of progesterone in the retina. Instead, we draw upon the growing evidence that progestins achieve their neuroprotective effects through activation of the non-classical progesterone receptors.[77,143–145]

Progesterone-mediated signalling through PGRMC1 is incredibly diverse. A small, complex signalling molecule, PGRMC1 has multiple signalling domains[145] and can bind several steroid hormones, although progesterone is bound with the highest affinity.[152] Steroid binding will initiate a wide variety of responses, not least because PGRMC1 has a number of signalling partner proteins. Preferential binding of these proteins enable regulation of cellular damage repair, drug and hormone metabolism, apoptosis suppression, membrane trafficking and repression of gene transcription, to name a few.[146,153–156] Although structurally similar to PGRMC1, little is known about PGRMC1-binding partner PGRMC2.[145,146,157,158] Indeed, the specific role that PGRMC2 plays in progesterone signalling is not yet clear, for PGRMC2 cannot bind progesterone directly.[145] Instead, PGRMC2 acts as a companion protein to PGRMC1, regulating cellular changes indirectly.

Finally, the membrane progesterone receptors (mPRα, β, γ) are non-classical membrane receptors. Making up part of the class II PAQR family of proteins (α: PAQR7, β: PAQR8; γ: PAQR5),[145,159,160] these receptors are not structurally nor functionally related to either PGRMC1 or PGRMC2. The mPRs are instead predicted to be G-protein coupled receptors,[144,161] which would instead link them with the alkaline ceramidase protein family.[159] mPR mRNA has been isolated from several vertebrate species, including human and mouse[90,94,148] and has been shown to be highly conserved in sequence and structural similarity. Expressed on the plasma membrane of cells, the mPRs bind progestins in a specific, displaceable, high-affinity and limited capacity manner.[148,162,163] Interestingly, mPRs are expressed in both PR A/B positive and negative cells[162,163] and so can mediate progestin's anti-apoptotic effects in both a PR A/B-dependent and -independent manner.[162]

Progesterone lacks oestrogen's phenol ring (Figure 8.1). Since it is this phenol ring that enables oestrogen to assimilate into cellular membranes,[133,134] progesterone has a limited capacity to integrate.[134] In addition, progesterone, unlike oestrogen,[122,132] has no proven antioxidant capabilities. However, progesterone and its derivatives have been shown to induce an upregulation in pro-survival growth factors such as basic fibroblast growth factor (bFGF)[45,89,164] and leukaemia inhibitory factor.[91] In fact it is the relationship between progesterone and bFGF that plays a central role in the regulation of granulosa cell survival and in the survival of stressed 661W cone photoreceptor-like cells *in vitro*.[89,92,119,165] We suggest therefore that it is progesterone's capability to promote this pro-survival signalling that will maintain cellular viability[165] and prevent the cleavage of pro-apoptotic PARP and caspase 3.[89,166,167] However, the direct link between these pro-survival signalling events and their associated progesterone receptor interaction has not yet been elucidated.

8.4 Discussion

In this review, we have demonstrated that both glucocorticoids and sex steroids provide protection to photoreceptor cells in certain models of retinal degeneration (Table 8.1). Outside of their structural differences, the main distinctions between these two steroidal groups lie in their clinical use and in their known signalling mechanisms (Figures 8.2 and 8.3).

Firstly, from examining glucocorticoid clinical use, it is obvious that steroids provide enhanced therapeutic potential when administered *via* a slow-release mechanism. Both DEX and FA benefited from incorporation into polymers, enabling enhanced retention within the vitreous.[30,31,37] Through the use of slow-release systems, fewer ophthalmological interventions are required by the patient, thereby reducing the probabilities of complications including cataract development or increased intraocular pressure. If the non-feminizing derivatives of oestrogen or the synthetic progestins are to be developed into clinical therapies for retinal degenerations, their potential would be improved through utilizing an equivalent delivery mechanism.

Secondly, even though there is evidence that DEX can protect photoreceptors in the light damage model of retinal degeneration,[18–22] this has been disputed[47] and possibly depends on the role of AP-1 in the cell death process (Figure 8.2). Oestradiols, in contrast, protect photoreceptors in various retinal degeneration models.[84–88] They possess four separate, well proven signalling mechanisms by which they protect neurons (Figure 8.3), three of which are proven to play a role in the retina.[85,86,88] So, while corticosteroids undoubtedly benefit the retina in inflammatory diseases, this may not translate into protection for retinal degenerations. Given the amount of pre-clinical data on non-feminizing oestradiols, it is probable that these steroids could be of significant therapeutic value in some retinal diseases, but clinical trials are needed.

Thirdly, and finally, the benefits of devoting some time to delineating the underlying signalling mechanisms must be stressed. Although the previous paragraph emphasizes that there already appears to be ample evidence to pursue a clinical trial of oestradiols in retinal degenerations, there are still many grey areas in steroidal pro-survival signalling. As an example it is appropriate to return to the role of DEX and AP-1 in photoreceptor degeneration. Without the follow-up studies by Gu *et al.*[47] and Hao *et al.*,[19] it appeared that DEX was an ideal candidate for a retinal degeneration therapy.[22] Through comparing the three studies (Figure 8.2), it became apparent that DEX would only be appropriate when photoreceptor death was reliant on the activation of AP-1. The studies into developing non-feminizing oestradiols exhibit a similar theme. Through extensive and detailed research, it was evident that the phenolic ring of oestrogen provides neuroprotection,[84,133,134] which was subsequently incorporated into effective derivatives. Whilst it is important to move forward with clinical trials using proven neuroprotective steroids, useful information may also be gained from continuing basic research into their signalling mechanisms.

In summary, the potential of glucocorticoids, oestradiols and progestins to be neuroprotectants in retinal degenerations is evident from the literature. However, there are valuable lessons about drug delivery to be ascertained from the administration of glucocorticoids in the treatment of inflammatory ocular conditions. These should be studied carefully and applied to the sex steroids oestrogen and progesterone, before commencement of large-scale clinical trials. While knowledge of the underlying mechanisms of cell survival will of course be beneficial to the design of future steroidal derivatives, the clinical testing of promising candidates should not be delayed. It is evident that there is still work to be done in explicating the exact mechanism of cell survival behind each of these compounds; however, in general, there appears to be a great deal of evidence supporting their use in retinal disease prevention. The current investigations into these steroids highlights that the health of the retina as a whole is essential to photoreceptor maintenance and each of the steroids accomplishes this through interacting with different retinal cell types. Therefore, we suggest that there is great potential in developing at least one of these steroids in the quest to find an effective, neuroprotective-based therapy for degenerative retinal diseases.

Acknowledgements

This work was funded by the Science Foundation of Ireland, the Health Research Board of Ireland and Fighting Blindness Ireland.

References

1. S. P. Daiger, L. S. Sullivan and S. J. Bowne, Genes and mutations causing retinitis pigmentosa, *Clin. Genet.*, 2013, **84**, 132–141.
2. S. M. Cashman, J. Gracias, M. Adhi and R. Kumar-Singh, Adenovirus mediated delivery of factor H attenuates complement C3 induced pathology in the murine retina: a potential gene therapy for AMD, *J. Gene Med.*, 2015, **17**, 229–243.
3. Z. Wu, *et al.*, A long-term efficacy study of gene replacement therapy for RPGR-associated retinal degeneration, *Hum. Mol. Genet.*, 2015, **24**, 3956–3970.
4. X. Guo, *et al.*, A short N-terminal domain of HDAC4 preserves photoreceptors and restores visual function in retinitis pigmentosa, *Nat. Commun.*, 2015, **6**, 8005.

5. J.-X. Zhang, N.-L. Wang and Q.-J. Lu, Development of gene and stem cell therapy for ocular neurodegeneration, *Int. J. Ophthalmol.*, 2015, **8**, 622–630.
6. D. Trifunović, *et al.*, Neuroprotective strategies for the treatment of inherited photoreceptor degeneration, *Curr. Mol. Med.*, 2012, **12**, 598–612.
7. F. Doonan and T. G. Cotter, Norgestrel may be a potential therapy for retinal degenerations, *Expert Opin. Invest. Drugs*, 2012, **21**, 579–581.
8. C. O'Driscoll, J. O'Connor, C. J. O'Brien and T. G. Cotter, Basic fibroblast growth factor-induced protection from light damage in the mouse retina in vivo, *J. Neurochem.*, 2008, **105**, 524–536.
9. A. J. Payne, S. Kaja, Y. Naumchuk, N. Kunjukunju and P. Koulen, Antioxidant drug therapy approaches for neuroprotection in chronic diseases of the retina, *Int. J. Mol. Sci.*, 2014, **15**, 1865–1886.
10. M. Ishikawa, T. Yoshitomi, C. F. Zorumski and Y. Izumi, Neurosteroids are endogenous neuroprotectants in an ex vivo glaucoma model, *Invest. Ophthalmol. Visual Sci.*, 2014, **55**, 8531–8541.
11. J. L. Edelman, Differentiating intraocular glucocorticoids, *Ophthalmologica*, 2010, **224**, 25–30.
12. C. D. Toran-Allerand, K. Hashimoto, W. T. Greenough and M. Saltarelli, Sex steroids and the development of the newborn mouse hypothalamus and preoptic area in vitro: III. Effects of estrogen on dendritic differentiation, *Brain Res.*, 1983, **283**, 97–101.
13. C. D. Toran-Allerand, Sex steroids and the development of the newborn mouse hypothalamus and preoptic area in vitro. II. Morphological correlates and hormonal specificity, *Brain Res.*, 1980, **189**, 413–427.
14. C. D. Toran-Allerand, Sex steroids and the development of the newborn mouse hypothalamus and preoptic area in vitro: implications for sexual differentiation, *Brain Res.*, 1976, **106**, 407–412.
15. L. Garay, *et al.*, Steroid protection in the experimental autoimmune encephalomyelitis model of multiple sclerosis, *Neuroimmunomodulation*, 2008, **15**, 76–83.
16. P. Mannella, A. M. Sanchez, M. S. Giretti, A. R. Genazzani and T. Simoncini, Oestrogen and progestins differently prevent glutamate toxicity in cortical neurons depending on prior hormonal exposure via the induction of neural nitric oxide synthase, *Steroids*, 2009, **74**, 650–656.
17. K. J. Gurling, Evaluation of an androgen, methylandrostenediol, in the treatment of diabetic retinopathy, *Br. J. Ophthalmol.*, 1955, **39**, 151–154.
18. J. Fu, T. T. Lam and M. O. Tso, Dexamethasone ameliorates retinal photic injury in albino rats, *Exp. Eye Res.*, 1992, **54**, 583–594.
19. W. Hao, *et al.*, Evidence for two apoptotic pathways in light-induced retinal degeneration, *Nat. Genet.*, 2002, **32**, 254–260.
20. H. She, *et al.*, Photoreceptor protection after photodynamic therapy using dexamethasone in a rat model of choroidal neovascularization, *Invest. Ophthalmol. Visual Sci.*, 2008, **49**, 5008–5014.

21. A. Wenzel, C. Grimm, M. Samardzija and C. E. Remé, The genetic modifier Rpe65Leu(450): effect on light damage susceptibility in c-Fos-deficient mice, *Invest. Ophthalmol. Visual Sci.*, 2003, **44**, 2798–2802.
22. A. Wenzel, *et al.*, Prevention of photoreceptor apoptosis by activation of the glucocorticoid receptor, *Invest. Ophthalmol. Visual Sci.*, 2001, **42**, 1653–1659.
23. R. B. Bhisitkul, *et al.*, Neuroprotective effect of intravitreal triamcinolone acetonide against photoreceptor apoptosis in a rabbit model of subretinal hemorrhage, *Invest. Ophthalmol. Visual Sci.*, 2008, **49**, 4071–4077.
24. H. J. Shin, S. H. Lee, H. Chung and H. C. Kim, Association between photoreceptor integrity and visual outcome in diabetic macular edema, *Graefe's Arch. Clin. Exp. Ophthalmol.*, 2012, **250**, 61–70.
25. I. V. Glybina, A. Kennedy, P. Ashton, G. W. Abrams and R. Iezzi, Photoreceptor neuroprotection in RCS rats via low-dose intravitreal sustained-delivery of fluocinolone acetonide, *Invest. Ophthalmol. Visual Sci.*, 2009, **50**, 4847–4857.
26. I. V. Glybina, A. Kennedy, P. Ashton, G. W. Abrams and R. Iezzi, Intravitreous delivery of the corticosteroid fluocinolone acetonide attenuates retinal degeneration in S334ter-4 rats, *Invest. Ophthalmol. Visual Sci.*, 2010, **51**, 4243–4252.
27. Z. Cebeci and N. Kir, Role of implants in the treatment of diabetic macular edema: focus on the dexamethasone intravitreal implant, *Diabetes, Metab. Syndr. Obes.*, 2015, **8**, 555–566.
28. C. M. Jermak, J. T. Dellacroce, J. Heffez and G. A. Peyman, Triamcinolone acetonide in ocular therapeutics, *Surv. Ophthalmol.*, 2007, **52**, 503–522.
29. H. W. Kwak and D. J. D'Amico, Evaluation of the retinal toxicity and pharmacokinetics of dexamethasone after intravitreal injection, *Arch. Ophthalmol.*, 1992, **110**, 259–266.
30. J. A. Haller, *et al.*, Randomized, sham-controlled trial of dexamethasone intravitreal implant in patients with macular edema due to retinal vein occlusion, *Ophthalmology*, 2010, **117**, 1134–1146.e3.
31. J. A. Haller, *et al.*, Dexamethasone intravitreal implant in patients with macular edema related to branch or central retinal vein occlusion twelve-month study results, *Ophthalmology*, 2011, **118**, 2453–2460.
32. E. Moisseiev, M. Goldstein, M. Waisbourd, A. Barak and A. Loewenstein, Long-term evaluation of patients treated with dexamethasone intravitreal implant for macular edema due to retinal vein occlusion, *Eye*, 2013, **27**, 65–71.
33. P. M. Beer, *et al.*, Intraocular concentration and pharmacokinetics of triamcinolone acetonide after a single intravitreal injection, *Ophthalmology*, 2003, **110**, 681–686.
34. M. J. Abraldes, M. Fernández and F. Gómez-Ulla, Intravitreal triamcinolone in diabetic retinopathy, *Curr. Diabetes Rev.*, 2009, **5**, 18–25.

35. P. U. Dugel, D. Eliott, H. L. Cantrill, T. Mahmoud, R. Avery and S. Erickson, I-vation TA: 24-month clinical results of the phase I safety and preliminary efficacy study, in *Proceedings of the Association for Research in Vision and Ophthalmology Annual Meeting* (2009).
36. Clinicaltrials.gov, A Study of MK0140 in Diabetic Patients with Macular Edema, Available at https://clinicaltrials.gov/ct2/archive/NCT00692614, Accessed 7 Jan 2016.
37. S. G. Schwartz and H. W. Flynn, Fluocinolone acetonide implantable device for diabetic retinopathy, *Curr. Pharm. Biotechnol.*, 2011, **12**, 347–351.
38. T. A. Macky, D. Helmy and N. El Shazly, Retinal toxicity of triamcinolone's vehicle (benzyl alcohol): an electrophysiologic and electron microscopic study, *Graefe's Arch. Clin. Exp. Ophthalmol.*, 2007, **245**, 817–824.
39. S.-Y. Yu, F. M. Damico, F. Viola, D. J. D'Amico and L. H. Young, Retinal toxicity of intravitreal triamcinolone acetonide: a morphological study, *Retina*, 2006, **26**, 531–536.
40. National Eye Institute, Facts About Macular Edema, National Institutes of Health, 2015, Available at https://nei.nih.gov/health/macular-edema/fact_sheet, Accessed 10 May 2016.
41. G. P. Chrousos and T. Kino, Intracellular glucocorticoid signaling: a formerly simple system turns stochastic, *Sci. STKE*, 2005, **2005**, pe48.
42. R. H. Oakley and J. A. Cidlowski, Cellular processing of the glucocorticoid receptor gene and protein: new mechanisms for generating tissue-specific actions of glucocorticoids, *J. Biol. Chem.*, 2011, **286**, 3177–3184.
43. N. E. Zanchi, *et al.*, Glucocorticoids: extensive physiological actions modulated through multiple mechanisms of gene regulation, *J. Cell. Physiol.*, 2010, **224**, 311–315.
44. F. L. Groeneweg, H. Karst, E. R. de Kloet and M. Joëls, Mineralocorticoid and glucocorticoid receptors at the neuronal membrane, regulators of nongenomic corticosteroid signalling, *Mol. Cell. Endocrinol.*, 2012, **350**, 299–309.
45. F. Doonan, C. O'Driscoll, P. Kenna and T. G. Cotter, Enhancing survival of photoreceptor cells in vivo using the synthetic progestin Norgestrel, *J. Neurochem.*, 2011, **118**, 915–927.
46. A. Wenzel, C. Grimm, M. Samardzija and C. E. Remé, Molecular mechanisms of light-induced photoreceptor apoptosis and neuroprotection for retinal degeneration, *Prog. Retinal Eye Res.*, 2005, **24**, 275–306.
47. D. Gu, *et al.*, Steroids do not prevent photoreceptor degeneration in the light-exposed T4R rhodopsin mutant dog retina irrespective of AP-1 inhibition, *Invest. Ophthalmol. Visual Sci.*, 2009, **50**, 3482–3494.
48. Q. Meng and Y. Xia, c-Jun, at the crossroad of the signaling network, *Protein Cell*, 2011, **2**, 889–898.
49. M. A. Cubilla, V. Bermúdez, M. D. Marquioni Ramella, T. P. Bachor and A. M. Suburo, Mifepristone, a blocker of glucocorticoid receptors, promotes photoreceptor death, *Invest. Ophthalmol. Visual Sci.*, 2013, **54**, 313–322.

50. S. Kfir-Erenfeld, R. V. Sionov, R. Spokoini, O. Cohen and E. Yefenof, Protein kinase networks regulating glucocorticoid-induced apoptosis of hematopoietic cancer cells: fundamental aspects and practical considerations, *Leuk. Lymphoma*, 2010, **51**, 1968–2005.
51. B. Peng, *et al.*, Suppression of microglial activation is neuroprotective in a mouse model of human retinitis pigmentosa, *J. Neurosci.*, 2014, **34**, 8139–8150.
52. H. van Attikum and S. M. Gasser, Crosstalk between histone modifications during the DNA damage response, *Trends Cell Biol.*, 2009, **19**, 207–217.
53. Z. Smit-McBride, *et al.*, In vivo gene expression profiling of retina postintravitreal injections of dexamethasone and triamcinolone at clinically relevant time points for patient care, *Invest. Ophthalmol. Visual Sci.*, 2011, **52**, 8965–8978.
54. M. Rehak, *et al.*, Effects of intravitreal triamcinolone acetonide on retinal gene expression in a rat model of central retinal vein occlusion, *Graefe's Arch. Clin. Exp. Ophthalmol.*, 2011, **249**, 1175–1183.
55. S. P. Ayalasomayajula, P. Ashton and U. B. Kompella, Fluocinolone inhibits VEGF expression via glucocorticoid receptor in human retinal pigment epithelial (ARPE-19) cells and TNF-alpha-induced angiogenesis in chick chorioallantoic membrane (CAM), *J. Ocul. Pharmacol. Ther.*, 2009, **25**, 97–103.
56. H. Chung, J. J. Hwang, J. Y. Koh, J.-G. Kim and Y. H. Yoon, Triamcinolone acetonide-mediated oxidative injury in retinal cell culture: comparison with dexamethasone, *Invest. Ophthalmol. Visual Sci.*, 2007, **48**, 5742–5749.
57. S. L. Roche, A. C. Wyse-Jackson, A. M. Byrne, A. M. Ruiz-Lopez and T. G. Cotter, Alterations to retinal architecture prior to photoreceptor loss in a mouse model of retinitis pigmentosa, *Int. J. Dev. Biol.*, 2016, **60**, 127–139.
58. L. Zhao, *et al.*, Microglial phagocytosis of living photoreceptors contributes to inherited retinal degeneration, *EMBO Mol. Med.*, 2015, 7, 1179–1197.
59. C. Martínez-Fernández de la Cámara, *et al.*, Infliximab reduces Zaprinast-induced retinal degeneration in cultures of porcine retina, *J. Neuroinflammation*, 2014, **11**, 172.
60. M. Zhang, G. Xu, W. Liu, Y. Ni and W. Zhou, Role of fractalkine/CX3CR1 interaction in light-induced photoreceptor degeneration through regulating retinal microglial activation and migration, *PLoS One*, 2012, 7, e35446.
61. L. Yang, X. Zhu and M. O. M. Tso, A possible mechanism of microglia-photoreceptor crosstalk, *Mol. Vision*, 2007, **13**, 2048–2057.
62. O. Uckermann, *et al.*, The glucocorticoid triamcinolone acetonide inhibits osmotic swelling of retinal glial cells via stimulation of endogenous adenosine signaling, *J. Pharmacol. Exp. Ther.*, 2005, **315**, 1036–1045.

63. S. Singhal, J. M. Lawrence, T. E. Salt, P. T. Khaw and G. A. Limb, Triamcinolone attenuates macrophage/microglia accumulation associated with NMDA-induced RGC death and facilitates survival of Müller stem cell grafts, *Exp. Eye Res.*, 2010, **90**, 308–315.
64. W. Shen, *et al.*, Effect of glucocorticoids on neuronal and vascular pathology in a transgenic model of selective Müller cell ablation, *Glia*, 2014, **62**, 1110–1124.
65. M. A. Dyer and C. L. Cepko, Control of Müller glial cell proliferation and activation following retinal injury, *Nat. Neurosci.*, 2000, **3**, 873–880.
66. A.-M. G. Psarra, M.-L. Bochaton-Piallat, G. Gabbiani, C. E. Sekeris and M. Tsacopoulos, Mitochondrial localization of glucocortocoid receptor in glial (Müller) cells in the salamander retina, *Glia*, 2003, **41**, 38–49.
67. M. Zhao, *et al.*, Differential regulations of AQP4 and Kir4.1 by triamcinolone acetonide and dexamethasone in the healthy and inflamed retina, *Invest. Ophthalmol. Visual Sci.*, 2011, **52**, 6340–6347.
68. B. J. Fan, D. Y. Wang, C. C. Y. Tham, D. S. C. Lam and C. P. Pang, Gene expression profiles of human trabecular meshwork cells induced by triamcinolone and dexamethasone, *Invest. Opthalmol. Visual Sci.*, 2008, **49**, 1886.
69. O.-Y. Tektas and E. Lütjen-Drecoll, Structural changes of the trabecular meshwork in different kinds of glaucoma, *Exp. Eye Res.*, 2009, **88**, 769–775.
70. N. A. Compagnone, Treatments for spinal cord injury: is there hope in neurosteroids? *J. Steroid Biochem. Mol. Biol.*, 2008, **109**, 307–313.
71. K. K. Borowicz, B. Piskorska, M. Banach and S. J. Czuczwar, Neuroprotective actions of neurosteroids, *Front. Endocrinol.*, 2011, **2**, 50.
72. M. C. Gonzalez Deniselle, *et al.*, Basis of progesterone protection in spinal cord neurodegeneration, *J. Steroid Biochem. Mol. Biol.*, 2002, **83**, 199–209.
73. M. C. Gonzalez Deniselle, *et al.*, Progesterone neuroprotection in the Wobbler mouse, a genetic model of spinal cord motor neuron disease, *Neurobiol. Dis.*, 2002, **11**, 457–468.
74. S. L. Gonzalez, *et al.*, Progesterone neuroprotection in spinal cord trauma involves up-regulation of brain-derived neurotrophic factor in motoneurons, *J. Steroid Biochem. Mol. Biol.*, 2005, **94**, 143–149.
75. C. Espinosa-García, A. Aguilar-Hernández, M. Cervantes and G. Moralí, Effects of progesterone on neurite growth inhibitors in the hippocampus following global cerebral ischemia, *Brain Res.*, 2014, **1545**, 23–34.
76. S. Yousuf, F. Atif, I. Sayeed, H. Tang and D. G. Stein, Progesterone in transient ischemic stroke: a dose-response study, *Psychopharmacology*, 2014, **231**, 3313–3323.
77. Y. Qin, *et al.*, Progesterone attenuates Aβ25-35-induced neuronal toxicity via JNK inactivation and progesterone receptor membrane component 1-dependent inhibition of mitochondrial apoptotic pathway, *J. Steroid Biochem. Mol. Biol.*, 2015, **154**, 1–10.

78. D. G. Stein, R. I. Geddes and E. A. Sribnick, Recent developments in clinical trials for the treatment of traumatic brain injury, *Handb. Clin. Neurol.*, 2015, **127**, 433–451.
79. D. G. Stein, A clinical/translational perspective: can a developmental hormone play a role in the treatment of traumatic brain injury? *Horm. Behav.*, 2013, **63**, 291–300.
80. D. G. Stein, Progesterone in the treatment of acute traumatic brain injury: a clinical perspective and update, *Neuroscience*, 2011, **191**, 101–106.
81. Z. Liu, S. Mu and X. Wang, Effect of estrogen on recovering the injured nervous system, *Pak. J. Pharm. Sci.*, 2015, **28**, 1471–1475.
82. M. Stanojlović, *et al.*, Repeated low-dose 17β-estradiol treatment prevents activation of apoptotic signaling both in the synaptosomal and cellular fraction in rat prefrontal cortex following cerebral ischemia, *Neurochem. Int.*, 2015, **83–84**, 1–8.
83. M. Chakrabarti, *et al.*, Estrogen receptor agonists for attenuation of neuroinflammation and neurodegeneration, *Brain Res. Bull.*, 2014, **109**, 22–31.
84. J. A. Dykens, *et al.*, Photoreceptor preservation in the S334ter model of retinitis pigmentosa by a novel estradiol analog, *Biochem. Pharmacol.*, 2004, **68**, 1971–1984.
85. X. Yu, *et al.*, Involvement of insulin/phosphoinositide 3-kinase/Akt signal pathway in 17 beta-estradiol-mediated neuroprotection, *J. Biol. Chem.*, 2004, **279**, 13086–13094.
86. E. Nixon and J. W. Simpkins, Neuroprotective effects of nonfeminizing estrogens in retinal photoreceptor neurons, *Invest. Opthalmol. Visual Sci.*, 2012, **53**, 4739.
87. C. Li, *et al.*, 17beta-estradiol (betaE2) protects human retinal Müller cell against oxidative stress in vitro: evaluation of its effects on gene expression by cDNA microarray, *Glia*, 2006, **53**, 392–400.
88. H. Li, *et al.*, 17β-estradiol impedes Bax-involved mitochondrial apoptosis of retinal nerve cells induced by oxidative damage via the phosphatidylinositol 3-kinase/Akt signal pathway, *J. Mol. Neurosci.*, 2013, **50**, 482–493.
89. A. C. Wyse-Jackson and T. G. Cotter, The synthetic progesterone 'Norgestrel' is neuroprotective in stressed photoreceptor-like cells and retinal explants, mediating its effects via basic fibroblast growth factor, protein kinase A and glycogen synthase kinase 3β signalling, *Eur. J. Neurosci.*, 2016, **43**, 899–911.
90. A. C. Wyse-Jackson, S. L. Roche, A. M. Byrne, A. M. Ruiz-Lopez and T. G. Cotter, Progesterone receptor signalling in retinal photoreceptor neuroprotection, *J. Neurochem.*, 2016, **136**, 63–77.
91. A. M. Byrne, S. L. Roche, A. M. Ruiz-Lopez, A. C. Wyse-Jackson and T. G. Cotter, The synthetic progestin norgestrel acts to increase LIF levels in the rd10 mouse model of retinitis pigmentosa, *Mol. Vision*, 2016, **22**, 264–274.

92. A. M. Ruiz Lopez, *et al.*, Pro-survival redox signalling in progesterone-mediated retinal neuroprotection, *Eur. J. Neurosci.*, 2017, **1**, 1–10.
93. S. L. Roche, A. C. Wyse-Jackson, A. M. Ruiz-Lopez, A. M. Byrne and T. G. Cotter, Fractalkine-CX3CR1 signaling is critical for progesterone-mediated neuroprotection in the retina, *Sci. Rep.*, 2017, **7**, 43067.
94. S. L. Roche, *et al.*, Progesterone attenuates microglial-driven retinal degeneration and stimulates protective fractalkine-CX3CR1 signaling, *PLoS One*, 2016, **11**, e0165197.
95. A. M. Byrne, *et al.*, The synthetic progestin norgestrel modulates Nrf2 signaling and acts as an antioxidant in a model of retinal degeneration, *Redox Biol.*, 2016, **10**, 128–139.
96. F. Neumann, *et al.*, Sex steroids inhibit osmotic swelling of retinal glial cells, *Neurochem. Res.*, 2010, **35**, 522–530.
97. R. S. Allen, *et al.*, Progesterone treatment in two rat models of ocular ischemia, *Invest. Ophthalmol. Visual Sci.*, 2015, **56**, 2880–2891.
98. D. Feskanich, E. Cho, D. A. Schaumberg, G. A. Colditz and S. E. Hankinson, Menopausal and reproductive factors and risk of age-related macular degeneration, *Arch. Ophthalmol.*, 2008, **126**, 519–524.
99. S. Laçin, *et al.*, Assessment of the effectiveness of postmenopausal tibolone therapy on neural functions by measuring visual evoked potentials: a placebo-controlled study, *Eur. J. Obstet. Gynecol. Reprod. Biol.*, 2001, **98**, 72–76.
100. F. F. Verit, H. Oguz, Y. Ozkul and O. Bozkurt, Long-term effects of tibolone on ocular functions in postmenopausal women, *Arch. Gynecol. Obstet.*, 2006, **275**, 255–261.
101. M. C. Deschênes, *et al.*, Postmenopausal hormone therapy increases retinal blood flow and protects the retinal nerve fiber layer, *Invest. Ophthalmol. Visual Sci.*, 2010, **51**, 2587–2600.
102. C. W. S. Pintzka and A. K. Håberg, Perimenopausal hormone therapy is associated with regional sparing of the CA1 subfield: a HUNT MRI study, *Neurobiol. Aging*, 2015, **36**, 2555–2562.
103. M. Bourque, D. E. Dluzen and T. Di Paolo, Neuroprotective actions of sex steroids in Parkinson's disease, *Front. Neuroendocrinol.*, 2009, **30**, 142–157.
104. G. E. Hale and C. L. Shufelt, Hormone therapy in menopause: an update on cardiovascular disease considerations, *Trends Cardiovasc. Med.*, 2015, **25**, 540–549.
105. A. Hayashi, A. W. Weinberger, H. C. Kim and E. de Juan, Genistein, a protein tyrosine kinase inhibitor, ameliorates retinal degeneration after ischemia-reperfusion injury in rat, *Invest. Ophthalmol. Visual Sci.*, 1997, **38**, 1193–1202.
106. C. Cascio, *et al.*, A caspase-3-dependent pathway is predominantly activated by the excitotoxin pregnenolone sulfate and requires early and late cytochrome c release and cell-specific caspase-2 activation in the retinal cell death, *J. Neurochem.*, 2002, **83**, 1358–1371.

107. X. Yu, *et al.*, Protection against hydrogen peroxide-induced cell death in cultured human retinal pigment epithelial cells by 17β-estradiol: a differential gene expression profile, *Mech. Ageing Dev.*, 2005, **126**, 1135–1145.
108. A. Giddabasappa, *et al.*, 17-β estradiol protects ARPE-19 cells from oxidative stress through estrogen receptor-β, *Invest. Opthalmol. Visual Sci.*, 2010, **51**, 5278.
109. X. Zhou, *et al.*, Retinal ganglion cell protection by 17-β-estradiol in a mouse model of inherited glaucoma, *Dev. Neurobiol.*, 2007, **67**, 603–616.
110. S. Fraser-Bell, J. Wu, R. Klein, S. P. Azen and R. Varma, Smoking, alcohol intake, estrogen use, and age-related macular degeneration in Latinos: the Los Angeles Latino Eye Study, *Am. J. Ophthalmol.*, 2006, **141**, 79–87.
111. M. N. Haan, Hormone therapy and age-related macular degeneration, *Arch. Ophthalmol.*, 2006, **124**, 988.
112. G. Scuderi, *et al.*, Effects of phytoestrogen supplementation in post-menopausal women with dry eye syndrome: a randomized clinical trial, *Can. J. Ophthalmol.*, 2012, **47**, 489–492.
113. G. J. Chader and L. Reif-Lehrer, Hormonal effects on the neural retina: corticoid uptake, specific binding and structural requirements for the induction of glutamine synthetase, *Biochim. Biophys. Acta*, 1972, **264**, 186–196.
114. S. Macaione, R. Campisi and G. Barberio, [Effects of progesterone, hydrocortisone and triiodothyronine on the GABA system of the rat retina during postnatal growth], *Boll. Soc. Ital. Biol. Sper.*, 1973, **49**, 908–914.
115. W. K. O'Steen, Ovarian steroid effects on light-induced retinal photoreceptor damage, *Exp. Eye Res.*, 1977, **25**, 361–369.
116. I. Káldi and A. Berta, Progesterone administration fails to protect albino male rats against photostress-induced retinal degeneration, *Eur. J. Ophthalmol.*, 2004, **14**, 306–314.
117. S. Swiatek-De Lange, A. Stampfl and S. Hauck, Membrane initiated effects of progesterone on calcium dependent signaling and activation of VEGF gene expression in retinal glial cells, *Glia*, 2007, **1073**, 1061–1073.
118. A. M. Ruiz Lopez, *et al.*, Pro-survival redox signalling in progesterone-mediated retinal neuroprotection, *Eur. J. Neurosci.*, 2017, **46**, 1663–1672.
119. A. C. Wyse-Jackson, *et al.*, Progesterone analogue protects stressed photoreceptors via bFGF-mediated calcium influx, *Eur. J. Neurosci.*, 2016, **44**, 3067–3079.
120. N. Lu, C. Li, Y. Cheng and A.-L. Du, Protective effects of progesterone against high intraocular pressure-induced retinal ischemia-reperfusion in rats, *J. South Med. Univ.*, 2008, **28**, 2026–2029.

121. D. Lebesgue, V. Chevaleyre, R. S. Zukin and A. M. Etgen, Estradiol rescues neurons from global ischemia-induced cell death: multiple cellular pathways of neuroprotection, *Steroids*, 2009, **74**, 555–561.
122. D. W. Brann, K. Dhandapani, C. Wakade, V. B. Mahesh and M. M. Khan, Neurotrophic and neuroprotective actions of estrogen: basic mechanisms and clinical implications, *Steroids*, 2007, **72**, 381–405.
123. L. Raz, M. M. Khan, V. B. Mahesh, R. K. Vadlamudi and D. W. Brann, Rapid estrogen signaling in the brain, *Neurosignals*, 2008, **16**, 140–153.
124. S. Elliot, *et al.*, Subtype specific estrogen receptor action protects against changes in MMP-2 activation in mouse retinal pigmented epithelial cells, *Exp. Eye Res.*, 2008, **86**, 653–660.
125. S. J. Elliot, *et al.*, Estrogen receptor beta protects against in vivo injury in RPE cells, *Exp. Eye Res.*, 2010, **90**, 10–16.
126. Y. Hayashi, *et al.*, Neuroprotective effect of 17beta-estradiol against N-methyl-D-aspartate-induced retinal neurotoxicity via p-ERK induction, *J. Neurosci. Res.*, 2007, **85**, 386–394.
127. R. Russo, *et al.*, 17Beta-estradiol prevents retinal ganglion cell loss induced by acute rise of intraocular pressure in rat, *Prog. Brain Res.*, 2008, **173**, 583–590.
128. D. M. Kumar, *et al.*, Role of nonfeminizing estrogen analogues in neuroprotection of rat retinal ganglion cells against glutamate-induced cytotoxicity, *Free Radicals Biol. Med.*, 2005, **38**, 1152–1163.
129. R. R. Krishnamoorthy, A. F. Clark, D. Daudt, J. K. Vishwanatha and T. Yorio, A forensic path to RGC-5 cell line identification: lessons learned, *Invest. Ophthalmol. Visual Sci.*, 2013, **54**, 5712–5719.
130. M. R. Al-Ubaidi, RGC-5: are they really 661W? The saga continues, *Exp. Eye Res.*, 2014, **119**, 115.
131. S. de Voogd, *et al.*, Estrogen receptors alpha and beta and the risk of open-angle glaucoma: the Rotterdam Study, *Arch. Ophthalmol.*, 2008, **126**, 110–114.
132. B. Moosmann and C. Behl, The antioxidant neuroprotective effects of estrogens and phenolic compounds are independent from their estrogenic properties, *Proc. Natl. Acad. Sci. U. S. A.*, 1999, **96**, 8867–8872.
133. A. Dicko, M. Morissette, S. Ben Ameur, M. Pézolet and T. Di Paolo, Effect of estradiol and tamoxifen on brain membranes: investigation by infrared and fluorescence spectroscopy, *Brain Res. Bull.*, 1999, **49**, 401–405.
134. Y. Liang, *et al.*, Membrane fluidity effects of estratrienes, *Brain Res. Bull.*, 2001, **54**, 661–668.
135. I. Roglio, *et al.*, Neuroprotective effects of dihydroprogesterone and progesterone in an experimental model of nerve crush injury, *Neuroscience*, 2008, **155**, 673–685.
136. C. Choksuchat, S. Zhao, T. D. Deutch, T. D. Kimble and D. F. Archer, Effects of progesterone, levonorgestrel and medroxyprogesterone acetate on apoptosis in human endometrial endothelial cells, *Contraception*, 2009, **79**, 139–145.

137. H.-J. Yu, *et al.*, Progesterone attenuates neurological behavioral deficits of experimental autoimmune encephalomyelitis through remyelination with nucleus-sublocalized Olig1 protein, *Neurosci. Lett.*, 2010, **476**, 42–45.
138. P. A. Friberg, D. G. J. Larsson and H. Billig, Dominant role of nuclear progesterone receptor in the control of rat periovulatory granulosa cell apoptosis, *Biol. Reprod.*, 2009, **80**, 1160–1167.
139. C. A. Lange, Integration of progesterone receptor action with rapid signaling events in breast cancer models, *J. Steroid Biochem. Mol. Biol.*, 2008, **108**, 203–212.
140. D. X. Wen, Y. F. Xu, D. E. Mais, M. E. Goldman and D. P. McDonnell, The A and B isoforms of the human progesterone receptor operate through distinct signaling pathways within target cells, *Mol. Cell. Biol.*, 1994, **14**, 8356–8364.
141. Y. C. Li, S. Hayes and A. P. Young, Steroid hormone receptors activate transcription in glial cells of intact retina but not in primary cultures of retinal glial cells, *J. Mol. Neurosci.*, 1997, **8**, 145–158.
142. R. N. Hanna, *et al.*, Characterization and expression of the nuclear progestin receptor in zebrafish gonads and brain, *Biol. Reprod.*, 2010, **82**, 112–122.
143. Y. Pang and P. Thomas, Progesterone signals through membrane progesterone receptors (mPRs) in MDA-MB-468 and mPR-transfected MDA-MB-231 breast cancer cells which lack full-length and N-terminally truncated isoforms of the nuclear progesterone receptor, *Steroids*, 2011, **76**, 921–928.
144. P. Moussatche and T. J. Lyons, Non-genomic progesterone signalling and its non-canonical receptor, *Biochem. Soc. Trans.*, 2012, **40**, 200–204.
145. S. L. Petersen, *et al.*, Nonclassical progesterone signalling molecules in the nervous system, *J. Neuroendocrinol.*, 2013, **25**, 991–1001.
146. M. a. Cahill, Progesterone receptor membrane component 1: an integrative review, *J. Steroid Biochem. Mol. Biol.*, 2007, **105**, 16–36.
147. R. M. Lösel, D. Besong, J. J. Peluso and M. Wehling, Progesterone receptor membrane component 1-many tasks for a versatile protein, *Steroids*, 2008, **73**, 929–934.
148. Y. Zhu, J. Bond and P. Thomas, Identification, classification, and partial characterization of genes in humans and other vertebrates homologous to a fish membrane progestin receptor, *Proc. Natl. Acad. Sci. U. S. A.*, 2003, **100**, 2237–2242.
149. R. D. Brinton, *et al.*, Progesterone receptors: form and function in brain, *Front. Neuroendocrinol.*, 2008, **29**, 313–339.
150. E. J. Faivre and C. A. Lange, Progesterone receptors upregulate Wnt-1 to induce epidermal growth factor receptor transactivation and c-Src-dependent sustained activation of Erk1/2 mitogen-activated protein kinase in breast cancer cells, *Mol. Cell. Biol.*, 2007, **27**, 466–480.

151. L. A. Wickham, *et al.*, Identification of androgen, estrogen and progesterone receptor mRNAs in the eye, *Acta Ophthalmol. Scand.*, 2000, **78**, 146–153.
152. C. Meyer, R. Schmid, P. C. Scriba and M. Wehling, Purification and partial sequencing of high-affinity progesterone-binding site(s) from porcine liver membranes, *Eur. J. Biochem.*, 1996, **239**, 726–731.
153. J. J. Peluso, A. Pappalardo, R. Losel and M. Wehling, Progesterone membrane receptor component 1 expression in the immature rat ovary and its role in mediating progesterone's antiapoptotic action, *Endocrinology*, 2006, **147**, 3133–3140.
154. J. J. Peluso, J. Romak and X. Liu, Progesterone receptor membrane component-1 (PGRMC1) is the mediator of progesterone's antiapoptotic action in spontaneously immortalized granulosa cells as revealed by PGRMC1 small interfering ribonucleic acid treatment and functional analysis of PGRMC1 mutations, *Endocrinology*, 2008, **149**, 534–543.
155. H. J. Rohe, I. S. Ahmed, K. E. Twist and R. J. Craven, PGRMC1 (progesterone receptor membrane component 1): a targetable protein with multiple functions in steroid signaling, P450 activation and drug binding, *Pharmacol. Ther.*, 2009, **121**, 14–19.
156. J. J. Peluso, J. DeCerbo and V. Lodde, Evidence for a genomic mechanism of action for progesterone receptor membrane component-1, *Steroids*, 2012, **77**, 1007–1012.
157. A. Wendler and M. Wehling, PGRMC2, a yet uncharacterized protein with potential as tumor suppressor, migration inhibitor, and regulator of cytochrome P450 enzyme activity, *Steroids*, 2013, **78**, 555–558.
158. K. A. Intlekofer and S. L. Petersen, 17β-estradiol and progesterone regulate multiple progestin signaling molecules in the anteroventral periventricular nucleus, ventromedial nucleus and sexually dimorphic nucleus of the preoptic area in female rats, *Neuroscience*, 2011, **176**, 86–92.
159. B. R. Kupchak, I. Garitaonandia, N. Y. Villa, J. L. Smith and T. J. Lyons, Antagonism of human adiponectin receptors and their membrane progesterone receptor paralogs by TNFalpha and a ceramidase inhibitor, *Biochemistry*, 2009, **48**, 5504–5506.
160. P. Thomas and Y. Pang, Membrane progesterone receptors: evidence for neuroprotective, neurosteroid signaling and neuroendocrine functions in neuronal cells, *Neuroendocrinology*, 2012, **96**, 162–171.
161. Y. Zhu, C. D. Rice, Y. Pang, M. Pace and P. Thomas, Cloning, expression, and characterization of a membrane progestin receptor and evidence it is an intermediary in meiotic maturation of fish oocytes, *Proc. Natl. Acad. Sci. U. S. A.*, 2003, **100**, 2231–2236.
162. P. Thomas, *et al.*, Steroid and G protein binding characteristics of the seatrout and human progestin membrane receptor alpha subtypes and their evolutionary origins, *Endocrinology*, 2007, **148**, 705–718.

163. G. E. Dressing, R. Alyea, Y. Pang and P. Thomas, Membrane progesterone receptors (mPRs) mediate progestin induced antimorbidity in breast cancer cells and are expressed in human breast tumors, *Horm. Cancer*, 2012, **3**, 101–112.
164. M. Guthridge, J. Bertolini, J. Cowling and M. T. Hearn, Localization of bFGF mRNA in cyclic rat ovary, diethylstilbesterol primed rat ovary, and cultured rat granulosa cells, *Growth Factors*, 1992, **7**, 15–25.
165. J. J. Peluso and A. Pappalardo, Progesterone maintains large rat granulosa cell viability indirectly by stimulating small granulosa cells to synthesize basic fibroblast growth factor, *Biol. Reprod.*, 1999, **60**, 290–296.
166. V. Gómez-Vicente, M. Donovan and T. G. Cotter, Multiple death pathways in retina-derived 661W cells following growth factor deprivation: crosstalk between caspases and calpains, *Cell Death Differ.*, 2005, **12**, 796–804.
167. T. J. Miller, *et al.*, Photoreceptor cell apoptosis induced by the 2-nitroimidazole radiosensitizer, CI-1010, is mediated by p53-linked activation of caspase-3, *Neurotoxicology*, 2006, **27**, 44–59.
168. Oestrogen Hormone Information (2) The Gender Centre Inc. Available at http://www.gendercentre.org.au/resources/fact-sheets/old/oestrogen-hormone-information-2.htm, Accessed 11 Jan 2016.

CHAPTER 9

Alternative Experimental Models of Ciliary Trafficking and Dysfunction in the Retina

STEPHEN P. CARTER[a,b], JANINA LEYK[c], OLIVER E. BLACQUE[a,b] AND BREANDÁN KENNEDY*[a,b]

[a]UCD School of Biomolecular and Biomedical Science, University College Dublin, Belfield, Dublin 4, Ireland; [b]UCD Conway Institute, University College Dublin, Belfield, Dublin 4, Ireland; [c]Department of Neuroscience, University of Oldenburg, Oldenburg, Germany
*E-mail: brendan.kennedy@ucd.ie

9.1 Introduction

9.1.1 Cilia and Retinal Biology

The molecular aetiology of inherited retinal degeneration (iRD) is extremely heterogeneous. To date, 256 genes have been associated with iRDs,[1] which, while individually rare, have an estimated overall incidence of 1 in 2000 individuals.[2] Collectively, iRD genes have diverse functions in retinal biology, from retina-specific roles in the visual cycle, to genes with general cellular functions such as protein degradation or mitochondrial function. However, by far the largest category of iRD-causative genes (~20%) are those with roles in cilium formation or trafficking.[3] This is, perhaps,

Drug Discovery Series No. 66
Therapies for Retinal Degeneration: Targeting Common Processes
Edited by Enrique J. de la Rosa and Thomas G. Cotter

Published by the Royal Society of Chemistry, www.rsc.org

unsurprising, as the outer segments of cone and rod photoreceptors are specialised primary cilia, modified to form an extensive surface area for efficient detection of light.

Primary cilia consist of a central ring of nine doublet microtubules called the axoneme, surrounded by a membrane which is continuous with the plasma membrane but compositionally distinct. The axoneme itself extends from the mother centriole which docks at the plasma membrane when cells are quiescent.

The ciliary axoneme is built by a microtubule motor-based trafficking process called intraflagellar transport (IFT). In this process, an IFT particle consisting of distinct IFT-A and IFT-B cargo-adaptor complexes is trafficked anterogradely (from base to tip) by heterotrimeric kinesin-II and retrogradely (from tip to base) by cytoplasmic dynein 2.[4] In addition to delivering tubulin subunits for axoneme extension, IFT also regulates the delivery and removal of ciliary receptors and other cargo involved in cilium function. In the case of the class A (rhodopsin-like) G-protein coupled receptors (GPCRs), import is mediated by the Tubby-family proteins TUB and TULP3.[5] Retrieval of receptors from cilia is performed by the BBSome, an eight-member ciliopathy protein complex which is bound to the IFT particle by the IFT25/27 adapter complex.[6,7] Mutations in both Tubby and BBS genes are causative of iRD in humans, mouse and zebrafish models.[8]

The most proximal portion of the ciliary axoneme is called the transition zone (the connecting cilium in photoreceptors), a region which acts as a diffusion barrier preventing the entry of non-ciliary proteins, thereby regulating the unique composition of the organelle. Numerous transition zone proteins are associated with human iRD, including CEP290 and CC2D2A. CEP290 is a scaffolding protein which associates with the ciliary base/transition zone and is required for both ciliogenesis in general and transition zone assembly specifically.[9] In photoreceptors, CC2D2A regulates the docking of rhodopsin-carrying vesicles at the ciliary base and, therefore, is necessary for the transport of rhodopsin into the outer segment.[10]

Multiple cellular trafficking pathways direct cargo to and from the cilium (see Figure 9.1), including clathrin-dependent endocytosis which regulates transforming growth factor-β signalling and membrane retrieval at the ciliary pocket,[11,12] and post-Golgi vesicular trafficking which delivers receptors including rhodopsin to the cilium.[13] The small GTPase RAB8 is a critical regulator of ciliary vesicle traffic, coating vesicles destined for the cilium and promoting their fusion with the periciliary membrane.[7] RAB8 activity is regulated by its guanine-nucleotide exchange factor Rabin8, which is itself regulated by RAB11. Several ciliary GPCRs, including rhodopsin and the three cone opsins possess ciliary targeting sequences,[14] although there is no one universal ciliary targeting sequence, probably reflecting the existence of multiple ciliary trafficking pathways. Peripheral membrane proteins are also trafficked to the cilium by a specific pathway composed of the lipid-binding proteins UNC119 (myristoylated proteins)[15] and PDE6D (prenylated proteins)[16] and another small GTPase,

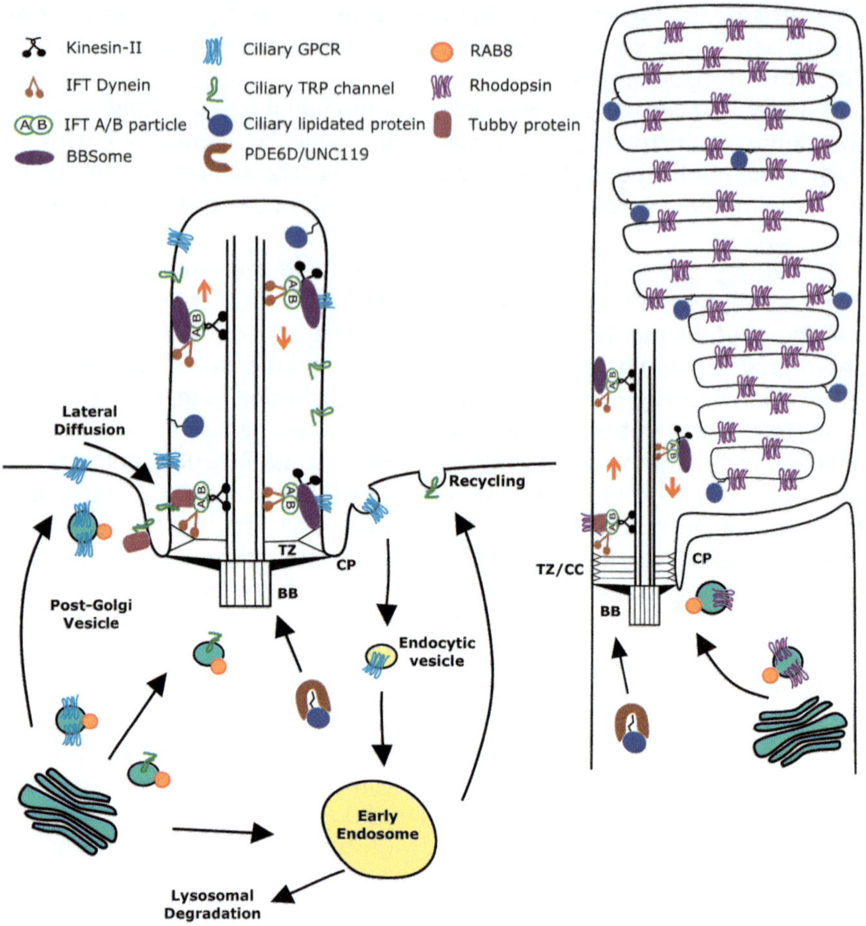

Figure 9.1 Schematic of a generic cilium and photoreceptor cilium (outer segment) and associated trafficking pathways (see text for details). IFT: intraflagellar transport; GPCR: G-protein coupled receptor; TRP: transient receptor potential; BB: basal body; TZ: transition zone; CC: connecting cilium; CP: ciliary pocket.

ARL3. For a summary of trafficking pathways to, within and from the cilium, see Figure 9.1.

9.1.2 Ciliopathies

A spectrum of human genetic disorders is associated with defective cilium formation or function, including Bardet–Biedl, Joubert, Meckel–Gruber and Senior–Løken syndromes, most of which present with retinal degeneration,[17] in addition to other phenotypes such as obesity, kidney cysts, polydactyly and mental retardation. Other syndromic ciliopathies include

Table 9.1 Summary of syndromic and non-syndromic ciliopathies which present with retinal dystrophy and selected representative genes.[a]

Presentation	Disease	Example causative genes
Syndromic	Bardet–Biedl syndrome	*BBS1-21*
	Alström syndrome	*ALMS1*
	Senior–Løken syndrome	*CEP290(BBS14)*, *NPHP4*
	Joubert syndrome	*CC2D2A*, *ARL13b*, *INPP5E*,
	Retinal dystrophy and obesity	*TUB*
Non-syndromic	Leber congenital amaurosis	*CEP290 (LCA10)*, *LCA5*, *RPGRIP1*
	Retinitis pigmentosa	*RPGR*, *RPGRIP1L*, *RP2*
	Cone–rod dystrophy	*RAB28*, *RPGRIP1*

[a]Data taken from Online Mendelian Inheritance in Man, http://www.omim.org/, last accessed April 2018.

a syndrome of rod–cone dystrophy and obesity caused by mutations in the gene *TUB*.[8] Clinical presentations also include non-syndromic ciliopathies in which the only phenotype is retinal degeneration, as exemplified by certain forms of Leber congenital amaurosis, retinitis pigmentosa and cone–rod dystrophy (see Table 9.1). Allelism is a common feature of ciliopathy genes, with different mutations in the same gene leading to different sets of phenotypes, even syndromic and non-syndromic. A primary example is *CEP290*, which is associated with Bardet–Biedl, Meckel–Gruber and Joubert syndromes, as well as non-syndromic Leber congenital amaurosis.[18] Interestingly, the penetrance of ciliopathy phenotypes in *CEP290* patients varies significantly between individuals. One possible explanation is pleiotropism of cilia genes, in which domain-specific functions of the encoded protein are affected differently by hypomorphic alleles. Another is the presence of background genetic modifiers, which significantly affects the penetrance of some ciliopathy phenotypes. For example, the A229T variant of RPGRIP1L increases the incidence of retinal degeneration in ciliopathies due to mutations in several other genes.[19] Finally, cell-specific differences in gene expression or function can modulate disease. For example, photoreceptors employ a high degree of alternative splicing.[20] Certain mutations (*e.g.* splice-site mutations) may impact photoreceptor-specific isoforms more severely than more broadly expressed variants, leading to non-syndromic phenotypes.[21] These factors can complicate the modelling of photoreceptor degeneration, due to species differences in the number, types and processing of isoforms. In summary, our understanding of the correlations between retinal ciliopathy genotypes and phenotypes requires innovative investigational approaches and models.

The pathomechanism of retinal degeneration due to cilia dysfunction in photoreceptors is most often ascribed to aberrant accumulation of proteins destined for the outer segment in the inner segment and/or leaking of inner segment proteins into the outer segment.[22] In other cells, the cilium is a highly privileged organelle and non-ciliary proteins are efficiently excluded.[23] However, the photoreceptor outer segment may act as a sink for membrane proteins, possibly due to its extremely high membrane content. Without exclusion of entry, the trafficking machinery of the outer segment must continuously work to remove non-constituent membrane proteins.[22,24] The high rate of traffic into and out of the outer segment means that even subtle disruption of the trafficking machinery could lead to membrane proteins normally present in other photoreceptor compartments accumulating in the outer segment and damaging the delicate membrane architecture. This makes photoreceptors particularly vulnerable to the effects of ciliary dysfunction and may account for both the prevalence of iRD amongst ciliopathy patients and the large number of ciliary genes linked to forms of iRD.

While the photoreceptor outer segment is the most well studied retinal cilium, other retinal cells also possess cilia, including amacrine and retinal ganglion cells, Müller glia and the retinal pigment epithelium (RPE).[25] The roles of these cilia in retinal biology and disease are poorly understood. Evidence from primary cultures links cilium-regulated sonic hedgehog signalling in Müller glia to their capacity to de-differentiate and re-differentiate into other retinal cell types.[26] As Müller glia can mediate retinal regeneration, understanding the role of ciliary signalling in this process could advance the development of therapies for iRD. In the RPE, the number of ciliated cells decreases postnatally in rodents for an unknown reason and the tight junction protein claudin-1 localises to some of these cilia, suggesting a link to cell–cell adhesion.[27] However, the role of cilia in RPE function remains unknown. Therefore, there is a scientific need to improve our understanding of the roles of cilia in other retinal cells and in blindness.

The cilium represents a cellular nexus for gene networks of critical importance to photoreceptor function and survival. Despite its human disease relevance (in particular disorders of the retina), limited therapeutics are developed for ciliopathies, representing a significant unmet clinical need. The generation of *in*/*ex vivo* and *in vitro* models of ciliary dysfunction that elucidate pathways amenable to therapeutic intervention is critical to the development of novel therapies. For example, such models can be used to screen for neurotrophic drugs preserving or restoring photoreceptor function and cell survival. Owing to their remarkable conservation, a diverse set of organisms from the unicellular eukaryotes *Chlamydomonas reinhardtii* and *Trypanosoma brucei* to human cell lines provide models of ciliary function and disease.[28] The advent of gene editing in multiple species has accelerated the generation of research models, particularly those harbouring patient alleles with the intention of more accurately recapitulating patient phenotypes. Each model comes with unique advantages and disadvantages, but all collectively provide valuable insight into the molecular and cellular basis of ciliary disease and retinal degeneration. Here, we critically review the amenability of selected model systems for elucidating ciliary trafficking and signalling pathways relevant to iRD and the development of therapies targeting these processes.

9.2 Rodent Models

Mouse models are the most widely used in the development of therapeutics for iRD. The genetic toolkit of mice includes tissue-specific and inducible knockouts, particularly useful in ciliopathies, as germline knockouts are often embryonic lethal. Inducible knockouts allow for investigation of the roles of genes in cilia maintenance and function in different spatio-temporal contexts. Knockout of *ARL3* is lethal in mice, but a retina-specific knockout has uncovered its postnatal role in outer segment formation, maintenance and function.[29] In addition, mice have been widely used in the development of gene- and cell-based therapies for iRD, including those resulting from ciliopathies.[30]

Disadvantages associated with mouse models include the relatively long (11 weeks) generation time and the fact that mice primarily avail of rod-mediated vision, unlike humans who rely more on cone-mediated vision. Additionally, mouse models do not always recapitulate the human phenotype and models with different alleles of the same gene may display distinct photoreceptor defects, as has been shown for retinitis pigmentosa 2 (*RP2*).[31] This could be due to species differences in gene usage, the above-mentioned allelism and genetic modifiers or any combination of these factors. While rodent models of ciliopathies and retinal degeneration have been informative, this review focuses on alternative, non-rodent metazoan models.

9.3 Induced Pluripotent Stem Cell-derived Retinal Organoids

9.3.1 Introduction

Since their discovery in 2006, induced pluripotent stem cells (iPSCs)[32] have flourished as a research model to study cellular aspects of disease, including iRD, in a human cell environment (for recent reviews see Yvon *et al.*[33] and Ovando-Roche *et al.*[34]). Patient-specific models can be developed by reprogramming somatic cells from iRD patients into pluripotent stem cells. These can be differentiated into target retinal cells for (1) studying pathological molecular pathways; (2) testing pharmacological or genetic therapies; and (3) cell replacement.

Human iPSC-derived retinal cells provide a significant alternative to animal models. One advantage is the more reliable genotype–phenotype relationship of inherited diseases due to human-specific splicing and expression mechanisms. Using iPSCs as a "disease-in-a-dish" system allows genetic, molecular and proteomic analyses of single retinal cell types, and pharmacological and gene-editing interventions are relatively simple. However, maintenance of iPSC cultures is time consuming and costly,[33] limiting their widespread use as a drug screening system. Moreover, there is high variability in cell growth and differentiation across iPSC lines, possibly due to genetic or epigenetic differences, which can confound interpretation of results.[34,35] Despite the progress in differentiating iPSCs into three-dimensional retinal organoids, such models only represent an isolated system and findings need to be validated *in vivo*.

So far, iPSCs have been successfully differentiated into RPE cells, ganglion cells[36] and photoreceptor-containing optic cups.[37] Patient-specific, iPSC-derived RPE cells are widely used as *in vitro* models of retinal diseases, such as age-related macular degeneration,[38] retinitis pigmentosa[39] and Best vitelliform macular dystrophy.[40] iPSC-derived ganglion cells represent novel models of glaucoma.[41] Pioneering work from Zhong *et al.*[37] showed iPSCs can form three-dimensional retinal organoids (or optic cups) with proper layering of neuroretinal cell types, including photoreceptors which develop connecting cilia and rudimentary outer segments. This breakthrough has prompted many laboratories to model iRDs or ciliopathies, using "mini retinas" to investigate pathogenic mechanisms in patient-specific photoreceptors (see Section 9.3.2 for examples).

Although disease-specific phenotypes can be recapitulated (see Section 9.3.2), major drawbacks of using retinal organoids to study iRDs are: (1) differences in size, shape and composition of optic cups derived from different cell lines; (2) unorganised and partially degraded inner retinal cells; and (3) inability of organoid photoreceptors to develop mature outer segments.[42] Addressing these issues, Wahlin *et al.*[42] reported a protocol whereby photoreceptors in retinal cups develop outer segments of normal length, with localisation of opsins. However, they still lack the organisation of outer segments *in vivo*, with only a few discs present at irregular angles to the ciliary axoneme.

9.3.2 iPSC Ciliopathy Models

As summarised in Table 9.2, a variety of animal models for ciliopathies with retinal phenotypes have generated profound knowledge about cilia biology in health and disease. However, the discrepancy between human and animal model phenotypes of *CEP290* and *RP2* mutations, for example, reveals pitfalls that need to be considered when using animals to model human ciliopathies.[31] iPSC technology has the potential to overcome this issue, as it allows the study of mutations in retinal cells derived from patients.

Mutations in *CEP290* are associated with a variety of ciliopathy phenotypes, including Leber congenital amaurosis (LCA10).[43] The reason for the pleiotropic manifestation of *CEP290* disease is controversial. Two recent studies addressed this using iPSC-derived retinal cells. Parfitt *et al.*[44] compared CEP290 levels in fibroblasts, iPSCs, iPSC-derived RPE and optic cups from *CEP290*-LCA patients to test for cell-type specific phenotypes. In all

Table 9.2 Summary of ciliopathy/inherited retinal degeneration models; advantages and disadvantages of each; and ciliopathies that have been modelled in each.[a]

Species/model	Advantages	Disadvantages	Diseases modelled
Mouse (*Mus musculus*)	Mammalian model	Species-specific transcript processing	Bardet–Biedl syndrome
	Well-established genetic tools	Generation/ development time	Joubert syndrome
	Rod-mediated vision	Cost	Retinitis pigmentosa
iPSC-derived retinal cells/ tissues	Human cells	*In vitro* model	Leber congenital amaurosis
	Human splicing mechanisms	Incomplete cell morphogenesis *e.g.* outer segment	Retinitis pigmentosa
	Chemical screens	Long culturing time	
		Cost	
Zebrafish (*Danio rerio*)	Fast development	Evolutionary distance from humans	Bardet–Biedl syndrome
	Relatively low cost	Redundancy between large numbers of paralogues	Joubert syndrome
	High-throughput genetic/chemical screens		
	Cone-mediated vision		
Caenorhabditis elegans	Extremely rapid development	Evolutionary distance from humans	Bardet–Biedl syndrome
	Ease of genetic manipulation	Biochemical studies are challenging	Cone–rod dystrophy
	Routine live microscopy		
	LOF mutants available for most genes		

[a]iPSC: induced pluripotent stem cells; LOF: loss-of-function.

cell types, CEP290 levels and ciliation were reduced. However, the authors demonstrated the highest levels of aberrant splicing and cilia defects in the optic cups, suggesting a cell-specific regulation of CEP290 expression that explains the retina-specific phenotype of LCA10. Shimada *et al.*[45] reported that fibroblasts of *CEP290*-LCA patients had reduced CEP290 protein levels, but ciliogenesis and cilia length were normal. However, in agreement with Parfitt *et al.*, the photoreceptors in iPSC-derived optic cups showed cilia defects, thereby supporting a tissue-specific requirement for CEP290 in cilia development.[45]

X-linked retinitis pigmentosa (XLRP) is predominantly caused by mutations in *RPGR* and *RP2*.[46] Both proteins are associated with cilia function, but their roles in the molecular pathologies of XLRP remain elusive. A recent study using iPSC-derived optic cups from individuals carrying *RPGR* mutations identified RPGR as a gelsolin-interacting protein.[47] Gelsolin regulates actin disassembly in the connecting cilium, which is an essential process for rhodopsin transport to photoreceptor outer segments. The *RPGR* mutations disturb this interaction. In support of the iPSC data, RPGR and gelsolin knockout mice show increased actin polymerisation and abnormal rhodopsin localisation. Hence, this iPSC study revealed insights into a novel molecular pathway affected by a disease-specific *RPGR* mutation.

RP2 localises to the ciliary base and is a GTPase activating protein for ARL3, indicating a role in ciliary trafficking.[48,49] Schwarz *et al.*[50] characterised fibroblasts and iPSC-derived RPE cells from a patient with a nonsense *RP2* mutation. Depletion of RP2 led to impaired IFT20 (intraflagellar transport 20) localisation, GB1 (transducin β-subunit) trafficking and Golgi cohesion; however, cilia incidence was unaffected. Interestingly, translational read-through-inducing drugs (TRIDs) rescue the phenotypes in both fibroblasts and iPSC-derived RPE, restoring up to 20% of endogenous RP2 protein. This was validated in a second study, wherein patient iPSCs were differentiated into optic cups.[51] Again, the *RP2* mutation did not reduce cilia number, but the cilia had reduced Kif7 at their tips, which was restored by treatment with TRIDs. In summary, both proof-of-concept studies demonstrate the potential of TRIDs as therapeutics for *RP2* nonsense mutations.

9.3.3 Future Prospects

For retinal organoids to prosper as *in vitro* models of retinal ciliopathies, it will be necessary to develop techniques allowing the development of morphologically normal and functionally intact outer segments. Co-incubation with RPE may help, as in neonatal retina cultures the RPE is needed for proper alignment of photoreceptors.[52] Wahlin *et al.*[42] proposed that the RPE is not essential, since they achieved outer segment with nearly normal length without direct RPE-photoreceptor contact. However, as the resulting outer segment developed only a few disorganised discs, the RPE may be necessary to support correct membrane architecture and long-term stability. Also of major importance is the use of appropriate controls. Patient-specific mutations can be corrected using gene-editing tools such as CRISPR/Cas9. These isogenic controls represent better controls than related healthy donors, whose different phenotypes may arise from genetic differences.[34] Furthermore, improvements in reprogramming and differentiation methods will reduce variability, time and costs, increasing the amenability of iPSC-derived retinal cells for high-throughput screening and personalised medicine.

9.4 Zebrafish

9.4.1 Introduction

Zebrafish (*Danio rerio*) is a widely used model in developmental biology and biomedical research, popular due to its advantages as a vertebrate with very rapid initial development, a relatively short life-cycle, large brood sizes and lower maintenance costs relative to mammals. The genetic toolbox of zebrafish includes germline chemical mutagenesis screens, transient morpholino knockdowns, Tol2 transposase mediated transgenesis and CRISPR/Cas9 gene editing. In vision research, zebrafish possess specific advantages over rodent models, most notably a retina layered with equivalent cell types to humans, in which photoreceptors are enriched to ~75% cones, and cone-mediated vision developing at 3–5 days post-fertilisation.[53] Like humans, zebrafish are primed for cone-mediated, photopic vision, in contrast to mice, which have evolved to primarily utilise rod-mediated, scotopic vision. In addition, it is relatively simple to measure visual function in zebrafish larvae using behavioural assays such as the optokinetic[54] and visual-motor response[55] assays.

Caution must be taken when interpreting data from zebrafish models of human disease as zebrafish are more evolutionarily distant from humans than rodents and rodent models themselves do not always accurately model human pathologies. Human phenotypes or syndromes are not always recapitulated in zebrafish models with mutations in the same gene. For example, mutations in *CC2D2A* cause Joubert and Meckel–Gruber syndromes, presenting with cerebellar ataxia, hypotonia and renal cysts, in addition to other developmental disorders. However, the zebrafish *sentinel* mutant, in which the *CC2D2A* orthologue is mutated, displays additional retinal degeneration, only seen in some *CC2D2A* patients.[56] This is possibly due to species differences in gene redundancy and/or the cell-specific roles of such genes.

A significant advantage of zebrafish is the ability to perform low- to high-throughput screens for novel compounds that rescue macro phenotypes. Such phenotype-based screens are molecular target agnostic, and with appropriate models they provide a powerful approach to drug discovery.[57] In such chemical screens the endpoint readouts can be suppression of a developmental phenotype with disease relevance (*e.g.* angiogenesis[58]) or rescue of a disease phenotype.[59] Phenotypes may be morphological, physiological or behavioural. Screening opportunities are enhanced by the development of high-throughput imaging pipelines for zebrafish.[60] One such screen for chemical modifiers of polycystic kidney disease, a common ciliopathy, identified histone deacetylase inhibitors as suppressors of the zebrafish *pkd2* mutant phenotype.[61]

Caveats associated with screens when randomised chemical libraries are applied include a lack of knowledge on the drug target, which may hinder further drug development. The hit rate of chemical screens in zebrafish has varied from as low as 0.02% to as high as 70%,[62] reflecting variability in methodology, number of compounds screened and the genetic background of zebrafish stocks used. These factors must be taken into careful consideration when designing a chemical screen. Care must also be taken when selecting a chemical library, which can introduce bias into the screen, as not all libraries are equally representative of the various classes of compound. A failure of a screen to uncover novel compounds could be due to under-sampling or sampling of the wrong chemical space for the biological process of interest. With the number of test drugs in the order of millions, it is impractical to perform a screen which samples the entirety of chemical space and so biologically active compounds with therapeutic potential will be missed.

9.4.2 Zebrafish Ciliopathy Models

As mentioned previously, one explanation for patients with mutations in the same gene displaying either syndromic or non-syndromic retinal degeneration could be cell- or tissue-specific isoforms of these genes. The importance of these isoforms is illustrated by work from Pretorius *et al.*,[63] who reported a retina-specific isoform of BBS3 (ARL6) that included an extra 13 bp exon between exons 6 and 7 (BBS3L). Morpholino knockdown of this isoform in zebrafish did not produce the classic BBS developmental phenotypes seen in zebrafish; however, it did result in decreased visual response and mislocalisation of green cone opsin. Interestingly, while the visual defects could be rescued by reintroduction of the BBS3L transcript, the "normal" isoform transcript could not. This indicates that the long isoform is necessary for normal visual function and opsin trafficking, and suggests that gene therapy to restore vision in BBS3 patients would need to include this splice-variant.

Zebrafish have also been used to uncover genetic modifiers of ciliopathies. The *sentinel* (*cc2d2a*) mutant displays shortened outer segment and opsin mislocalisation which is enhanced by knockdown of the gene *ninl*.[10] Both CC2D2A and NINL localise to the ciliary base in humans and zebrafish, where they regulate vesicle docking. Interestingly, the authors identified one patient with a *CC2D2A* mutation causing Joubert syndrome and a heterozygous mutation in *NINL*. The patient manifested a more severe form of Joubert syndrome, which included retinal degeneration and hearing loss, supporting *NINL* as a modifier of *CC2D2A*-associated Joubert syndrome, and that such interactions can be modelled in zebrafish.

The potential to uncover small-molecule therapeutics resolving ciliopathies is demonstrated by Jin *et al.*[64] who used zebrafish *leakytail* (*lkt*), a mutant of the ATP-binding cassette transporter ABCC4, which displays ciliogenesis defects. The authors linked these ciliary defects to reduced prostaglandin E_2 (PGE_2) signalling *via* the cilium-localised EP_4 receptor, the first time prostaglandin signalling was linked to ciliogenesis. Specifically, reduced PGE_2 secretion in the *lkt* mutant led to reduced cilium length in several tissues and consequent aberrant organogenesis. Pharmacological treatment of *lkt* fish with PGE_2 resulted in a partial rescue of ciliogenesis and associated morphological defects by increasing the rate of anterograde IFT. This innovative study uncovered drugs that promote PGE_2 secretion or signalling as candidate therapeutics to treat ciliopathies and holds promise that additional drugs can be identified in other models.

9.4.3 Future Prospects

CRISPR technology has revolutionised the genetic toolkit of zebrafish, allowing increasingly precise and complex genetic manipulations to be performed. For this reason zebrafish will continue to be a powerful model for investigating ciliary, retinal and other diseases. The feasibility of the aforementioned chemical screens also makes zebrafish an excellent model for the discovery of novel compounds with therapeutic potential in ciliopathies and retinal degeneration.

9.5 Caenorhabditis elegans

9.5.1 Introduction

Rapid development, feasibility of live microscopy and a wide array of genetic tools make *Caenorhabditis elegans* nematodes a powerful model for investigating the basic cell biology and molecular pathways underlying human disease. One of the greatest advantages of *C. elegans* is the ability to perform forward genetic screens with relative ease.[65] Such screens uncover novel genes and/or novel phenotypes of known genes in an unbiased fashion and have provided loss-of-function alleles of almost every *C. elegans* gene. Furthermore, forward genetic screens can generate hypomorphic alleles in which only part of the encoded protein is disrupted, allowing the investigation of domain-specific functions, and genes normally essential for life. The transparency of *C. elegans* facilitates microscopy, which when combined with transgenic fluorescent reporters allows for live imaging of protein trafficking and dynamics *in vivo*. Many of the genes underlying mammalian cilium biogenesis, function and disease are conserved in *C. elegans*,[66] allowing the nematode to serve as a general model of ciliary disease and provide insights into molecular pathomechanisms relevant to photoreceptor cilium dysfunction.

C. elegans possesses 60 ciliated cells, all sensory neurons which collectively mediate chemo-, osmo-, thermo- and mechanosensation.[67] Most cilia are contained within discrete environmentally exposed pores called sensilla, which are formed from the extended processes of glia-like support cells that wrap around the ciliated endings of the sensory neurons. *C. elegans* cilia come in many different shapes and sizes, ranging from rod-shape structures of varying lengths to elaborate multi-branched structures with highly expanded membranes.[66] Although nematode cilia broadly resemble those of their vertebrate counterparts, there are some notable differences, such as the absence of a stereotypical basal body, which partially degrades and remodels post-ciliogenesis. Given that all *C. elegans* cilia are on sensory neurons, it is interesting to note that many possess features reminiscent of vertebrate photoreceptors, such as expanded membranes and axonemal A-tubule extensions.[68]

9.5.2 *C. elegans* Ciliopathy Models

A number of genes associated with retinal dystrophies were first linked to cilia in *C. elegans*, with prominent first examples being the Bardet–Biedl syndrome genes BBS8,[69] BBS7[70] and ARL6 (BBS3).[71] Each of the BBS genes is exclusively expressed in the ciliated neurons of *C. elegans*, where they localise to the cilium and undergo IFT.[69] For these and other ciliopathy gene orthologues, *C. elegans* has been instrumental in dissecting their roles in regulating ciliary cargo sorting and providing a model for understanding the functions of these genes in photoreceptor biology and disease. For example, research in *C. elegans* has provided valuable insight into the roles of many ciliopathy proteins at the transition zone by uncovering a molecular hierarchy by which these proteins assemble into functional modules to regulate transition zone formation and diffusion barriers to ciliary protein entry and exit.[72]

The power of the forward genetics approach in *C. elegans* is illustrated by early screens for sensory mutants,[73] which generated large numbers of *che* (chemotaxis defective) and *osm* (osmotic-avoidance defective) strains, many of which were later discovered to harbour mutations in important ciliary genes. More recently, screens have been used to generate models in which ciliary modules such as the BBSome are disrupted, but cilia are otherwise normal. Wei *et al.* screened for ciliary mutants where IFT was compromised, but cilia length was relatively unaffected.[74] One such mutant generated in this screen was a G361R substitution within the WD40 domain of DYF-2, the worm orthologue of human IFT144. DYF-2 is normally required for IFT; however, the hypomorphic allele generated in this study showed only severely reduced retrograde IFT and a failure of the BBSome to enter cilia. Such an allele provides a unique opportunity to investigate the roles of the BBSome and retrograde IFT in ciliary trafficking.

Additionally, *C. elegans* can be used to investigate cilia genes linked exclusively to retinal phenotypes in humans. Several studies have shown mutations in RAB28, a Rab small GTPase, cause cone–rod dystrophy.[75–77] Subsequent work in *C. elegans* demonstrated that the worm orthologue, RAB-28, is expressed exclusively in ciliated cells, localises to the periciliary membrane and undergoes bidirectional IFT in a BBSome-dependent manner.[78] Functionally, RAB-28 regulates sensory pore formation, possibly by regulating signalling between the ciliated neurons and their glial cells. Assuming conservation of function between *C. elegans* and mammals, this raises the possibility that photoreceptor cilia, in addition to their role in phototransduction are involved in signalling to other nearby cells such as other photoreceptors or the RPE.

Again, as with all non-human models, care must be taken when extrapolating from *C. elegans* to humans. It is tempting to draw parallels between the distal segments of *C. elegans* cilia, where the microtubules extend as singlets and is thought to act as a specialised subcompartment for receptors, and the modified distal portions of some vertebrate cilia, such as those found in photoreceptors and olfactory neurons. In *C. elegans,* anterograde transport along the distal segment is performed exclusively by one of the two anterograde IFT kinesins, OSM-3.[79] However, the role of KIF17, the vertebrate orthologue of OSM-3, in photoreceptors remains controversial. While morpholino knockdown of *Kif17* results in severe outer segment length reduction,[80] knockout models display modest or no outer segment defects.[81,82]

9.5.3 Future Prospects

C. elegans is an excellent model for performing molecular genetic and cell biological studies, including those of cilia formation and function. As such, it is also a powerful model for dissecting the molecular basis of cilium-associated retinal degeneration. Again, CRISPR has enhanced the genetic tractability of *C. elegans*, making possible models with patient alleles and tagged endogenous genes where gene dosage can be more carefully controlled.

9.6 Conclusion

While cilia are remarkably well conserved organelles with the same pathways underlying their formation and function in organisms as diverse as single-celled algae, nematode worms and mammals, a number of important species differences exist, even between relatively closely related species such as mice and humans (see Table 9.2). Such differences are potential pitfalls to investigators using these species to model human disease and uncover druggable targets and therapeutic strategies. A holistic view which combines findings from non-human and "human" models such as iPSCs is therefore the best approach to uncovering the common processes which form the molecular basis of ciliopathies and retinal degeneration.

Acknowledgements

Research related to some of the topics discussed in this review was funded by an Irish Research Council grant (GOIPG/2014/683), the European Union's Horizon 2020 research and innovation programme under grant agreement No. 734907 (RISE/3D-NEONET project) and a Fighting Blindness-Medical Research Charities-Health Research Board project grant (MRCG 2014-3.b).

References

1. RetNet – Retinal Information Network.
2. M. M. Sohocki, S. P. Daiger, S. J. Bowne, J. A. Rodriquez, H. Northrup and J. R. Heckenlively, *et al.*, Prevalence of mutations causing retinitis pigmentosa and other inherited retinopathies, *Hum. Mutat.*, 2001, **17**, 42–51.

3. A. F. Wright, C. F. Chakarova, M. M. Abd El-Aziz and S. S. Bhattacharya, Photoreceptor degeneration: genetic and mechanistic dissection of a complex trait, *Nat. Rev. Genet.*, 2010, **11**, 273–284.
4. L. Hao and J. M. Scholey, Intraflagellar transport at a glance, *J. Cell Sci.*, 2009, **122**, 889–892.
5. H. B. Badgandi, S.-H. Hwang, I. S. Shimada, E. Loriot and S. Mukhopadhyay, Tubby family proteins are adapters for ciliary trafficking of integral membrane proteins, *J. Cell Biol.*, 2017, **216**, 743–760.
6. T. Eguether, J. T. San Agustin, B. T. Keady, J. A. Jonassen, Y. Liang and R. Francis, *et al.*, IFT27 links the BBSome to IFT for maintenance of the ciliary signaling compartment, *Dev. Cell*, 2014, **31**, 279–290.
7. M. V. Nachury, A. V. Loktev, Q. Zhang, C. J. Westlake, J. Peränen and A. Merdes, *et al.*, A core complex of BBS proteins cooperates with the GTPase Rab8 to promote ciliary membrane biogenesis, *Cell*, 2007, **129**, 1201–1213.
8. A. D. Borman, L. R. Pearce, D. S. Mackay, K. Nagel-Wolfrum, A. E. Davidson and R. Henderson, *et al.*, A homozygous mutation in the TUB gene associated with retinal dystrophy and obesity, *Hum. Mutat.*, 2014, **35**, 289–293.
9. C. Schouteden, D. Serwas, M. Palfy and A. Dammermann, The ciliary transition zone functions in cell adhesion but is dispensable for axoneme assembly in *C. elegans*, *J. Cell Biol.*, 2015, **210**, 35–44.
10. R. Bachmann-Gagescu, M. Dona, L. Hetterschijt, E. Tonnaer, T. Peters and E. de Vrieze, *et al.*, The ciliopathy protein CC2D2A associates with NINL and functions in RAB8-MICAL3-regulated vesicle trafficking, *PLoS Genet.*, 2015, **11**, e1005575.
11. C. A. Clement, K. D. Ajbro, K. Koefoed, M. L. Vestergaard, I. R. Veland and M. P. R. Henriques de Jesus, *et al.*, TGF-β signaling is associated with endocytosis at the pocket region of the primary cilium, *Cell Rep.*, 2013, **3**, 1806–1814.
12. O. I. Kaplan, D. B. Doroquez, S. Cevik, R. V. Bowie, L. Clarke and A. A. W. M. Sanders, *et al.*, Endocytosis genes facilitate protein and membrane transport in *C. elegans* sensory cilia, *Curr. Biol.*, 2012, **22**, 451–460.
13. I. Nemet, P. Ropelewski and Y. Imanishi, Rhodopsin trafficking and mistrafficking: signals, molecular components, and mechanisms, *Prog. Mol. Biol. Transl. Sci.*, 2015, **132**, 39–71.
14. N. F. Berbari, A. D. Johnson, J. S. Lewis, C. C. Askwith and K. Mykytyn, Identification of ciliary localization sequences within the third intracellular loop of G protein-coupled receptors, *Mol. Biol. Cell*, 2008, **19**, 1540–1547.
15. K. J. Wright, L. M. Baye, A. Olivier-Mason, S. Mukhopadhyay, L. Sang and M. Kwong, *et al.*, An ARL3-UNC119-RP2 GTPase cycle targets myristoylated NPHP3 to the primary cilium, *Genes Dev.*, 2011, **25**, 2347–2360.
16. M. C. Humbert, K. Weihbrecht, C. C. Searby, Y. Li, R. M. Pope and V. C. Sheffield, *et al.*, ARL13B, PDE6D, and CEP164 form a functional network for INPP5E ciliary targeting, *Proc. Natl. Acad. Sci. U. S. A.*, 2012, **109**, 19691–19696.

17. A. M. Waters and P. L. Beales, Ciliopathies: an expanding disease spectrum, *Pediatr. Nephrol.*, 2011, **26**, 1039–1056.
18. F. Coppieters, S. Lefever, B. P. Leroy and E. De Baere, CEP290, a gene with many faces: mutation overview and presentation of CEP290base, *Hum. Mutat.*, 2010, **31**, 1097–1108.
19. H. Khanna, E. E. Davis, C. A. Murga-Zamalloa, A. Estrada-Cuzcano, I. Lopez and A. I. den Hollander, *et al.*, A common allele in RPGRIP1L is a modifier of retinal degeneration in ciliopathies, *Nat. Genet.*, 2009, **41**, 739–745.
20. D. Murphy, B. Cieply, R. Carstens, V. Ramamurthy and P. Stoilov, The musashi 1 controls the splicing of photoreceptor-specific exons in the vertebrate retina, *PLoS Genet.*, 2016, **12**, e1006256.
21. D. Murphy, R. Singh, S. Kolandaivelu, V. Ramamurthy and P. Stoilov, Alternative splicing shapes the phenotype of a mutation in BBS8 to cause nonsyndromic retinitis pigmentosa, *Mol. Cell. Biol.*, 2015, **35**, 1860–1870.
22. P. Datta, C. Allamargot, J. S. Hudson, E. K. Andersen, S. Bhattarai and A. V. Drack, *et al.*, Accumulation of non-outer segment proteins in the outer segment underlies photoreceptor degeneration in Bardet–Biedl syndrome, *Proc. Natl. Acad. Sci. U. S. A.*, 2015, **112**, E4400–E4409.
23. J. F. Reiter, O. E. Blacque and M. R. Leroux, The base of the cilium: roles for transition fibres and the transition zone in ciliary formation, maintenance and compartmentalization, *EMBO Rep.*, 2012, **13**, 608–618.
24. S. A. Baker, M. Haeri, P. Yoo, S. M. Gospe 3rd, N. P. Skiba and B. E. Knox, *et al.*, The outer segment serves as a default destination for the trafficking of membrane proteins in photoreceptors, *J. Cell Biol.*, 2008, **183**, 485–498.
25. H. May-Simera, K. Nagel-Wolfrum and U. Wolfrum, Cilia – The sensory antennae in the eye, *Prog. Retinal Eye Res.*, 2017, **60**, 144–180.
26. S. Ferraro, A. I. Gomez-Montalvo, R. Olmos, M. Ramirez and M. Lamas, Primary cilia in rat mature Müller glia: downregulation of IFT20 expression reduces sonic hedgehog-mediated proliferation and dedifferentiation potential of Müller glia primary cultures, *Cell. Mol. Neurobiol.*, 2015, **35**, 533–542.
27. K. Nishiyama, H. Sakaguchi, J. G. Hu, D. Bok and J. G. Hollyfield, Claudin localization in cilia of the retinal pigment epithelium, *Anat. Rec.*, 2002, **267**, 196–203.
28. L. Vincensini, T. Blisnick and P. Bastin, 1001 model organisms to study cilia and flagella, *Biol. Cell.*, 2011, **103**, 109–130.
29. C. Hanke-Gogokhia, Z. Wu, C. D. Gerstner, J. M. Frederick, H. Zhang and W. Baehr, Arf-like protein 3 (ARL3) regulates protein trafficking and ciliogenesis in mouse photoreceptors, *J. Biol. Chem.*, 2016, **291**, 7142–7155.
30. D. L. Simons, S. L. Boye, W. W. Hauswirth and S. M. Wu, Gene therapy prevents photoreceptor death and preserves retinal function in a Bardet–Biedl syndrome mouse model, *Proc. Natl. Acad. Sci. U. S. A.*, 2011, **108**, 6276–6281.

31. D. A. Parfitt, A. Lane, C. Ramsden, K. Jovanovic, P. J. Coffey and A. J. Hardcastle, *et al.*, Using induced pluripotent stem cells to understand retinal ciliopathy disease mechanisms and develop therapies, *Biochem. Soc. Trans.*, 2016, **44**, 1245–1251.
32. K. Takahashi and S. Yamanaka, Induction of pluripotent stem cells from mouse embryonic and adult fibroblast cultures by defined factors, *Cell*, 2006, **126**, 663–676.
33. C. Yvon, C. M. Ramsden, A. Lane, M. B. Powner, L. da Cruz and P. J. Coffey, *et al.*, Using stem cells to model diseases of the outer retina, *Comput. Struct. Biotechnol. J.*, 2015, **13**, 382–389.
34. P. Ovando-Roche, A. Georgiadis, A. J. Smith, R. A. Pearson and R. R. Ali, Harnessing the potential of human pluripotent stem cells and gene editing for the treatment of retinal degeneration, *Curr. Stem Cell Rep.*, 2017, **3**, 112–123.
35. M. Grzybek, A. Golonko, M. Walczak and P. Lisowski, Epigenetics of cell fate reprogramming and its implications for neurological disorders modelling, *Neurobiol. Dis.*, 2017, **99**, 84–120.
36. P. Teotia, D. A. Chopra, S. M. Dravid, M. J. Van Hook, F. Qiu and J. Morrison, *et al.*, Generation of functional human retinal ganglion cells with target specificity from pluripotent stem cells by chemically defined recapitulation of developmental mechanism, *Stem Cells*, 2017, **35**, 572–585.
37. X. Zhong, C. Gutierrez, T. Xue, C. Hampton, M. N. Vergara and L.-H. Cao, *et al.*, Generation of three-dimensional retinal tissue with functional photoreceptors from human iPSCs, *Nat. Commun.*, 2014, **5**, 4047.
38. J. S. Saini, B. Corneo, J. D. Miller, T. R. Kiehl, Q. Wang and N. C. Boles, *et al.*, Nicotinamide ameliorates disease phenotypes in a human iPSC model of age-related macular degeneration, *Cell Stem Cell*, 2017, **20**, 635–647.e7.
39. C. M. Ramsden, B. Nommiste, A. R. Lane, A.-J. F. Carr, M. B. Powner and J. K. M. Smart, *et al.*, Rescue of the MERTK phagocytic defect in a human iPSC disease model using translational read-through inducing drugs, *Sci. Rep.*, 2017, **7**, 51.
40. R. Singh, W. Shen, D. Kuai, J. M. Martin, X. Guo and M. A. Smith, *et al.*, iPS cell modeling of best disease: insights into the pathophysiology of an inherited macular degeneration, *Hum. Mol. Genet.*, 2013, **22**, 593–607.
41. P. Teotia, M. J. Van Hook, C. S. Wichman, R. R. Allingham, M. A. Hauser and I. Ahmad, Modeling glaucoma: retinal ganglion cells generated from induced pluripotent stem cells of patients with six6 risk allele show developmental abnormalities, *Stem Cells*, 2017, **35**, 2239–2252.
42. K. J. Wahlin, J. A. Maruotti, S. R. Sripathi, J. Ball, J. M. Angueyra and C. Kim, *et al.*, Photoreceptor outer segment-like structures in long-term 3D retinas from human pluripotent stem cells, *Sci. Rep.*, 2017, **7**, 766.
43. A. I. den Hollander, R. K. Koenekoop, S. Yzer, I. Lopez, M. L. Arends and K. E. J. Voesenek, *et al.*, Mutations in the CEP290 (NPHP6) gene are a frequent cause of Leber congenital amaurosis, *Am. J. Hum. Genet.*, 2006, **79**, 556–561.

44. D. A. Parfitt, A. Lane, C. M. Ramsden, A.-J. F. Carr, P. M. Munro and K. Jovanovic, *et al.*, Identification and correction of mechanisms underlying inherited blindness in human iPSC-derived optic cups, *Cell Stem Cell*, 2016, **18**, 769–781.
45. H. Shimada, Q. Lu, C. Insinna-Kettenhofen, K. Nagashima, M. A. English and E. M. Semler, *et al.*, *In vitro* modeling using ciliopathy-patient-derived cells reveals distinct cilia dysfunctions caused by CEP290 mutations, *Cell Rep.*, 2017, **20**, 384–396.
46. J.-A. Sahel, K. Marazova and I. Audo, Clinical characteristics and current therapies for inherited retinal degenerations, *Cold Spring Harbor Perspect. Med.*, 2014, **5**, a017111.
47. R. Megaw, H. Abu-Arafeh, M. Jungnickel, C. Mellough, C. Gurniak and W. Witke, *et al.*, Gelsolin dysfunction causes photoreceptor loss in induced pluripotent cell and animal retinitis pigmentosa models, *Nat. Commun.*, 2017, **8**, 271.
48. S. Veltel, R. Gasper, E. Eisenacher and A. Wittinghofer, The retinitis pigmentosa 2 gene product is a GTPase-activating protein for Arf-like 3, *Nat. Struct. Mol. Biol.*, 2008, **15**, 373–380.
49. R. J. Evans, N. Schwarz, K. Nagel-Wolfrum, U. Wolfrum, A. J. Hardcastle and M. E. Cheetham, The retinitis pigmentosa protein RP2 links pericentriolar vesicle transport between the Golgi and the primary cilium, *Hum. Mol. Genet.*, 2010, **19**, 1358–1367.
50. N. Schwarz, A.-J. Carr, A. Lane, F. Moeller, L. L. Chen and M. Aguilà, *et al.*, Translational read-through of the RP2 Arg120stop mutation in patient iPSC-derived retinal pigment epithelium cells, *Hum. Mol. Genet.*, 2015, **24**, 972–986.
51. N. Schwarz, A. Lane, K. Jovanovic, D. A. Parfitt, M. Aguila and C. L. Thompson, *et al.*, Arl3 and RP2 regulate the trafficking of ciliary tip kinesins, *Hum. Mol. Genet.*, 2017, **26**, 2480–2492.
52. G. Pinzón-Duarte, K. Kohler, B. Arango-González and E. Guenther, Cell differentiation, synaptogenesis, and influence of the retinal pigment epithelium in a rat neonatal organotypic retina culture, *Vis. Res.*, 2000, **40**, 3455–3465.
53. J. Fadool and J. Dowling, Zebrafish: A model system for the study of eye genetics, *Prog. Retinal Eye Res.*, 2008, **27**, 89–110.
54. S. E. Brockerhoff, Measuring the optokinetic response of zebrafish larvae, *Nat. Protoc.*, 2006, **1**, 2448–2451.
55. F. Emran, J. Rihel and J. E. Dowling, A behavioral assay to measure responsiveness of zebrafish to changes in light intensities, *J. Visualized Exp.*, 2008, **20**, e923.
56. N. T. Gorden, H. H. Arts, M. A. Parisi, K. L. M. Coene, S. J. F. Letteboer and S. E. C. van Beersum, *et al.*, CC2D2A is mutated in Joubert syndrome and interacts with the ciliopathy-associated basal body protein CEP290, *Am. J. Hum. Genet.*, 2008, **83**, 559–571.
57. C. A. MacRae and R. T. Peterson, Zebrafish as tools for drug discovery, *Nat. Rev. Drug Discovery*, 2015, **14**, 721–731.

58. A. L. Reynolds, Y. Alvarez, T. Sasore, N. Waghorne, C. T. Butler and C. Kilty, *et al.*, Phenotype-based discovery of 2-[(E)-2-(Quinolin-2-yl)vinyl] phenol as a novel regulator of ocular angiogenesis, *J. Biol. Chem.*, 2016, **291**, 7242–7255.
59. A. Asimaki, S. Kapoor, E. Plovie, A. Karin Arndt, E. Adams and Z. Liu, *et al.*, Identification of a new modulator of the intercalated disc in a zebrafish model of arrhythmogenic cardiomyopathy, *Sci. Transl. Med.*, 2014, **6**, 240ra74.
60. J. H. Westhoff, S. Giselbrecht, M. Schmidts, S. Schindler, P. L. Beales and B. Tönshoff, *et al.*, Development of an automated imaging pipeline for the analysis of the zebrafish larval kidney, *PLoS One*, 2013, **8**, e82137.
61. Y. Cao, N. Semanchik, S. H. Lee, S. Somlo, P. E. Barbano and R. Coifman, *et al.*, Chemical modifier screen identifies HDAC inhibitors as suppressors of PKD models, *Proc. Natl. Acad. Sci. U. S. A.*, 2009, **106**, 21819–21824.
62. J. L. Tan and L. I. Zon, Chemical screening in zebrafish for novel biological and therapeutic discovery, *Methods Cell Biol.*, 2011, **105**, 493–516.
63. P. R. Pretorius, L. M. Baye, D. Y. Nishimura, C. C. Searby, K. Bugge and B. Yang, *et al.*, Identification and functional analysis of the vision-specific BBS3 (ARL6) long isoform, *PLoS Genet.*, 2010, **6**, e1000884.
64. D. Jin, T. T. Ni, J. Sun, H. Wan, J. D. Amack and G. Yu, *et al.*, Prostaglandin signalling regulates ciliogenesis by modulating intraflagellar transport, *Nat. Cell Biol.*, 2014, **16**, 841–851.
65. E. M. Jorgensen and S. E. Mango, The art and design of genetic screens: *Caenorhabditis elegans*, *Nat. Rev. Genet.*, 2002, **3**, 356–369.
66. N. J. Lambacher, A.-L. Bruel, T. J. P. van Dam, K. Szymańska, G. G. Slaats and S. Kuhns, *et al.*, TMEM107 recruits ciliopathy proteins to subdomains of the ciliary transition zone and causes Joubert syndrome, *Nat. Cell Biol.*, 2016, **18**, 122–131.
67. P. Inglis, The sensory cilia of *Caenorhabditis elegans*, *WormBook*, 2006, , DOI: 10.1895/wormbook.1.126.1.
68. M. A. Silverman and M. R. Leroux, Intraflagellar transport and the generation of dynamic, structurally and functionally diverse cilia, *Trends Cell Biol.*, 2009, **19**, 306–316.
69. S. J. Ansley, J. L. Badano, O. E. Blacque, J. Hill, B. E. Hoskins and C. C. Leitch, *et al.*, Basal body dysfunction is a likely cause of pleiotropic Bardet–Biedl syndrome, *Nature*, 2003, **425**, 628–633.
70. O. E. Blacque, M. J. Reardon, C. Li, J. McCarthy, M. R. Mahjoub and S. J. Ansley, *et al.*, Loss of *C. elegans* BBS-7 and BBS-8 protein function results in cilia defects and compromised intraflagellar transport, *Genes Dev.*, 2004, **18**, 1630–1642.
71. Y. Fan, M. A. Esmail, S. J. Ansley, O. E. Blacque, K. Boroevich and A. J. Ross, *et al.*, Mutations in a member of the Ras superfamily of small GTP-binding proteins causes Bardet–Biedl syndrome, *Nat. Genet.*, 2004, **36**, 989–993.
72. C. L. Williams, C. Li, K. Kida, P. N. Inglis, S. Mohan and L. Semenec, *et al.*, MKS and NPHP modules cooperate to establish basal body/

transition zone membrane associations and ciliary gate function during ciliogenesis, *J. Cell Biol.*, 2011, **192**, 1023–1041.
73. L. A. Perkins, E. M. Hedgecock, J. N. Thomson and J. G. Culotti, Mutant sensory cilia in the nematode *Caenorhabditis elegans*, *Dev. Biol.*, 1986, **117**, 456–487.
74. Q. Wei, Y. Zhang, Y. Li, Q. Zhang, K. Ling and J. Hu, The BBSome controls IFT assembly and turnaround in cilia, *Nat. Cell Biol.*, 2012, **14**, 950–957.
75. S. Roosing, K. Rohrschneider, A. Beryozkin, D. Sharon, N. Weisschuh and J. Staller, *et al.*, Mutations in RAB28, encoding a farnesylated small GTPase, are associated with autosomal-recessive cone-rod dystrophy, *Am. J. Hum. Genet.*, 2013, **93**, 110–117.
76. R. Riveiro-Álvarez, Y. A. Xie, M.-Á. López-Martínez, T. Gambin, R. Pérez-Carro and A. Ávila-Fernández, *et al.*, New mutations in the RAB28 gene in 2 Spanish families with cone-rod dystrophy, *JAMA Ophthalmol.*, 2015, **133**, 133–139.
77. G.-I. Lee, C. Lee, S. Subramanian, N. K. D. Kim, C.-S. Ki and W.-Y. Park, *et al.*, A novel likely pathogenic variant in the RAB28 gene in a Korean patient with cone-rod dystrophy, *Ophthalmic Genet.*, 2017, 1–3.
78. V. L. Jensen, S. Carter, A. A. W. M. Sanders, C. Li, J. Kennedy and T. A. Timbers, *et al.*, Whole-organism developmental expression profiling identifies RAB-28 as a novel ciliary GTPase associated with the BBsome and intraflagellar transport, *PLoS Genet.*, 2016, **12**, e1006469.
79. J. J. Snow, G. Ou, A. L. Gunnarson, M. R. S. Walker, H. Mimi Zhou and I. Brust-Mascher, *et al.*, Two anterograde intraflagellar transport motors cooperate to build sensory cilia on *C. elegans* neurons, *Nat. Cell Biol.*, 2004, **6**, 1109–1113.
80. C. Insinna, N. Pathak, B. Perkins, I. Drummond and J. C. Besharse, The homodimeric kinesin, Kif17, is essential for vertebrate photoreceptor sensory outer segment development, *Dev. Biol.*, 2008, **316**, 160–170.
81. L. Jiang, B. M. Tam, G. Ying, S. Wu, W. W. Hauswirth and J. M. Frederick, *et al.*, Kinesin family 17 (osmotic avoidance abnormal-3) is dispensable for photoreceptor morphology and function, *FASEB J.*, 2015, **29**, 4866–4880.
82. T. R. Lewis, S. R. Kundinger, A. L. Pavlovich, J. R. Bostrom, B. A. Link and J. C. Besharse, Cos2/Kif7 and Osm-3/Kif17 regulate onset of outer segment development in zebrafish photoreceptors through distinct mechanisms, *Dev. Biol.*, 2017, **425**, 176–190.

CHAPTER 10

Drug Delivery Systems for the Treatment of Diseases Affecting the Retina and Optic Nerve

I. BRAVO-OSUNA[†a,b], V. ANDRÉS-GUERRERO[†a,b], IRENE T. MOLINA-MARTÍNEZ[a,b] AND R. HERRERO-VANRELL*[a,b]

[a]Complutense University, Pharmaceutical Innovation in Ophthalmology Research Group, UCM 920415, Department of Pharmaceutics and Food Technology, Faculty of Pharmacy, Plaza Ramón y Cajal s/n, Madrid, 28040, Spain; [b]Ocular Pathology National Net (OFTARED) of the Institute of Health Carlos III, Spain, Health Research Institute of the San Carlos Clinical Hospital (IdISSC), Madrid, Spain
*E-mail: rociohv@ucm.es

10.1 Introduction

Diseases affecting the retina and the optic nerve are usually chronic and multifactorial, and are the major causes of irreversible blindness. Some of them are related to ageing, and others are inherited. In these pathologies, the active therapeutic substance must reach the posterior segment of the eye. However, the efficacy of treatment is conditioned for the poor

[†]Both authors contributed equally to the present work and should therefore be considered as first authors.

Drug Discovery Series No. 66
Therapies for Retinal Degeneration: Targeting Common Processes
Edited by Enrique J. de la Rosa and Thomas G. Cotter

Published by the Royal Society of Chemistry, www.rsc.org

access of the drug to the retinal target tissues. In fact, effective biological barriers limit the access of the drug to the retina and surroundings areas when using systemic or topical ophthalmic administration. For these reasons, the treatment of diseases of the back of the eye often requires a direct deposit of the drug in an anatomical place close to the target site. Intravitreal injections bypass the blood–ocular barriers and allow high intraocular concentrations of the active substances.[1] Despite to their advantages, this route of administration is not well tolerated by patients and it is associated with secondary effects such as vitreous hemorrhages, increase of intraocular pressure, optic atrophy, retinal detachment, cataracts and endophthalmitis, among others. Furthermore, the risk of side effects increases with the number of injections.

Successful therapy of retinal pathologies requires the development of pharmaceutical formulations able to release the active molecule near to the target tissue for as long as possible. Intraocular drug delivery systems (IDDS) are emerging therapeutic tools in the treatment of diseases affecting the posterior segment, as they are able to provide effective concentrations of the drug for a long period, thus avoiding successive injections. Depending on their size, IDDS are classified in implants (>1 mm), microsystems (1–1000 μm) and nanosystems (1–1000 nm)[2] (Figure 10.1).

Several nanosystems have gained attention in the past decades as effective drug delivery systems to target retinal tissues, especially nanoparticles, liposomes and more recently, niosomes and dendrimers. The small size of these systems makes them suitable to be taken up by cells. Additionally, they can protect labile molecules against the external environment, increasing their half-life. As consequence, nanosystems are considered very useful for targeting specific molecules or genetic material into the cells, revealing the utility of these systems to treat retinal disorders. Implants and microparticles are able to release the drug for longer periods of time compared to nanoparticles, being useful for the treatment of chronic pathologies. The main advantage of microparticles is that they can be injected using small needles. Furthermore, different amounts of particles can be administered depending on the patient's need, allowing a personalized therapy. In addition, they can

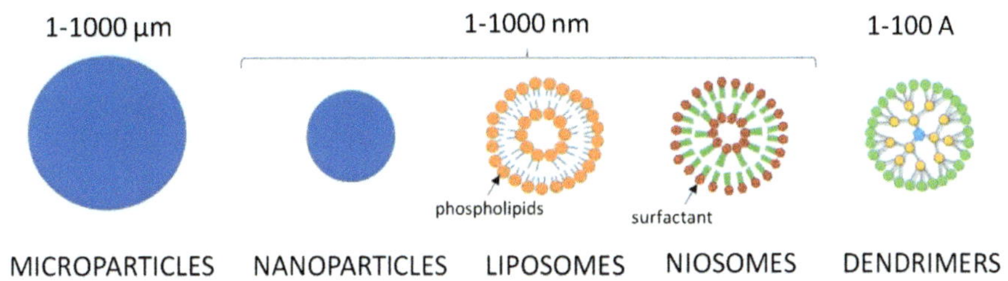

Figure 10.1 Main microsystems and nanosystems under development for ocular drug delivery: microparticles, nanoparticles, liposomes, niosomes and dendrimers.

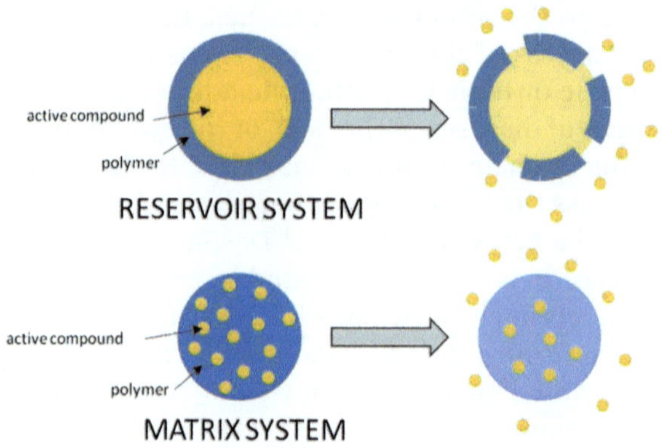

Figure 10.2 Structure of reservoir and matrix systems.

be loaded with more than one active substance, and are thus of interest in the treatment of multifactorial retinal pathologies.

Attending to their structure, the devices can be classified as reservoir (the drug is surrounded by a polymeric membrane) and matrix systems (the active substance is dispersed in the polymeric network). Particles are named microcapsules and nanocapsules for reservoir structures and microspheres and nanospheres in matrix structures (Figure 10.2). IDDS can be prepared from biodegradable and non-biodegradable polymers. Biodegradable biomaterials have the advantage of disappearing from the site of injection after releasing the active substance, avoiding the need of surgery to remove the device. Among the biodegradable polymers used in the preparation of IDDS are the homopolymer poly(lactic) acid (PLA) and the copolymer poly(lactic-*co*-glycolic) acid (PLGA).[3]

This chapter covers a description of the IDDS most used for the treatment of diseases affecting the retina and the optic nerve.

10.2 Drug Delivery Systems

A certain number of products are already in clinical development, racing to be the first to market as a sustained delivery for the treatment of diseases affecting the retina and optic nerve. Some of these products have met challenges in the approach of delivery, such as long-term device retention or the insertion process, which have limited their final use in ophthalmology. Ocular drug delivery systems that have been active in phases I to III of clinical trials within the past 10 years are summarized in Table 10.1.

10.2.1 Implants

To date, several devices >1 mm in size have been prepared for the treatment of posterior segment diseases. They are named according to their shape: implants, scleral plugs, pellets, discs, film and rods (Figure 10.3). Several

Table 10.1 Ocular drug delivery systems that have been active in phases I to III of clinical trials within the past 10 years.[a]

Date	Clinical trial	Product	Drug	Company	Description	Indication
2006	Phase I/II	RETAAC	Triamcinolone	Federal University of Sao Paulo	Intravitreal triamcinolone acetonide microspheres	Diffuse diabetic macular edema unresponsive to laser photocoagulation
2010	Phase II	Brimo PS DDS	Brimonidine	Allergan	Posterior segment delivery system	Age-related macular degeneration
2012	Phase II/III	DE-102	Betamethasone	Santen Pharmaceutical	Betamethasone-loaded microspheres, intravitreal	DME, macular edema following branch retinal vein occlusion
2013	Phase II	L-PPDS	Latanoprost	Mati Therapeutics	Punctal plug	POAG or OH
2014	Phase I/II	Latanoprost SR	Latanoprost	Pfizer	Subconjunctival controlled-release insert	POAG and OH
2015	Phase I/II	VS101 insert	Latanoprost	ViSci Ltd	Subconjunctival insert	POAG and OH
2016	Phase II	Glaukos Travoprost	Travoprost	Glaukos	Trabecular implant	POAG
2016	Phase III	Bimatoprost SR	Bimatoprost	Allergan	Intracameral implant	POAG and OH
2016	Phase IIb	OTX-TP	Travoprost	Ocular Therapeutics	Intracanalicular depot	OH
2016	Phase II	RPDS	Ranibizumab	Genetech, Inc	Refillable intravitreal implant	Subfoveal neovascular age-related macular degeneration
2017	Phase II	ProDex	Dexamethasone	Taiwan Liposome Company	Lipid-based technology	Macular edema due to retinal vein occlusion
2017	Phase II	ENV515-XR	Travoprost	Envisia	Intracameral print particle formulation	OH

[a]POAG: primary open angle glaucoma; OH: ocular hypertension; DME: diabetic macular edema.

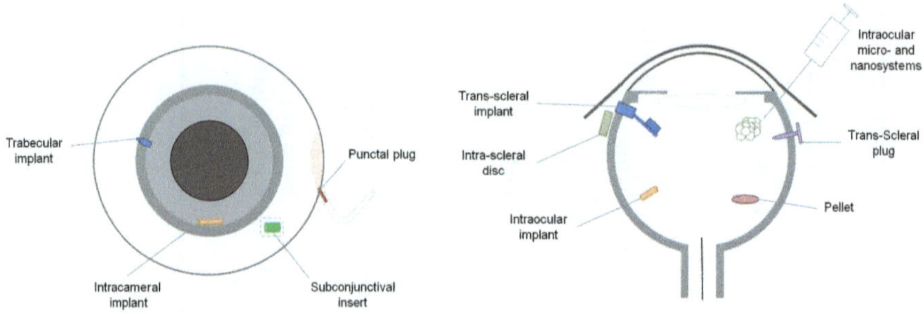

Figure 10.3 Drug delivery systems with different administration routes for the treatment of diseases affecting the retina and optic nerve.

non-biodegradable intraocular implants (Vitrasert®, Retisert® and Iluvien®) have been approved for clinical use. The non-biodegradable devices remain inside the eye during the rest of the patient's lifetime or may require surgery to remove them. There is one biodegradable implant in the market (Ozurdex®). This biodegradable system disappears from the site of administration after delivering the drug.

Several implants (biodegradable and non-biodegradable) have undergone under clinical trials in recent years. In 2009, Allergan sponsored a study to evaluate the safety and efficacy of a brimonidine (400 µg or 200 µg) posterior segment drug delivery system in 44 subjects with previous rhegmatogenous macula-off retinal detachment. The study, completed in 2013, showed 6.7% of patients with at least a 15-letter increase from baseline in best corrected visual acuity in both study groups in comparison to control (no implant) (NCT00650702).

Different implants loaded with latanoprost have been evaluated. In 2011 pSivida Corp developed a long-term, sustained-release implant to deliver latanoprost for patients with ocular hypertension and glaucoma. The proposed device (bioerodible version of Durasert technology) was designed to be injected into the subconjunctival space. In 2014, the implant entered phase I/II clinical trials for ocular hypertension and primary open angle glaucoma (POAG) in the USA (NCT01180062). A phase II study to assess the efficacy of latanoprost punctal plug delivery system, sponsored by Mati Therapeutics in 2013, finished in 2016 with 86 subjects diagnosed with bilateral open angle-glaucoma or ocular hypertension and controlled intra-ocular pressure (IOP) (<21 mmHg). However, final results have not been posted to the clinical study register (NCT00845299). The safety and efficacy of three dose levels of the VS101 subconjunctival latanoprost insert is being evaluated in a phase I/II multicenter, randomized, controlled dose-finding study. The study, sponsored by ViSci Ltd is ongoing, but not recruiting participants, compares the slow-release formulation with topical administration of commercially available 0.005% latanoprost up to 12 weeks (NCT02129673).

Envisia's lead product candidate ENV515 is a fully biodegradable PRINT (particle replication in non-wetting templates) particle formulation of travoprost with the potential to lower IOP for more than 6 months through an intracameral single administration. The safety and efficacy of ENV515 travoprost extended release in patients with bilateral ocular hypertension or POAG will be investigated in a phase II study that is currently recruiting participants (NCT02371746).

In May 2017, Glaukos Corporation announced the completion of patient enrollment in a phase II study of its travoprost intraocular implant with the iDose delivery system in patients with glaucoma. The titanium implant is filled with travoprost and capped with a membrane designed to allow continuous controlled drug delivery into the anterior chamber. Once depleted, it can be removed and replaced in a subsequent procedure. An ongoing clinical trial will assess the safety and preliminary efficacy of two models of this system with different travoprost elution rates, which will be compared to topical 0.5% timolol maleate ophthalmic solution (NCT02754596).

The bimatoprost biodegradable implant from Allergan (Bimatoprost SR) has been designed to release bimatoprost over 6 months. The device is placed intracamerally in the eye using a prefilled single-use applicator. The safety and IOP-lowering effect of Bimatoprost SR (6 µg, 10 µg, 15 µg, or 20 µg) was evaluated in a phase I/II clinical trial that enrolled 75 patients diagnosed with POAG. Mean IOP reduction from baseline through week 16 in study eyes resulted in 7.2, 7.4, 8.1 and 9.5 mmHg with the 6 µg, 10 µg, 15 µg and 20 µg dose strengths of implant, respectively, *vs.* 8.4 mmHg in topical bimatoprost-treated pooled fellow eyes. Adverse events were reported within 2 days after the injection procedure and were transient, demonstrating a favorable efficacy and safety of this system over 6 months, in comparison to topical bimatoprost in overall IOP reduction through week 16.[4]

Ocular Therapeutix designed OTX-TP, a polyethylene glycol-based hydrogel punctal plug, to be implanted into the distal aspect of the inferior or superior canaliculus, to deliver travoprost for up to 90 days. In 2016, the company began enrollment in the first phase III clinical trial for the treatment of ocular hypertension and glaucoma. The study will compare OTX-TP with placebo inserts and is expected to enroll approximately 550 patients with POAG or ocular hypertension at 50 clinical sites (NCT02914509).

Genetech has pursued the development of the ranibizumab non-biodegradable port delivery system that is implanted in the eye with a 3.2 mm scleral incision, which allows the surgeon to refill the device as needed while delivering ranibizumab into the vitreous over an extended period of time. The first-in-humans results of the refillable implant in patients with neovascular age-related macular degeneration (AMD) were presented in 2012 at the American Academy of Ophthalmology and recently presented at the World Ophthalmology Conference in Tokyo.

10.2.2 Microtechnologies

The use of microparticles as injectable devices for the treatment of posterior segment diseases has become more popular over the past few years. In the case of chronic diseases biodegradable microspheres are of special interest. The main advantages of these pharmaceutical formulations are that they can release the drug for long periods of time with only one injection, thus avoiding frequent administrations. Furthermore, the administration can be made through a small incision by using 30–34 G needles. For the treatment of retinal pathologies particles can be administered by different routes (intravitreal, periocular or suprachoroidal injections). Microspheres are good candidates for personalized medicine, as different doses of the drug are administered depending on the amount of particles.

Over the last decades, a large variety of active substances have been included in microspheres. For the treatment of posterior segment diseases, antiproliferative, steroidal and non-steroidal, anti-inflammatory agents, immunosuppresants, antibiotics, antioxidants and even biotechnological products, among others, have been encapsulated in microparticles.

Biodegradable polymers have been widely employed to elaborate microparticles (*i.e.* gelatin, albumin, polyanhidrides, polyesthers and polyorthoesters). For ophthalmic purposes, PLA and PLGA are most employed. These biomaterials are widely used in US Food and Drug Administration (FDA)- and European Medicines Agency-approved drug products.[5] The degradation rate depends on the crystallinity grade of the polymer as well as on its molecular weight. Previous experience with the use of PLA and PLGA polymers has shown that low and medium molecular PLGA 50 : 50 (50% lactide and 50% glycolide) present short half-life *in vivo* degradation. Once administered, the copolymer is converted to lactic and glycolic acid, which are then eliminated from the body after their metabolism to carbon dioxide and water.

Tolerance of microspheres has been tested *in vitro* and *in vivo*. *In vitro* tolerance studies can be performed by using immortalized cell lines and primary cultures. These experiments reduce the use of animals and are used as a platform for the first screening. Cytotoxicity studies give information regarding to the response of cells to drug delivery systems (loaded and unloaded) as well as to the biomaterial employed in the formulation. Despite their usefulness, cell cultures do not represent the whole tissue and *in vivo* behavior cannot be predicted. In order to solve this gap between animal models and dissociated cell cultures, the development of organotypic cultures has emerged as a useful tool for closer *in vivo* conditions for screening drugs, biomaterials and devices.[6]

In vivo ocular tolerance of drug delivery systems depends on the device (size, shape and pharmaceutical composition) and the site of administration. Local administration of microspheres are mainly performed by intravitreal and periocular (sub-Tenon, subconjunctival and posterior juxtaescleral) injections.[7] Recently, suprachoroidal administration using microneedles has been assayed. Injection volumes for intravitreal administration are usually between 50 and 100 microliters,[8] although periocular sites allow higher volumes. PLGA microspheres have the tendency to aggregate, forming a whitish deposit.[9] Among the clinical signs, inflammation is the most frequent reaction after intravitreal injection of PLGA microspheres in rabbit eyes.[10] The reaction is similar to that observed for sutures made of PLGA after intramuscular administration in rabbits,[11] disappearing 2–4 weeks after administration. In any case, no retinal or choroidal damage were reported 1 month after injection.[12] On the contrary, no inflammatory intraocular reaction was associated to PLGA microspheres after periocular injection. In this case, only mild conjunctival congestion was reported. Similar results were observed in the case of suprachoroidal administration with normal fundus and absence of inflammation.[13]

Although special concern regarding visual impairment after intravitreal injection of microspheres, preliminary tolerance studies in humans showed the opposite[14] because of the tendency of the particles to aggregate at the site of administration.

Several parameters influence the degradation behaviour of biodegradable microspheres. Size, physical structure, drug loading, the amount of injected particles and polymer properties are the most relevant.[5]

As cited earlier, depending on the disease, PLGA microspheres have been loaded with different active substances. Adriamycin and retinoic acid microspheres have been tested for the treatment of proliferative retinopathy; anti-inflammatory and immunosuppressant agents (dexamethasone and cyclosporine) for uveitis; and other low molecular-weight drugs such as triamcinolone acetonide for macular edema, acyclovir for herpes infection, budesonide and celecoxib for diabetic retinopathy and ganciclovir for cytomegalovirus retinitis. Biotechnological products such as neurotrophic factors (glial cell line-derived neurotrophic factor (GDNF) for neuroprotection in glaucoma), anti-vascular endothelial growth factor (anti-VEGF) agent for AMD included in microspheres have demonstrated their long-term efficacy.[15]

Adryamicine PLA (3400 g mol^{-1}) and retinoic acid PLGA (50 : 50; 0.2 dl g^{-1}) microspheres (~50 μm and 60 μm, respectively) effectively decrease retinal detachment in animal models of proliferative vitreoretinopathy.[16,17]

Posterior subconjunctival injection of PLGA (85 : 15) microspheres loaded with budesonide (1.1 μm) in diabetic rats was useful to inhibit diabetes-induced elevations in VEGF and blood–retinal barrier leakage.

The use of intravitreal triamcinolone-loaded microspheres in volunteer patients suffering from diffuse macular edema was reported by Cardillo and co-workers in 2006.[14] The efficacy of PLGA microspheres was compared to conventional triamcinolone injections. Interestingly, the eyes treated with triamcinolone microspheres showed a marked decrease in retinal thickness as well as an improved visual acuity for approximately 12 months.[14]

Ganciclovir-loaded microspheres (10 mg) prepared from PLGA 50 : 50 controlled the progression of cytomegalovirus retinitis in an animal model of the disease (rabbits).[10] Also the periocular administration of the kinase inhibitor PKC412 delivered from PLGA microspheres (42.5 μm) decreased the areas of choroidal neovascularisation induced by laser photocoagulation.[18]

The usefulness of biodegradable microspheres (19.1 μm) loaded with GDNF to protect the retinal ganglion cells from degeneration induced by high IOP was tested in rats. In this work the authors confirmed the hypothesis that the long-term delivery of low amounts of the neurotrophic factor GDNF released from the PLGA microspheres resulted effective in a chronic degenerative disease such as glaucoma.[15]

Long-term delivery microparticulate platforms resulted effective in retinal inherent diseases such as retinitis pigmentosa. This is the case of the bile acid tauroursodeoxycholic acid released from PLGA (50:50, MW 35 000 g mol^{-1}) microspheres were intravitreally injected in homozygous P23H rats and slowed retinal degeneration and vision loss.[19]

Microparticles can be used in regenerative therapy. In fact, an improvement of retinal repair was observed with the co-transplantation of microspheres loaded with matrix metalloproteinase 2 with retinal progenitor cells thanks to the inhibition of the cell adhesion molecules.[20]

In 2012, the efficacy and safety of betamethasone microspheres (DE-102) for macular edema was evaluated in a phase II/III study sponsored by Santen Pharmaceutical in Japan (NCT01512901). Primary outcome measures were defined as change from baseline of best corrected visual acuity in letters score. Change in retinal thickness from baseline was also evaluated. The study was halted, unfortunately, because steroid-releasing microspheres could not exhibit superiority to crystalline triamcinolone acetonide, which has been already used clinically. The laboratories are currently investigating the use of these microspheres for drugs other than steroids for ophthalmic therapy (*i.e.* pegtaptanib, an aptamer with affinity to VEGF).[21]

10.2.3 Nanotechnologies

Nanoparticles are generally defined as solid submicron-sized drug carriers. While nanospheres can load solid active compounds, nanocapsules can entrap solid and liquid active molecules. Furthermore, in both cases the active compound can be loaded by simple adsorption on the particle surface. In most cases they are composed of biocompatible and biodegradable polymers, similar to those employed to prepare microparticles. However, metal and lipid materials can also be used in their preparation.

Several studies have demonstrated the usefulness of nanoparticles for the effective treatment of retinal and optic nerve diseases following different routes of intraocular administration. For example, tetra-iodothyroacetic acid, an antiangiogenic active compound, was encapsulated in PLGA nanoparticles and evaluated after intravitreal injection in mouse eyes. The administration of a single injection of loaded nanoparticles significantly reduced retinal neovascularization by 31% 3 days after injection.[22] In another study, an expression plasmid of plasminogen "kringle K5", a natural angiogenic inhibitor, was encapsulated in nanoparticles prepared by combination of PLGA and chitosan. K5-nanoparticles were able to attenuate VEGF and reduced leukostasis and vascular leakage in a rat model of diabetic retinopathy at least 4 weeks after intravitreal administration.[23]

Advances have been achieved in the subretinal administration of nanoparticles. For example, Cai *et al.*[24] determined the utility of nanoparticles using a therapeutic gene in a mouse model of retinitis pigmentosa compacted with CK30PEG, a peptide with a high transfection activity. Local delivery of nanoparticles transfected nearly all of the photoreceptor population and did not produce irreversible defects in retinal function. Additionally, nanoparticles were able to generate a cone function restoration to almost normal levels, and an improvement in rod function.[24]

Some efforts have been made to reach the retina after periocular administration of nanoparticles. However, the effectiveness of nanoparticles for sustained transcleral drug delivery to the retina is limited by their size. According to Amrite *et al.* (2008)[25], smaller nanoparticles (≤20 nm) are washed very fast by the systemic or lymphatic circulation, so only larger nanoparticles (≥200 nm) are appropriate carriers.[25] Budesonide-loaded nanoparticles (345 nm) were prepared with polylactide polymer and injected into the subconjuntival space in rat eyes. 7 days after injection, budesonide concentration in cornea and vitreous was higher in the nanoparticle-treated group than in animals that received the budesonide control solution.[26]

Liposomes are spherical vesicles composed of an aqueous core enclosed by concentric lipid bilayers. According to their size and the number of lipid bilayers, liposomes are classified into small unillamelar vesicles (SUVs), large unillamelar vesicles and multillamelar vesicles if more than one bilayer is present. While hydrophilic drugs can be entrapped in the aqueous core, lipophilic compounds can be incorporated into the lipid bilayers. The basic components of the lipid bilayers are phospholipids, cholesterol and in many cases, vitamin E as hydrophobic antioxidant agent. Liposomes are considered very promising nanocarriers for ocular administration because they are biocompatible, biodegradable and relatively non-toxic. Phospholipids can be chosen to lead with neutral, anionic or cationic liposomes. Positively charged liposomes have high efficiency for entrapping DNA or RNA, due to electrostatic interactions forming complexes named "lipoplexes", frequently used as gene carriers to treat ocular pathologies.[27,28] Furthermore, liposomes can be easily coated with different polymers conferring additional characteristics to the systems. However, problems related to stability in biological fluids and storage which limited drug capacity, and difficulties in sterilization make the development of liposomal formulations difficult to perform. Several studies have demonstrated that drug–liposome complexes diminish drug toxicity because of a limited amount of drug is in contact with tissues. For example, it has been shown that AmpB liposomal formulation (SUVs) composed by phosphatidil choline, cholesterol and tocopherol succinate produced less toxic effects than an AmpB marketed solution after a single intravitreal injection in healthy pigmented rabbits.[29]

Some studies have also evaluated the efficacy of lipoplexes for treatment of retinal disease. The first liposomal formulation for intravitreal injection marketed was Vitravene™ (formivisen, Isis Pharmaceuticals, Carlsbad, CA, USA) (approved by the US FDA in 1998). Vitraven™ is composed of liposomes encapsulating an antisense oligonucleotide for the treatment of cytomegalovirus retinitis in AIDS patients.[30] When using these formulations, special care must be taken to avoid aggregation in the vitreous and DNA displacement from the complex due to the presence of polyanionic glycoaminoglycans in the vitreous. Improvements on lipoplexes formulation are currently under study to minimize these problems, for example by using pegylated lipoplexes.[31]

Niosomes are vesicular nanosystems composed by amphiphilic compounds (surfactants) with the ability to self-assemble and form bilayers. Niosomes can entrap both hydrophilic and lipophilic actives either in an aqueous layer or in vesicular membrane, like liposomes, but offer better chemical stability. Similar to liposomes, niosomes with cationic surfaces can form electrostatic complexes with genetic material leading to "nioplexes". The nioplexes have been demonstrated to transfect efficiently the outer segments of retina, showing their potential as non-viral vectors to treat retinal diseases.[32]

In 2017, Taiwan Liposome Company sponsored a phase-II study to evaluate the safety and efficacy of a single intravitreal dose of TLC399 in subjects diagnosed with macular edema (NTC03093701). This formulation consists of a lipid-based technology that carries dexamethasone.

Dendrimers are synthetic polymers with a high potential in drug delivery. Although as yet, the use of dendrimers in drug delivery is less developed than other drug nanocarrier systems, they present high potential for ocular drug delivery and have demonstrated uses in different ophthalmic administration routes.

A typical dendrimer consists of three main structural components: (1) a multifunctional central core, (2) branched units and (3) surface groups. Dendrimers have semi-globular to globular structures of a few nanometres of diameter with a high density of functionalities on the surface. These pendant groups can be functionalized with targeting ligands able to direct the dendrimer to the desired tissue. In addition, cationic dendrimers have the ability to form stable complex with oligonucleotides, forming "dendriplexes" that are currently considered promising as non-viral vectors for gene therapy in inherited retinal diseases. This is the case of lipid–lysine dendrimers able to deliver ODN-1, a sense oligonucleotide with anti-VEGF activity, into the nuclei of retinal cells.[33] The transfection potential of the formed dendriplexes was confirmed in retinal pigmented epithelium cell cultures. These types of dendrimers were then used for *in vivo* studies showing a reduction in VEGF expression after intravitreal injection in choroidal neovascularisation animal model (rats) up to 4 months.[34] Polyamidoamine (PAMAM) dendrimers with OH terminal groups have an intrinsic ability to selectively localize microglia, and can deliver drugs inside these cells for a sustained period. Authors prepared conjugates of PAMAM-OH with fluocinolone acetonide and demonstrated that a single intravitreal injection in a retinal degeneration rat model was able to stop retinal degeneration, preserve photoreceptors and attenuate activated microglia for an entire month.[35] The subconjunctival route has been also tested. Kang and co-workers prepared PAMAM-carboplatin nanoparticles (258 nm) and observed that the mean tumour mass in eyes of retinoblastoma animal models (LHβ-Tag mice) was significantly reduced after their subconjunctival injection.[36]

10.3 Conclusion

Biodegradable and non-biodegradable implants are approved for clinical uses in the treatment of posterior segment diseases.

Among the IDDS able to release active substances for long periods of time are the biodegradable microparticles. The main advantage of these formulations is that they do not need surgical procedures for their administration and can be injected as a conventional injection. Furthermore, if particles are biodegradable, they disappear from the site of administration after releasing the drug.

Co-delivery of active substances from microparticulate drug delivery systems represents a great improvement in the treatment of multifactorial diseases (*i.e.* glaucoma and AMD). Microparticulate systems loaded with several active agents will allow diminishing the amount of the polymer injected into the vitreous cavity.

Non-viral gene therapy can be performed thanks to the use of nanosystems. Gene therapy can get benefits from the concomitant use of drug delivery systems, and they can help to maintain the integrity of the retinal structures or slow their degradation.

Acknowledgements

Research Group UCM 920415 (InnOftal); FEDER-CICYT FIS-PI17/00079 and PI17/00466 Sanitary Research Institute of the San Carlos Clinical Hospital (IdISSC), MINECO/AEI/FEDER, UE (MAT2013-43217R; MAT2017-83858-C2) and the ISCII-FEDER RETICS (OFTARED) (RD16/0008).

References

1. A. Urtti, Challenges and obstacles of ocular pharmacokinetics and drug delivery, *Adv. Drug Delivery Rev.*, 2006, **58**(11), 1131–1135.
2. I. Bravo-Osuna, V. Andres-Guerrero, P. P. Abal, I. T. Molina-Martinez and R. Herrero-Vanrell, Pharmaceutical microscale and nanoscale approaches for efficient treatment of ocular diseases, *Drug Delivery Transl. Res.*, 2016, **6**(6), 686–707.

3. R. Herrero-Vanrell and I. T. Molina-Martinez, PLA and PLGA microparticles for intravitreal drug delivery: an overview, *J. Drug Delivery Sci. Technol.*, 2007, **17**(1), 11–17.
4. R. A. Lewis, W. C. Christie, D. G. Day, E. R. Craven, T. Walters, M. Bejanian, S. S. Lee, M. L. Goodkin, J. Zhang, S. M. Whitcup, M. R. Robinson and B. S. S. Group, Bimatoprost sustained-release implants for glaucoma therapy: 6-month results from a phase I/II clinical trial, *Am. J. Ophthalmol.*, 2017, **175**, 137–147.
5. R. Herrero-Vanrell and M. F. Refojo, Biodegradable microspheres for vitreoretinal drug delivery, *Adv. Drug Delivery Rev.*, 2001, **52**(1), 5–16.
6. B. Arango-Gonzalez, A. Szabo, G. Pinzon-Duarte, A. Lukats, E. Guenther and K. Kohler, *In vivo* and *in vitro* development of S- and M-cones in rat retina, *Invest. Ophthalmol. Visual Sci.*, 2010, **51**(10), 5320–5327.
7. R. Herrero-Vanrell, I. Bravo-Osuna, V. Andres-Guerrero, M. Vicario-de-la-Torre and I. T. Molina-Martinez, The potential of using biodegradable microspheres in retinal diseases and other intraocular pathologies, *Prog. Retinal Eye Res.*, 2014, **42**, 27–43.
8. C. Garcia-Caballero, E. Prieto-Calvo, P. Checa-Casalengua, E. Garcia-Martin, V. Polo-Llorens, J. Garcia-Feijoo, I. T. Molina-Martinez, I. Bravo-Osuna and R. Herrero-Vanrell, Six month delivery of GDNF from PLGA/vitamin E biodegradable microspheres after intravitreal injection in rabbits, *Eur. J. Pharm. Sci.*, 2017, **103**, 19–26.
9. B. Khoobehi, M. O. Stradtmann, G. A. Peyman and O. M. Aly, Clearance of sodium fluorescein incorporated into microspheres from the vitreous after intravitreal injection, *Ophthalmic Surg.*, 1991, **22**(3), 175–180.
10. A. A. Veloso Jr, Q. Zhu, R. Herrero-Vanrell and M. F. Refojo, Ganciclovir-loaded polymer microspheres in rabbit eyes inoculated with human cytomegalovirus, *Invest. Ophthalmol. Visual Sci.*, 1997, **38**(3), 665–675.
11. G. E. Visscher, R. L. Robison, H. V. Maulding, J. W. Fong, J. E. Pearson and G. J. Argentieri, Biodegradation of and tissue reaction to 50 : 50 poly(D,L-lactide-*co*-glycolide) microcapsules, *J. Biomed. Mater. Res.*, 1985, **19**(3), 349–365.
12. G. G. Giordano, P. Chevez-Barrios, M. F. Refojo and C. A. Garcia, Biodegradation and tissue reaction to intravitreous biodegradable poly(D,L-lactide-*co*-glycolide)acid microspheres, *Curr. Eye Res.*, 1995, **14**(9), 761–768.
13. S. R. Patel, D. E. Berezovsky, B. E. McCarey, V. Zarnitsyn, H. F. Edelhauser and M. R. Prausnitz, Targeted administration into the suprachoroidal space using a microneedle for drug delivery to the posterior segment of the eye, *Invest. Ophthalmol. Visual Sci.*, 2012, **53**(8), 4433–4441.
14. J. A. Cardillo, A. A. Souza-Filho and A. G. Oliveira, Intravitreal Bioerudivel sustained-release triamcinolone microspheres system (RETAAC). Preliminary report of its potential usefulnes for the treatment of diabetic macular edema, *Arch. Soc. Esp. Oftalmol.*, 2006, **81**(12), 675–677, 679–681.
15. P. Checa-Casalengua, C. H. Jiang, I. Bravo-Osuna, B. A. Tucker, I. T. Molina-Martinez, M. J. Young and R. Herrero-Vanrell, Retinal ganglion cells survival in a glaucoma model by GDNF/Vit E PLGA microspheres

prepared according to a novel microencapsulation procedure, *J. Controlled Release*, 2011, **156**(1), 92–100.
16. T. Moritera, Y. Ogura, N. Yoshimura, Y. Honda, R. Wada, S. H. Hyon and Y. Ikada, Biodegradable microspheres containing adriamycin in the treatment of proliferative vitreoretinopathy, *Invest. Ophthalmol. Visual Sci.*, 1992, **33**(11), 3125–3130.
17. G. G. Giordano, M. F. Refojo and M. H. Arroyo, Sustained delivery of retinoic acid from microspheres of biodegradable polymer in PVR, *Invest. Ophthalmol. Visual Sci.*, 1993, **34**(9), 2743–2751.
18. Y. Saishin, R. L. Silva, Y. Saishin, K. Callahan, C. Schoch, M. Ahlheim, H. Lai, F. Kane, R. K. Brazzell, D. Bodmer and P. A. Campochiaro, Periocular injection of microspheres containing PKC412 inhibits choroidal neovascularization in a porcine model, *Invest. Ophthalmol. Visual Sci.*, 2003, **44**(11), 4989–4993.
19. L. Fernandez-Sanchez, I. Bravo-Osuna, P. Lax, A. Arranz-Romera, V. Maneu, S. Esteban-Perez, I. Pinilla, M. D. Puebla-Gonzalez, R. Herrero-Vanrell and N. Cuenca, Controlled delivery of tauroursodeoxycholic acid from biodegradable microspheres slows retinal degeneration and vision loss in P23H rats, *PLoS One*, 2017, **12**(5), e0177998.
20. J. Yao, B. A. Tucker, X. Zhang, P. Checa-Casalengua, R. Herrero-Vanrell and M. J. Young, Robust cell integration from co-transplantation of biodegradable MMP2-PLGA microspheres with retinal progenitor cells, *Biomaterials*, 2011, **32**(4), 1041–1050.
21. T. Yasukawa, Biomaterials for intraocular sustained drug delivery, in *Biomaterials and Regenerative Medicine in Ophthalmology*, ed. C. A. Karkin, Elsevier, Cambridge, MA, USA, 2nd edn, 2016, pp. 131–143.
22. T. Yoshida, J. Gong, Z. Xu, Y. Wei and E. J. Duh, Inhibition of pathological retinal angiogenesis by the integrin alphavbeta3 antagonist tetraiodothyroacetic acid (tetrac), *Exp. Eye Res.*, 2012, **94**(1), 41–48.
23. K. Park, Y. Chen, Y. Hu, A. S. Mayo, U. B. Kompella, R. Longeras and J. X. Ma, Nanoparticle-mediated expression of an angiogenic inhibitor ameliorates ischemia-induced retinal neovascularization and diabetes-induced retinal vascular leakage, *Diabetes*, 2009, **58**(8), 1902–1913.
24. X. Cai, Z. Nash, S. M. Conley, S. J. Fliesler, M. J. Cooper and M. I. Naash, A partial structural and functional rescue of a retinitis pigmentosa model with compacted DNA nanoparticles, *PLoS One*, 2009, **4**(4), e5290.
25. A. C. Amrite, H. F. Edelhauser, S. R. Singh and U. B. Kompella, Effect of circulation on the disposition and ocular tissue distribution of 20 nm nanoparticles after periocular administration, *Mol. Vision*, 2008, **14**, 150–160.
26. U. B. Kompella, N. Bandi and S. P. Ayalasomayajula, Subconjunctival nano- and microparticles sustain retinal delivery of budesonide, a corticosteroid capable of inhibiting VEGF expression, *Invest. Ophthalmol. Visual Sci.*, 2003, **44**(3), 1192–1201.
27. K. Abul-Hassan, R. Walmsley and M. Boulton, Optimization of non-viral gene transfer to human primary retinal pigment epithelial cells, *Curr. Eye Res.*, 2000, **20**(5), 361–366.

28. R. Naik, A. Mukhopadhyay and M. Ganguli, Gene delivery to the retina: focus on non-viral approaches, *Drug Discovery Today*, 2009, **14**(5–6), 306–315.
29. C. Tremblay, M. Barza, F. Szoka, M. Lahav and J. Baum, Reduced toxicity of liposome-associated amphotericin B injected intravitreally in rabbits, *Invest. Ophthalmol. Visual Sci.*, 1985, **26**(5), 711–718.
30. E. R. Rayburn and R. Zhang, Antisense, RNAi, and gene silencing strategies for therapy: mission possible or impossible?, *Drug Discovery Today*, 2008, **13**(11–12), 513–521.
31. A. Bochot, E. Fattal, V. Boutet, J. R. Deverre, J. C. Jeanny, H. Chacun and P. Couvreur, Intravitreal delivery of oligonucleotides by sterically stabilized liposomes, *Invest. Ophthalmol. Visual Sci.*, 2002, **43**(1), 253–259.
32. M. Mashal, N. Attia, G. Puras, G. Martinez-Navarrete, E. Fernandez and J. L. Pedraz, Retinal gene delivery enhancement by lycopene incorporation into cationic niosomes based on DOTMA and polysorbate 60, *J. Controlled Release*, 2017, **254**, 55–64.
33. N. Wimmer, R. J. Marano, P. S. Kearns, E. P. Rakoczy and I. Toth, Syntheses of polycationic dendrimers on lipophilic peptide core for complexation and transport of oligonucleotides, *Bioorg. Med. Chem. Lett.*, 2002, **12**(18), 2635–2637.
34. R. J. Marano, I. Toth, N. Wimmer, M. Brankov and P. E. Rakoczy, Dendrimer delivery of an anti-VEGF oligonucleotide into the eye: a long-term study into inhibition of laser-induced CNV, distribution, uptake and toxicity, *Gene Ther.*, 2005, **12**(21), 1544–1550.
35. R. Iezzi, B. R. Guru, I. V. Glybina, M. K. Mishra, A. Kennedy and R. M. Kannan, Dendrimer-based targeted intravitreal therapy for sustained attenuation of neuroinflammation in retinal degeneration, *Biomaterials*, 2012, **33**(3), 979–988.
36. S. J. Kang, C. Durairaj, U. B. Kompella, J. M. O'Brien and H. E. Grossniklaus, Subconjunctival nanoparticle carboplatin in the treatment of murine retinoblastoma, *Arch. Ophthalmol.*, 2009, **127**(8), 1043–1047.

CHAPTER 11

Gene Therapies for Retinal Degenerations

L. PETIT*[a] AND V. KALATZIS[b]

[a]University of Massachusetts Medical School, Department of Ophthalmology & Horae Gene Therapy Center, 368 Plantation Street, Worcester, 01605-2324, USA; [b]Institute for Neurosciences of Montpellier, Inserm U1051, 80 rue Augustin Fliche, 34091, Montpellier, France
*E-mail: lolita.petit@umassmed.edu

11.1 The Goals of Gene Therapy

Gene therapy seeks to treat a disease by altering gene expression of a patient's cells. It can be carried out by two routes: (1) the transfer of one or more therapeutic nucleic acids to the cells (gene addition) or (2) the editing/correction of the existing patient's genome (gene editing). The objectives of gene therapy are to restore a missing gene function, repress the expression of a defective gene or direct the production of an advantageous substance to help the cells fight a disease mechanism. Hence, gene therapy has the potential to cure diseases that are treatable but not curable with conventional medications and to provide treatments for conditions that were previously untreatable. In particular, gene therapy promises to overcome several limitations associated with the administration of therapeutic proteins or small molecules. Since few protein drugs can efficiently cross cell membranes, due to their

Drug Discovery Series No. 66
Therapies for Retinal Degeneration: Targeting Common Processes
Edited by Enrique J. de la Rosa and Thomas G. Cotter

Published by the Royal Society of Chemistry, www.rsc.org

charge and size, only gene therapy can provide a remedy when the intracellular expression of a therapeutic protein is needed. Also, most proteins have short half-lives and must be thus re-administered frequently. In contrast, gene therapy could, in theory, be infinite, obviating the need for repeated injections. Finally, the combination of appropriate gene delivery vectors and cell-specific promoters gives the possibility for selective local treatment of affected cells, which is rarely possible with the use of classic pharmaceutical approaches.

11.2 Vectors and Gene Delivery in Retinal Gene Therapy

The retina offers several advantages compared to other tissues or organs for *in vivo* gene delivery. The most important challenge for gene therapy is to distribute exogenous genetic material to the desired target cells safely and efficiently. To facilitate gene transfer, nucleic acids are packaged into molecular "vehicles" called vectors that can be taken up by the target cells. In the retina, vectors can be delivered in immediate proximity to the target cells by either subretinal injection (resulting in exposure of the outer retina) or intravitreal injection (resulting in exposure of the inner retina).[1] These localized delivery methods minimize the required dose of vector – keeping the possibility of toxic reactions to a minimum – but allow for high local concentrations around the target cells.[1] Moreover, the compartmentalized anatomy of the eye and the presence of the blood–retina barrier limit the extent of vector leakage into the systemic circulation and contribute to reducing the severity of potentially damaging immune responses against the vector epitopes or the transgene product (for review see Willett and Bennett[2]). Furthermore, retinal cells are post-mitotic, making it possible to use non-integrative vector systems for long-term gene expression. Because random insertion of the vector genome into the host cell genome could disturb or inactivate the expression of oncogenes or tumor suppressor genes,[3,4] the use of non-integrating systems substantially reduces the possibility of malignant transformation of the targeted cells. In addition, the retina benefits from its relative accessibility for monitoring and imaging in live animals and patients. The ability to follow disease progression and vector expression in subjects directly at multiple time points is a significant advantage of retinal gene therapy. Lastly, the untreated contralateral eye is extremely useful as a control in evaluating gene therapy outcomes when the disease progresses symmetrically.

Importantly, the field has succeeded in developing many well-defined vector systems with which to deliver nucleic acids to the retina, including the use of physiochemical methods to compact the DNA and transport it across the cell membrane.[5,6] However, the best approach to target retinal cells is still to employ a vehicle that has entirely evolved to deliver foreign genes into cells: viruses. Viruses infect cells with excellent specificity. They transfer their genetic material into host cells and use the host's machinery to produce more viral particles. When it comes to gene therapy, it is possible to capitalize on this process by removing viral sequences and replacing them with a therapeutic gene.[7] There are several viral vectors capable of delivering therapeutic transgenes to retinal cells,[8] including those derived from adenovirus (Ad),[9–13] lentivirus (LV)[13,14] and adeno-associated virus (AAV).[15–17] Each vector is unique in its cellular specificity, carrying capacity, integration profile and immunogenicity. Human Ad-based vectors were the first vectors evaluated for retinal gene transfer.[9–12] Ad vectors target the retinal pigment epithelium (RPE) very efficiently after subretinal injection.[9,12] However, they can induce inflammatory and immune responses that may result in rapid loss of therapeutic gene expression.[10,12,18] In contrast, AAV-based vectors are well tolerated in the eye. In addition, they are the only vectors that can efficiently transduce both the photoreceptors and the RPE after subretinal delivery in the adult retina.[13] Furthermore, unlike LV genomes, AAV vector genomes do not integrate into the host's genome, but persist primarily in the nucleus of target cells as monomeric and concatemeric episomes.[19] AAV has the additional advantage of being small, which facilitate its diffusion across retinal tissue barriers.[20–24] However, limitations in the cloning capacity of AAV (4.8 kb)[25] is a challenge for delivering large genes to photoreceptors.[26–32] Despite this, AAV is currently the vector of choice for treating most retinal diseases. At least 12 natural serotypes of AAV have been described so far. They are termed AAV1–12 in accordance with their unique sequences. AAV2 was the first AAV serotype identified as a potentially useful capsid vector for gene transfer.[33,34]

11.3 Clinical Applications of Retinal Gene Therapy

With such advantages, it is no surprise that over the past years retinal gene therapy has been developed for a myriad disorders, with promising results obtained not only in animal models, but also in humans.[35,36] At the time of writing, 41 clinical trials of gene therapy have been initiated for the treatment of a dozen different human diseases, from retinitis pigmentosa to age-related macular degeneration (AMD) (http://www.clinicaltrials.gov).[37]

11.3.1 Gene Addition Therapy for *RPE65*-Leber Congenital Amaurosis

Much of the current enthusiasm for the development of retinal gene therapy products stems from preclinical and clinical studies of Leber's congenital amaurosis (LCA). This early-onset blindness with autosomal-recessive inheritance can result from mutations in any of 22 different genes (http://www.retnet.org).[38] For one of these, *RPE65* (a gene that encodes a protein required for recycling visual pigment in the RPE), the addition of a healthy copy of the gene *via* subretinal injection of AAV-*RPE65* corrected the visual deficit.[39–42] Initial evidence for the efficacy of *RPE65* gene therapy came from the demonstration that subretinal injection of AAV2-c*Rpe65* restored rod and cone function and visual-guided behavior in the *Rpe65*$^{-/-}$ Briard dog.[43] Consistent with this premise, five phase-I/II clinical trials confirmed that AAV vectors delivering *RPE65* are safe and effective.[39–42,44–49] More than a third of patients showed some improvements in their visual function, as judged by an increase in light sensitivity, improved pupillary light reflex, improved visual acuity and visual fields, or improved mobility in dim light (reviewed in MacLaren[50] and Pierce and Bennett[51]). One protocol was successfully expanded to a trial in children,[52] demonstrating that early intervention vastly improved the potential for restoring vision, with some young patients being now able to read books or play baseball.[51] A follow-on study reported the results of AAV re-administration to the contralateral eye in the first three patients.[53] There were no safety concerns and evidence of benefit was found, which prompted initiation in 2009 of the first-ever randomized, control phase-III clinical trial for an eye disease (sponsored by Spark Therapeutics, Philadelphia, PA, USA). In 2015, the trial met its primary endpoint (improved bilateral mobility), as well as two secondary endpoints (light sensitivity and mobility for the first treated eye) with high statistical significance.[54] AAV2-*RPE65* (trade name: LUXTURNA™) is now under prioritized review by the US Food and Drug Administration and should become both the first approved gene therapy product for a genetic disease in the United States and the first pharmacological treatment for a form of inherited retinal diseases in the world.

11.3.2 AAV2-Based Gene Therapy for Other Forms of Retinal Degeneration

The positive results obtained by different groups for *RPE65*-LCA has facilitated a rapid expansion in the number of gene therapy for other retinal diseases.[35] Building on prior *RPE65* studies, several trials are underway to test safety and efficacy of the same AAV2 technology in other RPE-based genetic retinal disorders. These include conditions such as (1) autosomal recessive retinitis pigmentosa caused by mutations in the RPE-specific gene *MERTK*[55] and (2) choroideremia, an X-linked inherited retinal disease caused by mutations in *CHM*.[56,57] In both programs, safety was confirmed, and there were significant improvements in visual acuity in some of the patients.[55–57]

AAV2 vectors are used for more complex genetic retinal diseases, such as *ND4*-Leber hereditary optic neuropathy (LHON).[58–60] LHON is caused by mitochondrial defects in retinal ganglion cells, which are efficiently transduced through intravitreal injection of AAV2. Since AAV deliver genes to the cell's nucleus, the *ND4* transgene needs to be tagged with a mitochondrial targeting sequence (MT), to allow the proteins to be imported into the mitochondrial compartment after synthesis. The additional challenge includes the fact that most patients are initially unaware that they are affected until they experience loss of vision in the first eye, and it is difficult to predict when vision loss will start in the second eye. The goal of LHON gene therapy is, for now, to prevent further vision loss after symptoms manifest in the first eye. Preliminary data from early-phase clinical trials support the safety and tolerability of AAV2-gene therapy, but the efficacy of the treatment still needs to be established.[58–60] Moreover, in one patient, clinical benefits were lost after re-administration of the same AAV2-*ND4*-MT vector in the partner eye,[60] indicating that the vitreous cavity may be less immune-privileged than the subretinal space.[2,61–63]

In parallel, the AAV2 technology has expanded to common, acquired diseases of the retina.[64–66] Here, the goal of gene therapy is not to correct a genetic mutation responsible for the disease, but rather to target a gene known to be involved in the downstream pathway. The neovascular form of AMD is characterized by abnormal choroidal vessel proliferation behind the macula. Over-secretion of vascular endothelial growth factor (VEGF) plays a crucial role in the pathogenesis of neovascular AMD, and the current standard of care includes the repeated intravitreal injections of anti-VEGF inhibitors in patients.[67] Lions Eye Institute (Perth, WA, Australia) and Genzyme (Cambridge, MA, USA) are sponsoring clinical trials assessing the safety of AAV2 vectors expressing a soluble decoy VEGF receptor (FLT-1) to achieve long-lived and stable drug delivery and to reduce complications of repeat injections. Each study uses a different delivery route: subretinal[65,66] or intravitreal injection,[64] respectively. Regardless of the delivery route, both approaches have shown relatively good safety, but have yet failed to generate proof of efficacy where a positive outcome was highly expected. Research has now centered on novel vectors, including LV, as well as genes encoding other anti-angiogenic proteins.[68]

More recently, AAV2-based gene therapy has been explored in patients with end-stage retinal degeneration for its ability to render the retina light sensitive after most photoreceptors have degenerated.[69] In an ongoing phase I study, the channelrhodopsin 2 gene is delivered to ganglion cells to convert them into artificial photoreceptors. For the first time, gene therapy is not being used to slow or halt the progression of the disease, but to reverse blindness in end-stage retinas, independent of the nature of the disease.

11.3.3 Clinical Trials with Different Vectors for Improved Targeting or Delivery of Larger Genes

Despite the potential of AAV2-based applications for RPE and ganglion cell-based diseases, it was apparent from early studies seeking to rescue photoreceptors that improvements in AAV2 would be needed.[70–73] These observations stimulated the creation of next-generation AAV vectors that – while AAV2 is spreading into the clinic – could build on the existing platform and overcome current limitations. When serotypes 5, 8 and 9 were discovered, reports from multiple groups confirmed that they outperformed AAV2 for both photoreceptor and RPE transduction in small and large animals, demonstrating that serotype choice plays a significant role in successful transgene delivery.[17,74–76] Key advantages of novel serotypes include the ability to achieve a higher level of cone transduction[75,77–79] and a more rapid onset of transgene expression,[74,76] when compared to AAV2. Therapeutic efficacy of these novel AAV serotypes has been demonstrated in numerous animal models (for review see Petit *et al.*[35]), establishing a rich pipeline of gene therapy products that could be brought soon to the clinic,[35] such as AAV5-hRK1-*PDE6β* for the treatment of a recessive form of retinitis pigmentosa[80] and AAV8- and AAV5-hRK1-*GUCY2D* for LCA1.[77,81–83] One trial is already underway using AAV8-*CNGA3* for the treatment of achromatopsia. Although it is too early to adequately assess clinical benefit, publicly available information on visual benefits reported by the first patients is notable, given the low dose injected and the potentially damaging effect of retinal detachment (http://www.gesundheitsindustrie-bw.de).

Concurrent with an ongoing search for new AAV capsids, many groups have sought to engineer current serotypes to improve photoreceptor transduction. These methods focus mostly on modifications of AAV2 capsid to prevent vector ubiquitination and consequent degradation.[84–86] One engineered AAV2 serotype (AAV2tYP) that is better than AAV5 for photoreceptor transduction is currently being administered subretinally in ongoing clinical trials for X-linked retinoschisis, and for *CNGB3*-achromatopsia.[87] It should be tested soon for two other diseases: *CNGA3*-achromatopsia[88] and *RPGR*-X-linked retinitis pigmentosa.[89]

In parallel, two trials are ongoing using a non-human LV vector encoding ABCA4[90] and *MYO7A*[91,92] for the treatment of Stargardt and Usher Type 1B diseases, respectively. These genes are too large to be packaged in AAV,[93] yet can fit into an LV, which has a higher cargo capacity than AAV. A concern with LV remains the low transduction of mature photoreceptors[23] and the risk due to its integration into the host genome. At this point, results of the two trials remain unknown. Nonetheless, these results will provide valuable information not only regarding the efficacy of LV in the retina but also to determine whether LV may serve as a safe alternative vector for RPE diseases in which high level of transgene expression is required.[94–97]

11.4 Where Do We Go from Here?

Ongoing clinical trials provide evidence for the long-sought potential of gene therapy for people with selected retinal degenerative diseases. Importantly, in addition they shed light on areas that will require subsequent translational research to improve on efficacy and efficiency.

11.4.1 Maximizing Retinal Cell Transduction

As an example, RPE65 studies offered only modest benefits to patients compared to the remarkable gene therapy disease rescue obtained in the *Rpe65*$^{-/-}$ dog.[39,41,42,44,45] Additionally, in two of the RPE65 trials, the duration of benefits was limited by continued retinal degeneration.[98,99] It could be concluded that intervention had come "too late" in patients: photoreceptors had accumulated too many changes at the time of treatment and photoreceptor death was inevitable, even following reconstitution of RPE65 activity in RPE cells.[98] Consistent with that, relative to lifespan, the rate of retinal degeneration is higher in human than in dog models. However, in animals and patients, improvements in function and duration of treatment were not correlated with retinal thickness at the time of intervention.[99,100] The findings that even older dogs with advanced photoreceptor degeneration could show some improvements in retinal vision also indicate that the therapeutic window is not restricted to early stages of the disease.[100] This observation leads to a second hypothesis: the number of photoreceptors that were correctly rescued by the therapy at this stage of retinal degeneration was "too small", and treated cells were eventually overcome by non-cell autonomous changes in the tissue.[99,101] The low number of rescued cells in treated humans remains unexplained. However, what is clear from the data is the effects of *RPE65* replacement are dose-dependent.[99] In dogs, only higher doses have been associated with protection against progressive degeneration.[99] In humans, the highest vector dose has achieved improvements in visual sensitivity in a higher proportion of patients. Moreover, demand for RPE65 is 2.5 times higher in human *versus* dog retina.[99] Taken together, these findings suggest that next-generation vectors will be required for this RPE disease. This has prompted the development of a new gene therapy trial for *RPE65*-LCA using an AAV5 in combination with an optimized *RPE65* expression cassette.[102]

Equally important, these discussions highlight in our view the need to better consider the impact of the diseases on the efficacy of retinal gene therapy. To date, most of the vectors used for gene transfer in the retina have been studied in non-pathological tissues. However, retinal degeneration is accompanied by profound modifications of retinal cell homeostasis that very likely influence the efficacy of gene delivery and/or expression in retinal cells. For example, the presence of intact rod outer segments was recently shown to be important for AAV transduction, and this dependence might be limiting in advanced disease.[35] It is also becoming evident that endocytic pathways can become dysregulated in sick photoreceptors,[103] which has the potential to impact AAV intracellular trafficking significantly. Deciphering how retinal cells die, and how vectors bind, enter and traffic into both normal and sick retinal cells will be crucial before the field can further develop more efficient retinal gene therapy. This will require not only the convergence of the retinal biology and gene delivery fields, but also the use of appropriate systems that allow easy manipulation of the cell homeostasis and observation of alterations due to the retinal gene transfer. Induced pluripotent stem cell (iPSc)-derived retinal cells generated from fibroblasts of patients are intrinsically well suited for these studies.[104,105] They can detail the cell-based pathophysiological mechanisms of a disorder and enable researchers to optimize parameters for successful gene transfer *in vitro*.[104–107] Choroideremia is one such disorder, where the mechanisms leading to RPE and photoreceptor death remains unclear. Human iPSc-derived RPE generated from patients with choroideremia demonstrate an under-prenylation defect, as observed in patients' cells. iPSc-RPE was used to test the efficacy of different AAV serotypes and demonstrated rescue of the prenylation defect after gene therapy using AAV5-CHM.[104]

Furthermore, to obtain the maximal therapeutic effect for any retinal disease, it is essential to continue to improve patient selection using more rational criteria. In-depth characterization of the natural history for retinal degeneration in each condition is expected to help identify the patients who are most likely to benefit from the treatment. Patient-derived retinal cells, which have the specific advantage of retaining the genetic background of the donor, could provide a better understanding of the effects of the patient's specific genetic lesions on gene therapy before inclusion in clinical trials.[108]

11.4.2 Expanding the Number of Indications

One particular challenge to achieving a substantial clinical impact in the field of retinal gene therapy is the hundreds of genes involved in the pathogenesis of inherited retinal diseases. The rare nature of most retinal dystrophies means that only a relatively small number of individuals might benefit from treatments targeting specific gene mutations. This raises issues regarding whether the costs of developing such advanced therapies can be recovered after market authorization, and how to encourage investment in treatments for rare diseases that involve only a single administration.

One strategy to address this challenge is to take advantage of expanding knowledge of retinal degenerations and develop "platform"-type gene therapies approaches based on the commonalities between diseases and similar vectors.[109,110] Thus it has been proposed that extensive preclinical testing in animal models will no longer be required: the most efficient serotype will be selected in non-human primates or iPSc-derived cells, and the potency of the therapeutic vector directly evaluated *in vitro* before toxicological studies.[15]

The second strategy is to develop a generic gene therapy for photoreceptor loss that works across genetic subtypes.[35,111] The idea for such an approach was based on the realization that in many forms of retinal diseases, the patterns of photoreceptor cell death are similar. For example, as rods malfunction and die, there is always a subsequent death of neighboring healthy cones. Interestingly, cone death always starts near the end of the rod death phase, independently of the rate of rod death,[112] indicating a common pathologic cause and process. Preclinical studies have shown that gene transfer of neurotrophic factors, such as glial cell line-derived neurotrophic factor, ciliary neurotrophic factor and brain-derived neurotrophic factor, or anti-apoptotic proteins, such as XIAP, may prolong rod, and consequently, cone photoreceptor survival (for review see Petit *et al.*[35]). However, it is worth noting that their mode of action is not always fully understood, which makes the clinical translation of these approaches challenging. Such therapies could have possible pleiotropic effects, depending on the dose, delivery mode and/or target cells, and thus be associated with undesirable side effects. Another concern is that significant rod degeneration often occurs in many patients before they first visit an ophthalmologist, resulting in missed opportunities to rescue the rods and indirectly prevent the loss of cones. Several groups are therefore working on the identification of the mechanisms of cone death with the goal to directly target them. It is now widely accepted that secondary cone death is related to lack of energy and the associated oxidative stress.[112–115] Gene therapies that enhance glucose uptake or utilization in cone photoreceptors are expected to have a particular impact on the treatment of various forms of retinal degenerations.[114–120]

Another important limitation of retinal gene therapy remains the lack of effective vectors for delivery of large genes to photoreceptors. This prevents the application of gene therapy to common and severe inherited retinal dystrophies such as Stargardt disease, Usher syndrome type 2A or *CEP290*-LCA, even if benefits have been recently obtained in animal models using mini genes[121] or dual AAV approaches.[26–32] An attractive alternative to classic gene therapy for these large genes is the emerging field of therapeutic genome editing.[122] This relies on the introduction of a DNA break around the mutational site by highly specific endonucleases, which is then repaired by the cell's repair machinery using a DNA repair template, resulting in correction of the mutation *in situ*. However, several challenges need to be overcome before clinical application, including the lack of efficient methods for transient expression of nucleases to terminally differentiated photoreceptors, as well as potential off-target mutagenesis.[123]

11.5 Conclusions

The advances in the understanding of the genetics and pathophysiology of retinal diseases have now established the fundamentals of gene-based therapies for several disorders that have not responded to conventional treatments. Clinical trial data have confirmed the safety of retinal gene therapy and provided evidence that it is possible to obtain clinically relevant visual improvements after gene therapy. With the recent developments in AAV vectors, new clinical trials should translate into more and more encouraging results in patients. Simultaneously, lessons learned from the biology of retinopathies should enable the development of viable therapeutic strategies to prolong vision independently of the disease-causing mutations, further encouraging transfer to the clinic.

Acknowledgements

We apologize to those authors whose works have not been cited due to space limitation. Recent findings are referenced; older findings are discussed in the cited reviews. We would also like to acknowledge the French-American Fulbright Organisation, the Fondation Monahan, the Fondation de France and the AFM-Telethon for research support.

References

1. G. A. Ochakovski, K. U. Bartz-Schmidt and M. D. Fischer, *Front. Neurosci.*, 2017, **11**, 174.
2. K. Willett and J. Bennett, *Front. Immunol.*, 2013, **4**, 261.
3. S. Hacein-Bey-Abina, A. Garrigue, G. P. Wang, J. Soulier, A. Lim, E. Morillon, E. Clappier, L. Caccavelli, E. Delabesse, K. Beldjord, V. Asnafi, E. MacIntyre, L. Dal Cortivo, I. Radford, N. Brousse, F. Sigaux, D. Moshous, J. Hauer, A. Borkhardt, B. H. Belohradsky, U. Wintergerst, M. C. Velez, L. Leiva, R. Sorensen, N. Wulffraat, S. Blanche, F. D. Bushman, A. Fischer and M. Cavazzana-Calvo, *J. Clin. Invest.*, 2008, **118**, 3132.
4. S. Hacein-Bey-Abina, C. Von Kalle, M. Schmidt, M. P. McCormack, N. Wulffraat, P. Leboulch, A. Lim, C. S. Osborne, R. Pawliuk, E. Morillon, R. Sorensen, A. Forster, P. Fraser, J. I. Cohen, G. de Saint Basile, I. Alexander, U. Wintergerst, T. Frebourg, A. Aurias, D. Stoppa-Lyonnet, S. Romana, I. Radford-Weiss, F. Gross, F. Valensi, E. Delabesse, E. Macintyre, F. Sigaux, J. Soulier, L. E. Leiva, M. Wissler, C. Prinz, T. H. Rabbitts, F. Le Deist, A. Fischer and M. Cavazzana-Calvo, *Science*, 2003, **302**, 415.
5. P. Charbel Issa and R. E. MacLaren, *Clin. Exp. Ophthalmol.*, 2012, **40**, 39.
6. A. Koirala, S. M. Conley and M. I. Naash, *Biomaterials*, 2013, **34**, 7158.
7. M. A. Kay, J. C. Glorioso and L. Naldini, *Nat. Med.*, 2001, **7**, 33.
8. I. Trapani, A. Puppo and A. Auricchio, *Prog. Retinal Eye Res.*, 2014, **43**, 108.
9. J. Bennett, J. Wilson, D. Sun, B. Forbes and A. Maguire, *Invest. Ophthalmol. Visual Sci.*, 1994, **35**, 2535.
10. B. Mashhour, D. Couton, M. Perricaudet and P. Briand, *Gene Ther.*, 1994, **1**, 122.
11. C. Jomary, T. A. Piper, G. Dickson, L. A. Couture, A. E. Smith, M. J. Neal and S. E. Jones, *FEBS Lett.*, 1994, **347**, 117.
12. T. Li, M. Adamian, D. J. Roof, E. L. Berson, T. P. Dryja, B. J. Roessler and B. L. Davidson, *Invest. Ophthalmol. Visual Sci.*, 1994, **35**, 2543.
13. R. Kumar-Singh, *Vis. Res.*, 2008, **48**, 1671.
14. K. S. Balaggan and R. R. Ali, *Gene Ther.*, 2012, **19**, 145.
15. L. H. Vandenberghe and A. Auricchio, *Gene Ther.*, 2012, **19**, 162.
16. T. P. Day, L. C. Byrne, D. V. Schaffer and J. G. Flannery, *Adv. Exp. Med. Biol.*, 2014, **801**, 687.
17. E. M. Surace and A. Auricchio, *Vis. Res.*, 2008, **48**, 353.
18. J. Bennett, T. Tanabe, D. Sun, Y. Zeng, H. Kjeldbye, P. Gouras and A. M. Maguire, *Nat. Med.*, 1996, **2**, 649.
19. B. C. Schnepp, R. L. Jensen, C. L. Chen, P. R. Johnson and K. R. Clark, *J. Virol.*, 2005, **79**, 14793.
20. D. Dalkara, K. D. Kolstad, N. Caporale, M. Visel, R. R. Klimczak, D. V. Schaffer and J. G. Flannery, *Mol. Ther.*, 2009, **17**, 2096.
21. K. D. Kolstad, D. Dalkara, K. Guerin, M. Visel, N. Hoffmann, D. V. Schaffer and J. G. Flannery, *Hum. Gene Ther.*, 2010, **21**, 571.
22. A. Puppo, G. Cesi, E. Marrocco, P. Piccolo, S. Jacca, D. M. Shayakhmetov, R. J. Parks, B. L. Davidson, S. Colloca, N. Brunetti-Pierri, P. Ng, G. Donofrio and A. Auricchio, *Gene Ther.*, 2014, **21**, 855.

23. O. Gruter, C. Kostic, S. V. Crippa, M. T. Perez, L. Zografos, D. F. Schorderet, F. L. Munier and Y. Arsenijevic, *Gene Ther.*, 2005, **12**, 942.
24. M. Calame, M. Cachafeiro, S. Philippe, K. Schouwey, M. Tekaya, D. Wanner, C. Sarkis, C. Kostic and Y. Arsenijevic, *PLoS One*, 2011, **6**, e23782.
25. Z. Wu, H. Yang and P. Colosi, *Mol. Ther.*, 2010, **18**, 80.
26. P. Colella, I. Trapani, G. Cesi, A. Sommella, A. Manfredi, A. Puppo, C. Iodice, S. Rossi, F. Simonelli, M. Giunti, M. L. Bacci and A. Auricchio, *Gene Ther.*, 2014, **21**, 450.
27. I. Trapani, P. Colella, A. Sommella, C. Iodice, G. Cesi, S. de Simone, E. Marrocco, S. Rossi, M. Giunti, A. Palfi, G. J. Farrar, R. Polishchuk and A. Auricchio, *EMBO Mol. Med.*, 2014, **6**, 194.
28. I. Trapani, E. Toriello, S. de Simone, P. Colella, C. Iodice, E. V. Polishchuk, A. Sommella, L. Colecchi, S. Rossi, F. Simonelli, M. Giunti, M. L. Bacci, R. S. Polishchuk and A. Auricchio, *Hum. Mol. Genet.*, 2015, **24**, 6811.
29. L. S. Carvalho, H. T. Turunen, S. J. Wassmer, M. V. Luna-Velez, R. Xiao, J. Bennett and L. H. Vandenberghe, *Front. Neurosci.*, 2017, **11**, 503.
30. F. M. Dyka, S. L. Boye, V. A. Chiodo, W. W. Hauswirth and S. E. Boye, *Hum. Gene Ther: Methods.*, 2014, **25**, 166.
31. K. Chamberlain, J. M. Riyad and T. Weber, *Hum. Gene Ther, Methods.*, 2016, **27**, 1.
32. V. S. Lopes, S. E. Boye, C. M. Louie, S. Boye, F. Dyka, V. Chiodo, H. Fofo, W. W. Hauswirth and D. S. Williams, *Gene Ther.*, 2013, **20**, 824.
33. R. J. Samulski, K. I. Berns, M. Tan and N. Muzyczka, *Proc. Natl. Acad. Sci. U. S. A.*, 1982, **79**, 2077.
34. P. L. Hermonat and N. Muzyczka, *Proc. Natl. Acad. Sci. U. S. A.*, 1984, **81**, 6466–6470.
35. L. Petit, H. Khanna and C. Punzo, *Hum. Gene Ther.*, 2016, **27**, 563.
36. S. E. Boye, S. L. Boye, A. S. Lewin and W. W. Hauswirth, *Mol. Ther.*, 2013, **21**, 509.
37. A. Auricchio, A. J. Smith and R. R. Ali, *Hum. Gene Ther.*, 2017, **28**, 982.
38. O. F. Chacon-Camacho and J. C. Zenteno, *World J. Clin. Cases*, 2015, **3**, 112.
39. J. W. Bainbridge, A. J. Smith, S. S. Barker, S. Robbie, R. Henderson, K. Balaggan, A. Viswanathan, G. E. Holder, A. Stockman, N. Tyler, S. Petersen-Jones, S. S. Bhattacharya, A. J. Thrasher, F. W. Fitzke, B. J. Carter, G. S. Rubin, A. T. Moore and R. R. Ali, *N. Engl. J. Med.*, 2008, **358**, 2231.
40. A. V. Cideciyan, W. W. Hauswirth, T. S. Aleman, S. Kaushal, S. B. Schwartz, S. L. Boye, E. A. Windsor, T. J. Conlon, A. Sumaroka, J. J. Pang, A. J. Roman, B. J. Byrne and S. G. Jacobson, *Hum. Gene Ther.*, 2009, **20**, 999.
41. A. M. Maguire, F. Simonelli, E. A. Pierce, E. N. Pugh Jr, F. Mingozzi, J. Bennicelli, S. Banfi, K. A. Marshall, F. Testa, E. M. Surace, S. Rossi, A. Lyubarsky, V. R. Arruda, B. Konkle, E. Stone, J. Sun, J. Jacobs, L. Dell'Osso, R. Hertle, J. X. Ma, T. M. Redmond, X. Zhu, B. Hauck, O. Zelenaia, K. S. Shindler, M. G. Maguire, J. F. Wright, N. J. Volpe, J. W. McDonnell, A. Auricchio, K. A. High and J. Bennett, *N. Engl. J. Med.*, 2008, **358**, 2240.
42. R. G. Weleber, M. E. Pennesi, D. J. Wilson, S. Kaushal, L. R. Erker, L. Jensen, M. T. McBride, T. R. Flotte, M. Humphries, R. Calcedo, W. W. Hauswirth, J. D. Chulay and J. T. Stout, *Ophthalmology*, 2016, **123**, 1606.

43. G. M. Acland, G. D. Aguirre, J. Ray, Q. Zhang, T. S. Aleman, A. V. Cideciyan, S. E. Pearce-Kelling, V. Anand, Y. Zeng, A. M. Maguire, S. G. Jacobson, W. W. Hauswirth and J. Bennett, *Nat. Genet.*, 2001, **28**, 92.
44. A. V. Cideciyan, T. S. Aleman, S. L. Boye, S. B. Schwartz, S. Kaushal, A. J. Roman, J. J. Pang, A. Sumaroka, E. A. Windsor, J. M. Wilson, T. R. Flotte, G. A. Fishman, E. Heon, E. M. Stone, B. J. Byrne, S. G. Jacobson and W. W. Hauswirth, *Proc. Natl. Acad. Sci. U. S. A.*, 2008, **105**, 15112.
45. W. W. Hauswirth, T. S. Aleman, S. Kaushal, A. V. Cideciyan, S. B. Schwartz, L. Wang, T. J. Conlon, S. L. Boye, T. R. Flotte, B. J. Byrne and S. G. Jacobson, *Hum. Gene Ther.*, 2008, **19**, 979.
46. S. G. Jacobson, A. V. Cideciyan, R. Ratnakaram, E. Heon, S. B. Schwartz, A. J. Roman, M. C. Peden, T. S. Aleman, S. L. Boye, A. Sumaroka, T. J. Conlon, R. Calcedo, J. J. Pang, K. E. Erger, M. B. Olivares, C. L. Mullins, M. Swider, S. Kaushal, W. J. Feuer, A. Iannaccone, G. A. Fishman, E. M. Stone, B. J. Byrne and W. W. Hauswirth, *Arch. Ophthalmol.*, 2012, **130**, 9.
47. F. Simonelli, A. M. Maguire, F. Testa, E. A. Pierce, F. Mingozzi, J. L. Bennicelli, S. Rossi, K. Marshall, S. Banfi, E. M. Surace, J. Sun, T. M. Redmond, X. Zhu, K. S. Shindler, G. S. Ying, C. Ziviello, C. Acerra, J. F. Wright, J. W. McDonnell, K. A. High, J. Bennett and A. Auricchio, *Mol. Ther.*, 2010, **18**, 643.
48. F. Testa, A. M. Maguire, S. Rossi, E. A. Pierce, P. Melillo, K. Marshall, S. Banfi, E. M. Surace, J. Sun, C. Acerra, J. F. Wright, J. Wellman, K. A. High, A. Auricchio, J. Bennett and F. Simonelli, *Ophthalmology*, 2013, **120**, 1283.
49. J. Bennett, J. Wellman, K. A. Marshall, S. McCague, M. Ashtari, J. DiStefano-Pappas, O. U. Elci, D. C. Chung, J. Sun, J. F. Wright, D. R. Cross, P. Aravand, L. L. Cyckowski, J. L. Bennicelli, F. Mingozzi, A. Auricchio, E. A. Pierce, J. Ruggiero, B. P. Leroy, F. Simonelli, K. A. High and A. M. Maguire, *Lancet*, 2016, **388**, 661.
50. R. E. MacLaren, *Curr. Opin. Mol. Ther.*, 2009, **11**, 540.
51. E. A. Pierce and J. Bennett, *Cold Spring Harbor Perspect. Med.*, 2015, **5**, a017285.
52. A. M. Maguire, K. A. High, A. Auricchio, J. F. Wright, E. A. Pierce, F. Testa, F. Mingozzi, J. L. Bennicelli, G. S. Ying, S. Rossi, A. Fulton, K. A. Marshall, S. Banfi, D. C. Chung, J. I. Morgan, B. Hauck, O. Zelenaia, X. Zhu, L. Raffini, F. Coppieters, E. De Baere, K. S. Shindler, N. J. Volpe, E. M. Surace, C. Acerra, A. Lyubarsky, T. M. Redmond, E. Stone, J. Sun, J. W. McDonnell, B. P. Leroy, F. Simonelli and J. Bennett, *Lancet*, 2009, **374**, 1597.
53. J. Bennett, M. Ashtari, J. Wellman, K. A. Marshall, L. L. Cyckowski, D. C. Chung, S. McCague, E. A. Pierce, Y. Chen, J. L. Bennicelli, X. Zhu, G. S. Ying, J. Sun, J. F. Wright, A. Auricchio, F. Simonelli, K. S. Shindler, F. Mingozzi, K. A. High and A. M. Maguire, *Sci. Transl. Med.*, 2012, **4**, 120ra115.
54. S. Russell, J. Bennett, J. A. Wellman, D. C. Chung, Z. F. Yu, A. Tillman, J. Wittes, J. Pappas, O. Elci, S. McCague, D. Cross, K. A. Marshall, J. Walshire, T. L. Kehoe, H. Reichert, M. Davis, L. Raffini, L. A. George, F.

P. Hudson, L. Dingfield, X. Zhu, J. A. Haller, E. H. Sohn, V. B. Mahajan, W. Pfeifer, M. Weckmann, C. Johnson, D. Gewaily, A. Drack, E. Stone, K. Wachtel, F. Simonelli, B. P. Leroy, J. F. Wright, K. A. High and A. M. Maguire, *Lancet*, 2017, **390**, 849.

55. N. G. Ghazi, E. B. Abboud, S. R. Nowilaty, H. Alkuraya, A. Alhommadi, H. Cai, R. Hou, W. T. Deng, S. L. Boye, A. Almaghamsi, F. Al Saikhan, H. Al-Dhibi, D. Birch, C. Chung, D. Colak, M. M. LaVail, D. Vollrath, K. Erger, W. Wang, T. Conlon, K. Zhang, W. Hauswirth and F. S. Alkuraya, *Hum. Genet.*, 2016, **135**, 327.
56. R. E. MacLaren, M. Groppe, A. R. Barnard, C. L. Cottriall, T. Tolmachova, L. Seymour, K. R. Clark, M. J. During, F. P. Cremers, G. C. Black, A. J. Lotery, S. M. Downes, A. R. Webster and M. C. Seabra, *Lancet*, 2014, **383**, 1129.
57. T. L. Edwards, J. K. Jolly, M. Groppe, A. R. Barnard, C. L. Cottriall, T. Tolmachova, G. C. Black, A. R. Webster, A. J. Lotery, G. E. Holder, K. Xue, S. M. Downes, M. P. Simunovic, M. C. Seabra and R. E. MacLaren, *N. Engl. J. Med.*, 2016, **374**, 1996–1998.
58. X. Wan, H. Pei, M. J. Zhao, S. Yang, W. K. Hu, H. He, S. Q. Ma, G. Zhang, X. Y. Dong, C. Chen, D. W. Wang and B. Li, *Sci. Rep.*, 2016, **6**, 21587.
59. W. J. Feuer, J. C. Schiffman, J. L. Davis, V. Porciatti, P. Gonzalez, R. D. Koilkonda, H. Yuan, A. Lalwani, B. L. Lam and J. Guy, *Ophthalmology*, 2016, **123**, 558–570.
60. S. Yang, S. Q. Ma, X. Wan, H. He, H. Pei, M. J. Zhao, C. Chen, D. W. Wang, X. Y. Dong, J. J. Yuan and B. Li, *EBioMedicine*, 2016, **10**, 258.
61. Q. Li, R. Miller, P. Y. Han, J. Pang, A. Dinculescu, V. Chiodo and W. W. Hauswirth, *Mol. Vision*, 2008, **14**, 1760.
62. M. A. Kotterman, L. Yin, J. M. Strazzeri, J. G. Flannery, W. H. Merigan and D. V. Schaffer, *Gene Ther.*, 2015, **22**, 116.
63. R. F. Boyd, S. L. Boye, T. J. Conlon, K. E. Erger, D. G. Sledge, I. M. Langohr, W. W. Hauswirth, A. M. Komaromy, S. E. Boye, S. M. Petersen-Jones and J. T. Bartoe, *Gene Ther.*, 2016, **23**, 548.
64. J. S. Heier, S. Kherani, S. Desai, P. Dugel, S. Kaushal, S. H. Cheng, C. Delacono, A. Purvis, S. Richards, A. Le-Halpere, J. Connelly, S. C. Wadsworth, R. Varona, R. Buggage, A. Scaria and P. A. Campochiaro, *Lancet*, 2017, **390**, 50.
65. I. J. Constable, C. M. Pierce, C. M. Lai, A. L. Magno, M. A. Degli-Esposti, M. A. French, I. L. McAllister, S. Butler, S. B. Barone, S. D. Schwartz, M. S. Blumenkranz and E. P. Rakoczy, *EBioMedicine*, 2016, **14**, 168.
66. E. P. Rakoczy, C. M. Lai, A. L. Magno, M. E. Wikstrom, M. A. French, C. M. Pierce, S. D. Schwartz, M. S. Blumenkranz, T. W. Chalberg, M. A. Degli-Esposti and I. J. Constable, *Lancet*, 2015, **386**, 2395.
67. J. Ba, R. S. Peng, D. Xu, Y. H. Li, H. Shi, Q. Wang and J. Yu, *Drug Des., Dev. Ther.*, 2015, **9**, 5397.
68. P. A. Campochiaro, A. K. Lauer, E. H. Sohn, T. A. Mir, S. Naylor, M. C. Anderton, M. Kelleher, R. Harrop, S. Ellis and K. A. Mitrophanous, *Hum. Gene Ther.*, 2017, **28**, 99.

69. V. Busskamp and B. Roska, *Curr. Opin. Neurobiol.*, 2011, **21**, 942.
70. C. Jomary, K. A. Vincent, J. Grist, M. J. Neal and S. E. Jones, *Gene Ther.*, 1997, **4**, 683.
71. M. H. Tan, A. J. Smith, B. Pawlyk, X. Xu, X. Liu, J. B. Bainbridge, M. Basche, J. McIntosh, H. V. Tran, A. Nathwani, T. Li and R. R. Ali, *Hum. Mol. Genet.*, 2009, **18**, 2099.
72. R. R. Ali, G. M. Sarra, C. Stephens, M. D. Alwis, J. W. Bainbridge, P. M. Munro, S. Fauser, M. B. Reichel, C. Kinnon, D. M. Hunt, S. S. Bhattacharya and A. J. Thrasher, *Nat. Genet.*, 2000, **25**, 306.
73. G. M. Sarra, C. Stephens, M. de Alwis, J. W. Bainbridge, A. J. Smith, A. J. Thrasher and R. R. Ali, *Hum. Mol. Genet.*, 2001, **10**, 2353.
74. M. Allocca, C. Mussolino, M. Garcia-Hoyos, D. Sanges, C. Iodice, M. Petrillo, L. H. Vandenberghe, J. M. Wilson, V. Marigo, E. M. Surace and A. Auricchio, *J. Virol.*, 2007, **81**, 11372.
75. L. H. Vandenberghe, P. Bell, A. M. Maguire, C. N. Cearley, R. Xiao, R. Calcedo, L. Wang, M. J. Castle, A. C. Maguire, R. Grant, J. H. Wolfe, J. M. Wilson and J. Bennett, *Sci. Transl. Med.*, 2011, **3**, 88ra54.
76. M. Natkunarajah, P. Trittibach, J. McIntosh, Y. Duran, S. E. Barker, A. J. Smith, A. C. Nathwani and R. R. Ali, *Gene Ther.*, 2008, **15**, 463.
77. A. Manfredi, E. Marrocco, A. Puppo, G. Cesi, A. Sommella, M. Della Corte, S. Rossi, M. Giunti, C. M. Craft, M. L. Bacci, F. Simonelli, E. M. Surace and A. Auricchio, *Hum. Gene Ther.*, 2013, **24**, 982.
78. C. Mussolino, M. della Corte, S. Rossi, F. Viola, U. Di Vicino, E. Marrocco, S. Neglia, M. Doria, F. Testa, R. Giovannoni, M. Crasta, M. Giunti, E. Villani, M. Lavitrano, M. L. Bacci, R. Ratiglia, F. Simonelli, A. Auricchio and E. M. Surace, *Gene Ther.*, 2011, **18**, 637.
79. L. H. Vandenberghe, P. Bell, A. M. Maguire, R. Xiao, T. B. Hopkins, R. Grant, J. Bennett and J. M. Wilson, *PLoS One*, 2013, **8**, e53463.
80. L. Petit, E. Lheriteau, M. Weber, G. Le Meur, J. Y. Deschamps, N. Provost, A. Mendes-Madeira, L. Libeau, C. Guihal, M. A. Colle, P. Moullier and F. Rolling, *Mol. Ther.*, 2012, **20**, 2019.
81. S. L. Boye, T. Conlon, K. Erger, R. Ryals, A. Neeley, T. Cossette, J. Pang, F. M. Dyka, W. W. Hauswirth and S. E. Boye, *Invest. Ophthalmol. Visual Sci.*, 2011, **52**, 7098.
82. S. L. Boye, I. V. Peshenko, W. C. Huang, S. H. Min, I. McDoom, C. N. Kay, X. Liu, F. M. Dyka, T. C. Foster, Y. Umino, S. Karan, S. G. Jacobson, W. Baehr, A. Dizhoor, W. W. Hauswirth and S. E. Boye, *Hum. Gene Ther.*, 2013, **24**, 189.
83. S. L. Boye, J. J. Peterson, S. Choudhury, S. H. Min, Q. Ruan, K. T. McCullough, Z. Zhang, E. V. Olshevskaya, I. V. Peshenko, W. W. Hauswirth, X. Q. Ding, A. M. Dizhoor and S. E. Boye, *Hum. Gene Ther.*, 2015, **26**, 575.
84. H. Petrs-Silva, A. Dinculescu, Q. Li, S. H. Min, V. Chiodo, J. J. Pang, L. Zhong, S. Zolotukhin, A. Srivastava, A. S. Lewin and W. W. Hauswirth, *Mol. Ther.*, 2009, **17**, 463.

85. H. Petrs-Silva, A. Dinculescu, Q. Li, W. T. Deng, J. J. Pang, S. H. Min, V. Chiodo, A. W. Neeley, L. Govindasamy, A. Bennett, M. Agbandje-McKenna, L. Zhong, B. Li, G. R. Jayandharan, A. Srivastava, A. S. Lewin and W. W. Hauswirth, *Mol. Ther.*, 2011, **19**, 293.
86. F. M. Mowat, K. R. Gornik, A. Dinculescu, S. L. Boye, W. W. Hauswirth, S. M. Petersen-Jones and J. T. Bartoe, *Gene Ther.*, 2014, **21**, 96.
87. G. J. Ye, E. Budzynski, P. Sonnentag, T. M. Nork, N. Sheibani, Z. Gurel, S. L. Boye, J. J. Peterson, S. E. Boye, W. W. Hauswirth and J. D. Chulay, *Hum. Gene Ther.*, 2016, **27**, 72.
88. E. Gootwine, R. Ofri, E. Banin, A. Obolensky, E. Averbukh, R. Ezra-Elia, M. Ross, H. Honig, A. Rosov, E. Yamin, G. J. Ye, D. R. Knop, P. M. Robinson, J. D. Chulay and M. S. Shearman, *Hum. Gene Ther. Clin. Dev.*, 2017, **28**, 96.
89. W. A. Beltran, A. V. Cideciyan, S. E. Boye, G. J. Ye, S. Iwabe, V. L. Dufour, L. F. Marinho, M. Swider, M. S. Kosyk, J. Sha, S. L. Boye, J. J. Peterson, C. D. Witherspoon, J. J. Alexander, G. S. Ying, M. S. Shearman, J. D. Chulay, W. W. Hauswirth, P. D. Gamlin, S. G. Jacobson and G. D. Aguirre, *Mol. Ther.*, 2017, **25**, 1866.
90. J. Kong, S. R. Kim, K. Binley, I. Pata, K. Doi, J. Mannik, J. Zernant-Rajang, O. Kan, S. Iqball, S. Naylor, J. R. Sparrow, P. Gouras and R. Allikmets, *Gene Ther.*, 2008, **15**, 1311.
91. M. Zallocchi, K. Binley, Y. Lad, S. Ellis, P. Widdowson, S. Iqball, V. Scripps, M. Kelleher, J. Loader, J. Miskin, Y. W. Peng, W. M. Wang, L. Cheung, D. Delimont, K. A. Mitrophanous and D. Cosgrove, *PLoS One*, 2014, **9**, e94272.
92. T. Hashimoto, D. Gibbs, C. Lillo, S. M. Azarian, E. Legacki, X. M. Zhang, X. J. Yang and D. S. Williams, *Gene Ther.*, 2007, **14**, 584.
93. Y. Lai, Y. Yue and D. Duan, *Mol. Ther.*, 2010, **18**, 75.
94. A. Matet, C. Kostic, A. P. Bemelmans, A. Moulin, S. G. Rosolen, S. Martin, F. Mavilio, V. Amirjanians, K. Stieger, B. Lorenz, F. Behar-Cohen and Y. Arsenijevic, *Transl. Res.*, 2017, **188**, 40–57, e4.
95. A. P. Bemelmans, C. Kostic, S. V. Crippa, W. W. Hauswirth, J. Lem, F. L. Munier, M. W. Seeliger, A. Wenzel and Y. Arsenijevic, *PLoS Med.*, 2006, **3**, e347.
96. C. Kostic, S. V. Crippa, V. Pignat, A. P. Bemelmans, M. Samardzija, C. Grimm, A. Wenzel and Y. Arsenijevic, *PLoS One*, 2011, **6**, e16588.
97. M. Tschernutter, F. C. Schlichtenbrede, S. Howe, K. S. Balaggan, P. M. Munro, J. W. Bainbridge, A. J. Thrasher, A. J. Smith and R. R. Ali, *Gene Ther.*, 2005, **12**, 694.
98. A. V. Cideciyan, S. G. Jacobson, W. A. Beltran, A. Sumaroka, M. Swider, S. Iwabe, A. J. Roman, M. B. Olivares, S. B. Schwartz, A. M. Komaromy, W. W. Hauswirth and G. D. Aguirre, *Proc. Natl. Acad. Sci. U. S. A.*, 2013, **110**, E517.
99. J. W. Bainbridge, M. S. Mehat, V. Sundaram, S. J. Robbie, S. E. Barker, C. Ripamonti, A. Georgiadis, F. M. Mowat, S. G. Beattie, P. J. Gardner, K. L.

Feathers, V. A. Luong, S. Yzer, K. Balaggan, A. Viswanathan, T. J. de Ravel, I. Casteels, G. E. Holder, N. Tyler, F. W. Fitzke, R. G. Weleber, M. Nardini, A. T. Moore, D. A. Thompson, S. M. Petersen-Jones, M. Michaelides, L. I. van den Born, A. Stockman, A. J. Smith, G. Rubin and R. R. Ali, *N. Engl. J. Med.*, 2015, **372**, 1887.
100. M. J. Annear, F. M. Mowat, J. T. Bartoe, J. Querubin, S. A. Azam, M. Basche, P. G. Curran, A. J. Smith, J. W. Bainbridge, R. R. Ali and S. M. Petersen-Jones, *Hum. Gene Ther.*, 2013, **24**, 883.
101. C. L. Cepko and L. H. Vandenberghe, *Hum. Gene Ther.*, 2013, **24**, 242.
102. A. Georgiadis, Y. Duran, J. Ribeiro, L. Abelleira-Hervas, S. J. Robbie, B. Sunkel-Laing, S. Fourali, A. Gonzalez-Cordero, E. Cristante, M. Michaelides, J. W. Bainbridge, A. J. Smith and R. R. Ali, *Gene Ther.*, 2016, **23**, 857.
103. J. Z. Chuang, C. Vega, W. Jun and C. H. Sung, *J. Clin. Invest.*, 2004, **114**, 131.
104. N. Cereso, M. O. Pequignot, L. Robert, F. Becker, V. De Luca, N. Nabholz, V. Rigau, J. De Vos, C. P. Hamel and V. Kalatzis, *Mol. Ther. Methods Clin. Dev.*, 2014, **1**, 14011.
105. Y. Li, L. Chan, H. V. Nguyen and S. H. Tsang, *Adv. Exp. Med. Biol.*, 2016, **854**, 549.
106. Y. Li, W. H. Wu, C. W. Hsu, H. V. Nguyen, Y. T. Tsai, L. Chan, T. Nagasaki, I. H. Maumenee, L. A. Yannuzzi, Q. V. Hoang, H. Hua, D. Egli and S. H. Tsang, *Mol. Ther.*, 2014, **22**, 1688.
107. N. Schwarz, A. J. Carr, A. Lane, F. Moeller, L. L. Chen, M. Aguila, B. Nommiste, M. N. Muthiah, N. Kanuga, U. Wolfrum, K. Nagel-Wolfrum, L. da Cruz, P. J. Coffey, M. E. Cheetham and A. J. Hardcastle, *Hum. Mol. Genet.*, 2015, **24**, 972.
108. S. Torriano, N. Erkilic, V. Faugere, K. Damodar, C. P. Hamel, A. F. Roux and V. Kalatzis, *Hum. Mol. Genet.*, 2017, **26**, 3573.
109. M. O'Reilly, D. B. Kohn, J. Bartlett, J. Benson, P. J. Brooks, B. J. Byrne, C. Camozzi, K. Cornetta, R. G. Crystal, Y. Fong, L. Gargiulo, R. Gopal-Srivastava, K. A. High, S. G. Jacobson, R. C. Jambou, M. Montgomery, E. Rosenthal, R. J. Samulski, S. I. Skarlatos, B. Sorrentino, J. M. Wilson, Y. Xie and J. Corrigan-Curay, *Hum. Gene Ther.*, 2013, **24**, 355.
110. L. H. Vandenberghe, *Cold Spring Harbor Perspect. Med.*, 2015, **5**, a017442.
111. L. Petit and C. Punzo, *Discov. Med.*, 2016, **22**, 221.
112. C. Punzo, K. Kornacker and C. L. Cepko, *Nat. Neurosci.*, 2009, **12**, 44.
113. C. Punzo, W. Xiong and C. L. Cepko, *J. Biol. Chem.*, 2012, **287**, 1642.
114. A. Venkatesh, S. Ma, Y. Z. Le, M. N. Hall, M. A. Ruegg and C. Punzo, *J. Clin. Invest.*, 2015, **125**, 1446.
115. C. Cepko and C. Punzo, *Nature*, 2015, **522**, 428.
116. W. Wang, S. J. Lee, P. A. Scott, X. Lu, D. Emery, Y. Liu, T. Ezashi, M. R. Roberts, J. W. Ross, H. J. Kaplan and D. C. Dean, *Cell Rep.*, 2016, **15**, 372.
117. M. A. Kanow, M. M. Giarmarco, C. S. Jankowski, K. Tsantilas, A. L. Engel, J. Du, J. D. Linton, C. C. Farnsworth, S. R. Sloat, A. Rountree, I. R. Sweet,

K. J. Lindsay, E. D. Parker, S. E. Brockerhoff, M. Sadilek, J. R. Chao and J. B. Hurley, *eLife*, 2017, **6**, e28899.
118. N. Ait-Ali, R. Fridlich, G. Millet-Puel, E. Clerin, F. Delalande, C. Jaillard, F. Blond, L. Perrocheau, S. Reichman, L. C. Byrne, A. Olivier-Bandini, J. Bellalou, E. Moyse, F. Bouillaud, X. Nicol, D. Dalkara, A. van Dorsselaer, J. A. Sahel and T. Leveillard, *Cell*, 2015, **161**, 817.
119. L. C. Byrne, D. Dalkara, G. Luna, S. K. Fisher, E. Clerin, J. A. Sahel, T. Leveillard and J. G. Flannery, *J. Clin. Invest.*, 2015, **125**, 105.
120. W. Xiong, A. E. MacColl Garfinkel, Y. Li, L. I. Benowitz and C. L. Cepko, *J. Clin. Invest.*, 2015, **125**, 1433.
121. W. Zhang, L. Li, Q. Su, G. Gao and H. Khanna, *Hum. Gene Ther.*, 2018, **29**, 42.
122. D. B. Cox, R. J. Platt and F. Zhang, *Nat. Med.*, 2015, **21**, 121.
123. M. Yanik, B. Muller, F. Song, J. Gall, F. Wagner, W. Wende, B. Lorenz and K. Stieger, *Prog. Retinal Eye Res.*, 2017, **56**, 1.

CHAPTER 12

Stem Cell Therapies for Retinal Repair and Regeneration

KAREN EASTLAKE, WEIXIN WANG, WILLIAM D. B. LAMB AND G. ASTRID LIMB*

UCL Institute of Ophthalmology, NIHR Biomedical Research Centre at Moorfields Eye Hospital, 11-43 Bath Street, London EC1V 9EL, UK
*E-mail: g.limb@ucl.ac.uk

12.1 Introduction

New developments in the stem cell field have broadened the prospects of designing cell-based therapies to repair and restore visual function. Epidemiological studies of diseases such as age-related macular degeneration (AMD),[1] glaucoma[2] and diabetic retinopathy[3] suggest that blindness and severe visual impairment caused by these conditions are on the rise, therefore creating significant economic burden to healthcare systems across the world. Current treatments can control disease progression, but cannot restore lost function, for which new therapies are needed to repair the retina and restore visual function. Research undertaken in the stem cell field in recent years has raised hope for the use of these cells as therapeutic agents to prevent blindness and restore vision.

Most studies in the retina regenerative field have been aimed at replacing neurons by stem cell transplantation. However, neural replacement by cell grafting has not been successfully achieved, as this process requires

Drug Discovery Series No. 66
Therapies for Retinal Degeneration: Targeting Common Processes
Edited by Enrique J. de la Rosa and Thomas G. Cotter

Published by the Royal Society of Chemistry, www.rsc.org

integration of transplanted cells within major neural networks. Nonetheless, partial restoration of retinal function has been observed after experimental stem cell grafting, suggesting that stem cells may constitute an important source of neuroprotective factors to repair neural damage and this is a subject of current investigations.[4–6] Furthermore, following the discovery of the presence of Müller glia with stem cell characteristics in the adult human retina,[7,8] approaches to induce self-repair (endogenous regeneration) are at present being investigated by several groups. Current knowledge in the transplantation and endogenous regeneration research fields is discussed in the following sections.

12.1.1 Retinal Degeneration as a Cause of Visual Impairment and Blindness

Retina degenerative diseases are characterized by neural cell damage, which is the ultimate cause of visual impairment and blindness. The retina harbours six different types of neurons, namely cone and rod photoreceptors, bipolar, horizontal and amacrine cells and retinal ganglion cells (Figure 12.1). Two macroglia populations, comprising Müller glia and astrocytes, are also present in the retina, with Müller cells providing structural and metabolic support to all retinal neurons. Müller glia and retinal neurons have a common progenitor,[9] but the origin of retinal astrocytes has been debatable. Evidence has been presented that astrocyte migration along the already formed optic nerve appears to be dependent on the presence of vascular progenitors.[10]

During the visual process, rods and cones, the light-sensitive neurons, respond to light photons by producing neural impulses that trigger a cascade of neuronal processes. This leads to retinal ganglion cells synapsing with

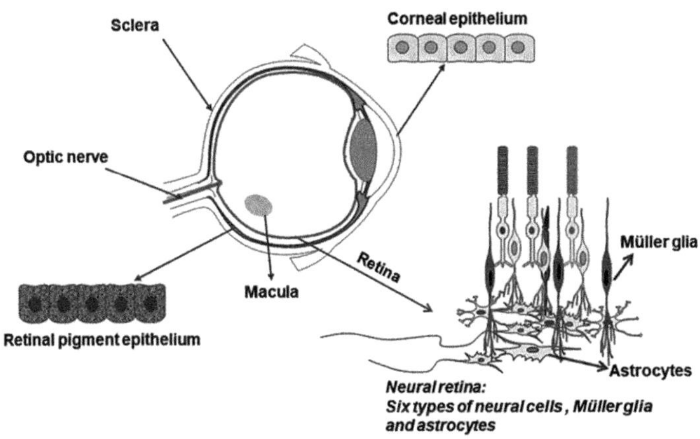

Figure 12.1 The neural retina: schematic representation of the anatomical localization of the retina and its neural and non-neural cellular components.

their targets in the brain to allow visual perception. Adjacent to the neural retina is the retinal pigment epithelium (RPE), which forms the blood–retinal barrier[11] and plays a very important role in the visual cycle and metabolic functions of retinal neurons.[12] When damage occurs to the RPE or selective populations of retinal neurons, a reorganization of the neural circuit ensues. This results in death of most neural cell populations in the retina, causing loss of visual function and consequently blindness. These pathological features are clearly illustrated during the development and progression of prevalent retinal degenerative diseases. For example, in AMD, initial damage is observed in the RPE, but during later stages photoreceptor cell death occurs due to impairment of their metabolic protection by RPE cells.[13,14] In other conditions, such as retinitis pigmentosa, degeneration of photoreceptors leads to retinal ganglion cell (RGC) loss due to axon compression and neural reorganization,[15] while damage to the optic nerve and RGC in glaucoma leads to photoreceptor and horizontal cell damage, with consequent blindness.[16]

Due to the complex network of the neural retina, methods to promote retina regeneration by stem cell transplantation are exceptionally challenging. During early retinal disease, regeneration of a single neural cell type may prove more feasible than in late disease when all neural cell types have been compromised and spread neural reorganization has occurred. Any approach to regenerate the retina will require not only a better understanding of the molecular mechanisms that promote adult stem cell differentiation into functional neurons and glia, but also of the requirements for successful migration, integration and survival of the transplanted cells. This is of special importance as a retina in need of repair may have lost the developmental cues that allow permissiveness for neural integration, and it is likely to exhibit pro-inflammatory and gliotic barriers that prevent successful regeneration by transplanted cells.[17,18]

12.2 Stem Cell Potential for Retina Regenerative Therapies

Regenerative medicine promises vast benefits for patients with progressive retinal disease, and stem cells provide an unlimited source of different cell types for potential therapies, as well as for drug screening and new drug development. Progress in the isolation and propagation of stem cells from various sources has opened the possibility of generating all types of retinal cells for therapeutic application. However, it is possible that stem cells obtained from different sources may differ in their ability to repair or restore function upon retinal transplantation, and understanding the potential of various stem cell sources for their clinical application may aid in the development of effective retinal therapies.

Pluripotent stem cells, including human embryonic stem cells (ESCs) and induced pluripotent stem cells (iPSCs) have been widely used to investigate the ability of these cells to repair or regenerate the retina in animal models of retina degeneration.[19–21] Other stem cells being explored for their potential to regenerate the retina include adult stem cells such as human Müller stem cells,[22–24] retinal progenitors,[25] which are closely related to Müller glia,[26] and mesenchymal stem cells derived from the umbilical cord and bone marrow.[27,28] Details on the use of these cell types in experimental studies to regenerate the retina will be described in the following section.

12.2.1 Embryonic Stem Cells

ESCs are derived from the inner cell mass of the blastocyst[29] and are capable of differentiating into most retinal cell types under specialized culture conditions *in vitro.* Because of their ability to self-renew and to differentiate into multiple neurons and glia,[30] ESCs have been widely recognized for their potential to developing retinal cell therapies. Efficient differentiation of ESCs into retinal progenitors has been obtained by inhibiting the Wnt and BMP pathways combined with the addition of IGF (insulin-like growth factor)-1 or a combination of Nodal and Wnt antagonists.[30] In addition, defined protocols to obtain differentiated RPE cells from human ESCs for clinical application have been successfully established by various investigators.[31,32] These protocols have been applied to the preparation of RPE cells that have been used in initial clinical trials to treat patients affected by Stargardt macular dystrophy and AMD.[33,34] Transplantation of retinal neurons derived from human ESCs has been widely explored in rodent models of retina degeneration[35–37] and normal non-human primate retina,[38] but clinical application of these cells has not yet been achieved.

12.2.2 Induced Pluripotent Stem Cells

Pluripotency of adult somatic cells such as keratinocytes,[39] hematopoietic stem cells[40] and dermal fibroblasts[41] can be induced by transfection with four transcription factors associated with the embryonic stem cell state: Oct4, Sox2, cMyc and Klf4[42]. However, because of the tumorigenic potential of c-Myc and the pathogenic risk of using viral vectors to generate iPSCs, various modified protocols have been designed to address these safety issues.[42,43] Despite being derived from adult tissue, these pluripotent cells share many features of ESCs, including DNA methylation, gene expression and chromatin state.[44] A major advantage of iPSCs over ESCs is their potential for autologous use, which could avoid the need for immunosuppressive therapies upon transplantation. Although these properties make these cells more desirable for regenerative medicine, there is evidence that iPSC lines can harbour different genetic and epigenetic signatures.[45] On this basis, extensive investigations to derive iPSC lines with stable genomic architecture are being undertaken for application in the clinic.

12.2.3 Adult Stem Cells

Ethical and tumorigenic concerns on the use of ESCs and iPSCs in cell therapies and limited methodology available to differentiate these cells into mature neural cell types have driven investigators to explore the potential application of adult stem cells in retinal regenerative medicine. Mesenchymal stem cells derived from bone marrow and umbilical cord blood[46–48] and adipose tissue,[49] as well as Müller glia and progenitor cells[22–25] have been induced to differentiate *in vitro* into retinal neurons and transplanted into the retina. RPE cells have been reported to *trans*-differentiate towards retinal neurons,[50] but convincing evidence that these cells have the potential to replace retinal neurons or RPE cells has not yet been presented. Adult neural stem cells derived from the adult hippocampus have also been transplanted into the eyes of experimental models of retinal disease with various outcomes,[51,52] and despite lack of evidence that mesenchymal or neural stem cells can regenerate the retina, these cells have been shown to be a potent source of trophic factors that aid in neural cell protection and recovery.[46,53] On this basis, their potential use may be likely restricted to cell-based neuroprotective strategies rather than neuronal replacement.

Retinal specific stem cells, known as Müller glia, have been identified in the neural retina of the human eye,[7,54] and they have been investigated for their potential use in retina regenerative therapies. Müller glial cells constitute the major glial cells of the neural retina. They expand across the whole width of the retina and provide structural and metabolic support to retinal neurons.[55,56] Müller glia are responsible for the spontaneous regeneration of the zebrafish observed throughout life[57] and are found in all vertebrate species including chicks[58] and small rodents such as mice[59] and rats,[60] but their endogenous regenerative ability is severely limited in chicks and small mammals.[61] Although Müller glial cells with stem cell characteristics are present in the adult human retina,[7] there is no evidence for their endogenous regenerative ability. However, human Müller with stem cell characteristics can be cultured indefinitely *in vitro* and display features of pluripotent progenitor cells in that they have the potential of unlimited cell renewal, express markers of stem cell and neural precursors and differentiate into cells with characteristic markers and functions of retinal neurons *in vitro*.[7,23,24,62] Experimental transplantation studies have shown that enriched populations of neural precursors derived from Müller stem cells significantly improve visual function in animal models of retinal ganglion cell depletion[22,24] and photoreceptor degeneration.[23]

12.3 Neural Stem Cells during Retinal Development and Adult Life

12.3.1 Embryonic Development of the Retina

During embryonic development, eye formation begins with evagination of the neural tube on either side of the developing forebrain. This leads to the formation of optic vesicles that further fold to form a double-layered optic cup. Further invagination of the optic cup drives differentiation of the optic stalk, neural retina and retinal pigment epithelium.[63] Cell lineage studies using tracers or retroviruses have shown that all neural retinal cells derive from a common multipotent retinal progenitor cell (RPC).[9,64] Current studies suggest that there is an order to which the retinal neurons and glia are generated; however, they do overlap with each other. RGCs are the first retinal neurons to develop, followed by horizontal cells, cone photoreceptors, amacrine cells, rod photoreceptors, bipolar and finally Müller glial cells.[64] Several models have been hypothesized to describe cell fate specification in the retina. During retinal cell differentiation progenitor cell division can therefore be asymmetric, symmetric mitotic or symmetric post-mitotic. The "competence model" of retinal cell fate determination suggests that each RPC goes through different states driven by extrinsic cues.[65] During these "competence" states, the progenitor cells can form only a particular subtype of retinal cell, but also have the ability to switch states at any given time. Another model suggests that early transcription factor expression in RPCs determines the production of future retinal cell types.[66] In addition, it has been suggested that all RPCs are comparable and extrinsic factors drive retinal fate specification.[67] It is therefore likely that a combination of these models may drive retinal cell fate determination, and this is the subject of investigations in the retinal development field.

Retinal progenitor cells are characterized by their expression of multiple transcription factors that drive their differentiation into mature retinal neurons through a sequence of activation driven by local and environmental signals.[68,69] Through activation of the downstream basic helix-loop-helix (bHLH) transcription factors Hes1 and Hes5, Notch pathway activation maintains the progenitor cell state. This prevents differentiation and promotes rapid proliferation which is thought to be required to complete the formation of all the retinal cell types.[70,71] To drive neural differentiation, Notch signalling is downregulated causing increased expression of bHLH (Math1/Math5/Math3/NeuroD/Ascl1) and homeodomain (Chx10/Pax6/Six3/Crx/Otx2/Prox1) activators.[72,73] However, Müller glia formation requires continual activation of Hes1 and Hes5 in the RPCs,[74,75] which is associated with the progenitor-like roles ascribed to Müller glia. Retinal ganglion cell development is driven by sonic hedgehog (Shh) signalling in a positive feedback loop, where already formed RGCs secrete Shh to promote further differentiation. RGC differentiation is accompanied by the expression of Atoh7, Pou4f1/2/3 (previously Brn3), Isl1 and NeuroD1, concomitant with a continued expression of the progenitor markers Pax6 and Math5.[70,76] Horizontal cells are generated by co-expression of Math3 and Pax6, and amacrine cells express these factors in addition to NeuroD. Photoreceptor generation is characterized by expression of the transcription factors Otx2 and Crx, where NR2E3 is specific to rod cell fate, and thyroid hormone receptor β2 determines the type of cones generated.[77,78] Although many studies have investigated the control of retinal cell fate determination, there is not a complete understanding of the interactions between intrinsic and extrinsic cues that direct neural differentiation in the human retina. It is therefore important for future studies to build upon the regulation of these factors and mechanisms involved, and new *in vitro* modelling of retinal organoid formation by ESCs and iPSCs may further confirm these mechanisms across different species.

12.3.2 Retinal Organoids Formed by ESCs and iPSCs *in vitro*

Recent advances in stem cell research have shown that through targeted differentiation, human iPSCs and ESCs can mimic early retinal development.[79–81] There has been much progress in the growth of 3D models of retina *in vitro*, which has not only facilitated investigations on the mechanisms of retinal neuron specification[79,82,83] and retinal lamination,[84] but also on the testing of new drugs for potential use in retinal therapies. Retinal organoid formation *in vitro* allows mapping of the molecular timings associated with retinal neuron development that are difficult to observe *in vivo*. Investigations in the field have shown that both mouse and human ESCs are able to form 3D optic cup-like structures following early inhibition of the Wnt signalling pathway, addition of matrigel, sonic hedgehog agonists and supplementation with foetal bovine serum.[79–81] Eiraku and Sasai have described four phases during retinal organoid development of mouse ESC that mimic retinal organogenesis *in vivo*. In the first stage, a spherical vesicle emerges from the embryoid body, which flattens in the second stage. During the third stage, the junction between neural retina, RPE and embryoid body narrows before invagination of the neural retina during the final stage.[83] The process of optic cup formation by human ESC occurs at approximately 24 days from the start of the differentiation process.[79] This coincides with the expression of Pax6-and Brn3-positive cells which are early markers of RGC differentiation. Lineage studies using a *venus* cDNA knock in gene for the photoreceptor marker Crx have shown that photoreceptors start emerging at days 28–34 and their number gradually increase over time, expressing recoverin at day 60 and rhodopsin and Nrl by day 126 after organoid initiation.[79,83,85] There is limited knowledge of the development of other neural retinal cells, including amacrine, bipolar and horizontal cells; however, markers such as ap2α (amacrine) Ptf1a, calretinin (horizontal/amacrine) and PCK, VSX2 (bipolar) positive cells have been observed in retinal organoids 7–18 weeks after initiation of organoid formation.[81,85] It has been generally accepted that Müller glia are the last retinal cells to form during retina development,[9] and in human ESC-derived retinal organoids, Müller glia staining for glutamine synthetase has been observed in 90 day-old organoids.[85] Genes associated with mature Müller glia markers, including DBI, GNAI2, GLUL and DKK3 have been also observed at 37 days following induction of organoid formation by human ESCs,[85] but it is not clear whether these are restricted to Müller glia or other cells present in the retinal organoids. A diagram illustrating the various stages observed during the development of retinal organoids from ESCs and iPSCs *in vitro* is shown in Figure 12.2.

Another important question in biological studies of optic cup formation *in vitro* is whether retinal organoids formed in the laboratory are able to exhibit retinal function. Using patch clamping, Zhong *et al.* have shown that organoids derived from human iPSCs not only contain photoreceptors with outer segment discs at 27 weeks after initiation of organoid formation, but that they respond to light stimulation in a comparable manner to rod photoreceptor

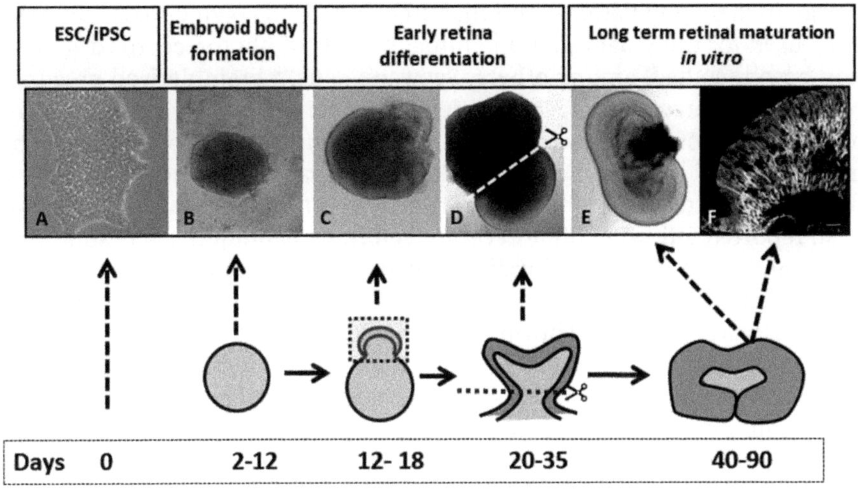

Figure 12.2 Schematic diagram illustrating the formation of retinal organoids from embryonic stem cells (ESCs) and induced pluripotent stem cells (iPSCs). (A) Undifferentiated ESC or iPSC; (B) embryoid body formation at days 2–12; (C) early protrusion of retinal organoid from embryoid body at days 12–18; (D) retinal organoid dissociated from undifferentiated cells at days 20–35; (E) matured organoid at days 40–90; (F) fluorescence microscopy image of retinal organoid section, stained for the Müller glia cell marker CRALBP.

cells *in vivo*.[86] Other groups have modified the original protocols developed by Nakano *et al.*[79] to encourage the development of organoids that favour the production of particular retinal cell types, such as photoreceptors[4] or RGCs,[87] which may have potential uses for retinal cell transplantation and drug screening.

12.4 Prospects for Retinal Repair and Regeneration Using Stem Cells

12.4.1 Stem Cell Transplantation

Knowledge gained in the stem cell field during recent years has widened the prospects of developing cell-based therapies to repair the retina that has been irreversibly damaged by disease or injury. Stem cell transplantation has been performed extensively in several experimental models of retinal degeneration, but cell therapies to regenerate the human neural retina are far from being implemented in the clinic, and still constitute a major challenge. Understandably, the majority of retinal transplantation studies published to date have been addressed to regenerate the retina in animal models of photoreceptor degeneration. This is possibly due to the fact that photoreceptor damage is the main feature of retinal degenerative diseases that prevail in large populations, and that include retinitis pigmentosa and AMD. Cells used for transplantation in animal models of photoreceptor degeneration have included, among others, Schwann cells,[88] brain-derived precursor cells,[89,90] photoreceptor precursors derived from the immature retina *in vivo*[91] or from pluripotent stem cells,[4] ESC-derived neural precursors,[92] mesenchymal stem cells[93] and Müller stem cells.[23] Successful outcomes in models of photoreceptor degeneration with partial restoration of visual function have been reported following photoreceptor cell transplantation in the subretinal space, and these effects were originally attributed to anatomical integration of the grafted cells into the neural circuitry.[91,94,95] However, recent observations from various groups have shown that transplanted photoreceptor cells do not integrate into the retina, but transfer intracellular material to the host cells. As a result, fluorescent proteins used to trace transplanted cells are transferred into adjacent host retinal cells, giving the appearance of transplant integration.[96–98] These studies have led investigators to reconsider whether improved visual function by cell transplantation should be aimed to replace neurons or to deliver neuroprotective factors to repair the degenerated retina. Other factors limiting the success of stem cell transplantation into the degenerated retina constitute the glia scarring and microglia accumulation, often occurring as a result of degenerative disease.[18,99] Addressing these problems may facilitate the migration and survival of transplanted stem cells.

In contrast with the high number of transplantation studies aimed at restoring photoreceptor cell function, stem cell studies to repair or regenerate the RCG layer and optic nerve have been limited. RGC degeneration occurs in conditions such as glaucoma, ischaemic optic neuropathies and inherited optic neuropathies. These constitute major causes of blindness for which stem cell-based therapies may be beneficial. Various sources of stem cells have been investigated for their ability to repair RGC function in animal models of glaucoma-like disease and include bone marrow mesenchymal stem cells,[100] oligodendrocyte precursors,[101] iPSC-derived trabecular meshwork cells,[102] olfactory ensheathing cells[103] and Müller stem cells.[22,24] As seen with transplantation involving photoreceptors, stem cell studies to repair or regenerate the glaucomatous-like retina have proved to have a beneficial effect, as demonstrated by improvement in RGC function. However, evidence for transplant cell integration with axon extension to the optic nerve and brain has not been achieved, for which it has been suggested that neurotrophic and neuroprotective factors released by the transplanted cells are responsible for the improvement of RGC function.[22,104]

Unlike stem cell therapies to regenerate the neural retina, therapies to replace the RPE, the pigmented cells that function as a retinal–blood barrier and metabolically support photoreceptor neurons,[105] have been taken to the clinic. These cells have been derived from iPSCs and ESCs, for which easy and reproducible protocols for RPE differentiation and transplantation have been established.[106–108] Initial clinical trials using ESC-derived RPE cells showed that transplanted RPE cells do not abnormally proliferate, form teratomas or trigger graft rejection or any other pathological reactions.[109]

Subsequent implementation of larger phase I/II clinical trials which have provided evidence of medium- to long-term safety and graft survival.[34] However, very little improvement in visual function has been achieved, for which it has been acknowledged that there is the need to implement more rigorous treatment strategies and more accurate clinical testing prior to applying these therapies in a large patient cohort.[110] RPE cells with neural and mesenchymal progenicity have been identified in the adult human retina,[50] and because of their reported ability to differentiate *in vitro* to form stable RPE monolayers, it has been suggested that they could be used in cell replacement therapies to treat macular disease.[111] However, this approach may not be feasible for autologous or allogeneic human transplantation, because of health and regulatory risks associated with the use of adult human tissue and the high costs of manufacturing these cells. In addition, the abundant sources of RPE cells easily obtained from pluripotent stem cells surpass the need for the use of cadaveric or autologous RPE cells.

As with any organ transplantation, both ESCs and iPSCs express major histocompatibility antigens,[112] capable of eliciting immunological reactivity in the recipient. Despite the eye being an immune-privileged site, issues with histocompatibility of the transplanted cells and the recipient need to be considered as this immunological barrier is most likely to be impaired in the degenerated retina. Avoidance of immunoreactivity can be achieved by the use of autologous cells, for which the use of iPSCs may provide a practical approach to transplantation. However, due to the high costs of manufacturing a cell line for a single application, the concept of "cell banks" to maximize the availability of histocompatible cells has gained support over the last few years, and the establishment of cell banks is being actively pursued. This approach will potentially provide cells that can be matched to the major HLA (human leukocyte antigen) loci of the transplant recipient, including A, B and DR, to avoid immune rejection.

12.5 Endogenous Regeneration – Targeting Factors with Potential to Induce Regeneration

To date, research in the retina regenerative field has focused on cell transplantation to replace damaged neurons in experimental models of retinal degeneration. Despite many types of adult and pluripotent stem cells being used, a consensus on the most appropriate cell sources for retinal therapies has not yet been achieved. If this research is to be translated into humans there is still uncertainty on the long-term cell survival of grafted cells, and concerns with efficacy and allogeneic reactivity are still in need of further investigation. In addition, true regeneration by transplantation has not yet been demonstrated. Based on evidence that quiescent Müller glia with stem cell characteristics are harboured in the adult human retina, a more practical approach to regeneration in preference to transplantation could involve the stimulation of neural cell renewal by resident Müller cells.

It has been known for a long time that fish and amphibians regenerate the neural retina throughout life;[113] however, it was originally thought that warm-blooded vertebrates and mammals were incapable of retinal regeneration. Investigations implicating Müller glia as a source of retinal neurons were first reported in the new-born chick.[58] Further studies in the rat retina showed that Müller glia could be stimulated to proliferate and differentiate into bipolar cells and rod photoreceptors in response to neurotoxic damage by *N*-methyl-D-aspartate (NMDA),[60] and that Müller glia in the adult mouse retina could be induced to proliferate and express markers of neural progenitors upon neurotoxic damage and stimulation with growth factors *in vivo*.[59] Subsequent investigations of retinal stem cells and their niches in the Tg(*gfap*: GFP) transgenic zebrafish provided compelling evidence for the role of Müller glia as the retinal cells responsible for the regeneration observed in this species.[57,114] These studies demonstrated that in the adult zebrafish Müller glia form the retinal stem cell niche, and that in response to local injury these cells partially and transiently dedifferentiate. During these events, Müller glia re-enter the cell cycle, undergo interkinetic nuclear migration and divide asymmetrically to generate retinal progenitors that give rise to new retinal neurons.[57,114] More recent studies have further shown that newly generated neurons in the zebrafish retina can synaptically re-connect their dendrites and axons, suggesting that new neurons may achieve functional integration in the retina. However, stereotypic synaptic distributions do not resemble those of the original neural population, indicating that developmental cues may be required to achieve true regeneration.[115]

Müller glia with stem cell characteristics have been identified in the adult human retina,[7,8] from where they can be isolated and cultured indefinitely *in vitro*.[7] As indicated earlier, this population of human Müller glia display stem cell characteristics, including unlimited cell renewal, expression of stem cell and neural precursor markers, as well as the potential to differentiate into various retinal neurons *in vitro*.[23,24] However, in contrast to lower vertebrates, these cells lack regenerative capacity *in vivo,* and it can be speculated that this is possibly due to inhibitory factors present in the developed retina that prevent uncontrolled proliferation and tumour formation. Factors produced as a result of inflammatory processes occurring during retina degeneration may also prevent Müller glia from exerting a regenerative function. It would be therefore important to identify these factors in order to develop strategies for the self-repair and regeneration of the diseased retina.

Most human retinal diseases that cause blindness are characterized by abnormal Müller glia proliferation, which leads not to repair, but to the formation of glial scarring (reactive gliosis).[116] Irrespective of aetiology, gliosis is accompanied by production of inflammatory cytokines such as tumour necrosis factor-α (TNFα), heparin-binding epidermal growth factor, vascular endothelium growth factor (VEGF) and transforming growth factor- β (TGF-β),[116] and Müller glia have been shown to constitute a major source of cytokines and inflammatory factors present in the gliotic retina.[117] Although some of these factors have been shown to stimulate the formation of multipotent Müller-derived progenitors in the uninjured retina zebrafish retina and to promote the regenerative response in this species,[118,119] there is no evidence that they have a regenerative role in the human retina. On the contrary, it is possible that these factors may have autocrine effects that prevent Müller cell progenicity. This is suggested by observations that TGF-β, a cytokine found to be highly upregulated in the gliotic retina, significantly downregulates the canonical Wnt signalling pathway in Müller glia *in vitro*, inhibiting their differentiation into photoreceptors.[120] On this basis, it is reasonable to hypothesize that interaction of pathways involved in neurogenesis and inflammation may affect the ability of Müller glia to endogenously regenerate the diseased retina. Current views favour investigations into the feasibility of activating the potential neurogenesis of these cells in order to regenerate the mammalian retina.[121–123]

Several genes expressed by retinal progenitors are known to be activated in the zebrafish retina in response to injury. One of these genes, the achaete-scute complex-like 1a (ascl1a) factor has been shown to be a key regulator in the generation of Müller glia-derived progenitors responsible for retina regeneration in this species.[124] While it has been demonstrated that TNF-α is produced by dying retinal neurons and that it promotes Ascl1a expression required for Müller glia proliferation in the zebrafish retina,[118] there is no suggestion that this cytokine may have the same regenerative potential in the mammalian retina. Ascl1 is not expressed in the mammalian Müller glia after injury, but its overexpression in mouse retinal explants *in vitro* and dissociated Müller glia, causes upregulation of retinal progenitor-specific genes and induces differentiation of these cells into functional retinal neurons that respond to neurotransmitters.[125] In addition, overexpression of Ascl1a in the NMDA-injured retina of young mice causes Müller glia to differentiate into amacrine and bipolar cells and photoreceptors *in vivo*.[126] Further studies in this field have indicated that despite Ascl1 overexpression, the neurogenic ability of the mouse retina is lost by postnatal day 16. In addition, lack of neurogenesis was shown to be accompanied by decreased chromatin accessibility in mature Müller glia, suggesting that epigenetic factors limit retina regeneration by these cells.[127] This was demonstrated by observations that overexpression of Ascl1 by Müller glia in the presence of a histone deacetylase inhibitor enabled Müller glia in adult mice to differentiate into cells expressing neural retinal markers, with the ability to synapse with host retinal neurons and respond to light.[127]

In view of the latest findings and evidence for the presence of Müller glia with potential neurogenic ability in the adult human retina, it may be possible that in the non-distant future, new approaches to induce endogenous regeneration of the human neural retina can be developed. This would constitute a preferable alternative to stem cell transplantation currently being investigated, and will transform the lives of millions of people currently affected by retinal degenerative disease.

12.6 Stem Cells as a Source of Neuroprotective Factors to Promote Neural Survival and Regeneration

In view of the data shown above, it is possible that stem cell transplantation to repair or regenerate the neural retina may not be aimed solely at replacing neurons, but that delivery of neuroprotective and reparative molecules by stem cells should be viewed as an important factor when developing such therapies. In addition, microglia control with anti-inflammatory agents may need to be considered as an adjunct to cell therapies in order to prolong cell survival and therefore the neurotrophic potential of grafted stem cells within the host retina.

Human neural progenitor cells have been shown to protect photoreceptors, maintain synaptic integrity and support horizontal cell survival when co-cultured with porcine retinal explants.[128] Similarly, co-culture of neural progenitors with rat and mice retina *in vitro* have been shown to delay photoreceptor degeneration,[129] and this effect has been attributed to their release of neurotrophic factors. In addition, bone marrow-derived mesenchymal stem cell (MSC)[130] and oligodendrocyte precursors[101] have been shown to exert a neuroprotective effect on RGCs upon transplantation into the eye of rats with experimental glaucoma. In view of the evidence that transplanted stem cells do not integrate into the retina, but that they induce recovery of retinal function, it is suggested that this effect is due to the neuroprotective ability of stem cells. Hence, strategies for cell transplantation as a source of neurotrophins or direct delivery of recombinant factors into the vitreous have gained much support.

Various approaches to directly deliver retinal neuroprotective agents have been used by many investigators, and have included direct injection of recombinant proteins into the vitreous,[131–133] non-viral or viral vector-based delivery of neurotrophin genes[134–136] or cell encapsulation in semipermeable biomaterials for delivery of neuroprotective factors.[137,138] However, the effects of neurotrophins delivered by these methods have been shown to be only transient, and problems can arise by repeated intraocular injections of these factors, as shown by observations that multiple applications of BDNF do not have an additive effect because continuous application of BDNF causes downregulation of its receptor.[139] It is not clear whether retinal-specific stem cells would be more efficient at promoting endogenous repair and or regeneration of neurons and axonal networks in the degenerated retina, or whether any type of cell with the ability to produce neurotrophins would suffice as a source of trophic factors for retinal repair, and this is subject of current investigations in the field.

While studies into the nature of neurotrophic factors released by MSCs have identified BDNF and β-nerve growth factor as the factors responsible for the neuroprotective effect of these cells when transplanted at sites of nerve injury,[140] approaches to enhance the neurotrophic effects of stem cells have also been used. Neural stem cells have been modified to secrete ciliary neurotrophic factor, resulting in a significant increase in RGC survival and axon regeneration upon intravitreal transplantation in a mouse model of glaucoma.[141] Rat and human bone marrow-derived MSCs induced to secrete high levels of BDNF, glial cell-derived nerve growth factor and VEGF exert a marked neuroprotective effect in rat eyes after optic nerve transection,[142] while MSCs transfected with retrovirus to express high levels of BDNF release high levels of this protein when transplanted into the axotomized rat retina.[143] Based on this evidence, justification for cell transplantation to repair or regenerate the retina may be mainly based on the assumption that cells do not integrate into the diseased tissue, but that neuroprotective factors produced by long-lived stem cells within the tissue can maintain and repair retinal function without achieving true regeneration.

12.7 Conclusions and Future Trends

Although progress on the understanding of the potential and limitations of stem cells has been made in recent years, much research is still needed before stem cell therapies can be implemented in the clinic. Encouraging data has demonstrated improvement of visual function after stem cell transplantation into experimental models of retina degeneration. However, neural cell replacement has not yet been achieved, with visual improvement being ascribed to neuroprotective factors released by transplanted cells. Furthermore, evidence that the adult human retina harbours a population Müller glia with stem cell characteristics opens exciting avenues for research into methodologies to induce endogenous regeneration of the retina, as an alternative to stem cell transplantation.

References

1. A. R. Rudnicka, Z. Jarrar, R. Wormald, D. G. Cook, A. Fletcher and C. G. Owen, *Ophthalmology*, 2012, **119**, 571.
2. R. R. Bourne, H. R. Taylor, S. R. Flaxman, J. Keeffe, J. Leasher and K. Naidoo, *et al.*, *PLoS One*, 2016, **11**, e0162229.
3. N. M. Holekamp, *Am. J. Manag. Care*, 2016, **22**, 300-s6.
4. A. Gonzalez-Cordero, E. L. West, R. A. Pearson, Y. Duran, L. S. Carvalho and C. J. Chu, *et al.*, *Nat. Biotechnol.*, 2013, **31**, 741.
5. C. M. Ramsden, M. B. Powner, A. J. Carr, M. J. Smart, L. da Cruz and P. J. Coffey, *Development*, 2013, **140**, 2576.
6. N. K. Wang, J. Tosi, J. M. Kasanuki, C. L. Chou, J. Kong and N. Parmalee, *et al.*, *Transplantation*, 2010, **89**, 911.
7. J. M. Lawrence, S. Singhal, B. Bhatia, D. J. Keegan, T. A. Reh and P. J. Luthert, *et al.*, *Stem Cells*, 2007, **25**, 2033.
8. B. Bhatia, S. Singhal, J. M. Lawrence, P. T. Khaw and G. A. Limb, *Exp. Eye Res.*, 2009, **89**, 373.

9. D. L. Turner and C. L. Cepko, *Nature*, 1987, **328**, 131.
10. J. Stone and Z. Dreher, *J. Comp. Neurol.*, 1987, **255**, 35.
11. A. Bennis, J. G. Jacobs, L. A. E. Catsburg, J. B. Ten Brink, C. Koster and R. O. Schlingemann, *et al.*, *Stem Cell Rev.*, 2017, **13**, 659.
12. C. B. Wright, T. M. Redmond and J. M. Nickerson, *Prog. Mol. Biol. Transl. Sci.*, 2015, **134**, 433.
13. P. Tornquist, A. Alm and A. Bill, *Eye*, 1990, **4**, 303.
14. A. Kauppinen, J. J. Paterno, J. Blasiak, A. Salminen and K. Kaarniranta, *Cell. Mol. Life Sci.*, 2016, **73**, 1765.
15. M. P. Villegas-Perez, J. M. Lawrence, M. Vidal-Sanz, M. M. Lavail and R. D. Lund, *J. Comp. Neurol.*, 1998, **392**, 58.
16. P. Janssen, R. Naskar, S. Moore, S. Thanos and H. J. Thiel, *Ger. J. Ophthalmol.*, 1996, **5**, 378.
17. R. A. Pearson, C. Hippert, A. B. Graca and A. C. Barber, *Vis. Neurosci.*, 2014, **31**, 333.
18. S. Singhal, J. M. Lawrence, B. Bhatia, J. S. Ellis, A. S. Kwan and A. Macneil, *et al.*, *Stem Cells*, 2008, **26**, 1074.
19. M. Riera, L. Fontrodona, S. Albert, D. M. Ramirez, A. Seriola and A. Salas, *et al.*, *Mol. Ther. Methods Clin. Dev.*, 2016, **3**, 16010.
20. H. Nazari, L. Zhang, D. Zhu, G. J. Chader, P. Falabella and F. Stefanini, *et al.*, *Prog. Retinal Eye Res.*, 2015, **48**, 1.
21. H. Shirai, M. Mandai, K. Matsushita, A. Kuwahara, S. Yonemura and T. Nakano, *et al.*, *Proc. Natl. Acad. Sci. U. S. A.*, 2016, **113**, E81.
22. S. Becker, K. Eastlake, H. Jayaram, M. F. Jones, R. A. Brown and G. J. McLellan, *et al.*, *Stem Cells Transl. Med.*, 2016, **5**, 192.
23. H. Jayaram, M. F. Jones, K. Eastlake, P. B. Cottrill, S. Becker and J. Wiseman, *et al.*, *Stem Cells Transl. Med.*, 2014, **3**, 323.
24. S. Singhal, B. Bhatia, H. Jayaram, S. Becker, M. F. Jones and P. B. Cottrill, *et al.*, *Stem Cells Transl. Med.*, 2012, **1**, 188.
25. M. Semo, N. Haamedi, L. Stevanato, D. Carter, G. Brooke and M. Young, *et al.*, *Trans. Vis. Sci. Technol.*, 2016, **5**, 6.
26. B. R. Nelson, Y. Ueki, S. Reardon, M. O. Karl, S. Georgi and B. H. Hartman, *et al.*, *PLoS One*, 2011, **6**, e22817.
27. R. D. Lund, S. Wang, B. Lu, S. Girman, T. Holmes and Y. Sauve, *et al.*, *Stem Cells*, 2007, **25**, 602.
28. S. S. Park, S. Caballero, G. Bauer, B. Shibata, A. Roth and P. G. Fitzgerald, *et al.*, *Invest. Ophthalmol. Visual Sci.*, 2012, **53**, 986.
29. M. J. Evans and M. H. Kaufman, *Nature*, 1981, **292**, 154.
30. D. A. Lamba, M. O. Karl, C. B. Ware and T. A. Reh, *Proc. Natl. Acad. Sci. U. S. A.*, 2006, **103**, 12769.
31. D. E. Buchholz, B. O. Pennington, R. H. Croze, C. R. Hinman, P. J. Coffey and D. O. Clegg, *Stem Cells Transl. Med.*, 2013, **2**, 384.
32. M. Idelson, R. Alper, A. Obolensky, E. Ben-Shushan, I. Hemo and N. Yachimovich-Cohen, *et al.*, *Cell Stem Cell*, 2009, **5**, 396.
33. B. Nommiste, K. Fynes, V. E. Tovell, C. Ramsden, L. da Cruz and P. Coffey, *Prog. Brain Res.*, 2017, **231**, 225.

34. S. D. Schwartz, C. D. Regillo, B. L. Lam, D. Eliott, P. J. Rosenfeld and N. Z. Gregori, *et al.*, *Lancet*, 2015, **385**, 509.
35. D. A. Lamba, J. Gust and T. A. Reh, *Cell Stem Cell*, 2009, **4**, 73.
36. S. Decembrini, U. Koch, F. Radtke, A. Moulin and Y. Arsenijevic, *Stem Cell Rep.*, 2014, **2**, 853.
37. E. L. West, A. Gonzalez-Cordero, C. Hippert, F. Osakada, J. P. Martinez-Barbera and R. A. Pearson, *et al.*, *Stem Cells*, 2012, **30**, 1424.
38. J. R. Chao, D. A. Lamba, T. R. Klesert, A. Torre, A. Hoshino and R. J. Taylor, *et al.*, *Trans. Vis. Sci. Technol.*, 2017, **6**, 4.
39. T. Aasen, A. Raya, M. J. Barrero, E. Garreta, A. Consiglio and F. Gonzalez, *et al.*, *Nat. Biotechnol.*, 2008, **26**, 1276.
40. S. Eminli, A. Foudi, M. Stadtfeld, N. Maherali, T. Ahfeldt and G. Mostoslavsky, *et al.*, *Nat. Genet.*, 2009, **41**, 968.
41. W. E. Lowry, L. Richter, R. Yachechko, A. D. Pyle, J. Tchieu and R. Sridharan, *et al.*, *Proc. Natl. Acad. Sci. U. S. A.*, 2008, **105**, 2883.
42. M. Nakagawa, N. Takizawa, M. Narita, T. Ichisaka and S. Yamanaka, *Proc. Natl. Acad. Sci. U. S. A.*, 2010, **107**, 14152.
43. K. Okita, M. Nakagawa, H. Hyenjong, T. Ichisaka and S. Yamanaka, *Science*, 2008, **322**, 949.
44. M. Wernig, A. Meissner, R. Foreman, T. Brambrink, M. Ku and K. Hochedlinger, *et al.*, *Nature*, 2007, **448**, 318.
45. A. Chhabra, *Front. Immunol.*, 2017, **8**, 993.
46. L. Junyi, L. Na and J. Yan, *J. Craniofac. Surg*, 2015, **26**, 548.
47. H. Salehi, N. Amirpour, S. Razavi, E. Esfandiari and R. Zavar, *Ann. Anat.*, 2017, **210**, 52.
48. W. Jin, Y. Q. Xing and A. H. Yang, *In Vitro Cell. Dev. Biol. Anim.*, 2009, **45**, 321.
49. H. Rezanejad, Z. S. Soheili, F. Haddad, M. M. Matin, S. Samiei and A. Manafi, *et al.*, *Cell Tissue Res.*, 2014, **356**, 65.
50. E. Salero, T. A. Blenkinsop, B. Corneo, A. Harris, D. Rabin and J. H. Stern, *et al.*, *Cell Stem Cell*, 2012, **10**, 88.
51. D. English, N. K. Sharma, K. Sharma and A. Anand, *J. Cell. Biochem.*, 2013, **114**, 764.
52. Y. Kurimoto, H. Shibuki, Y. Kaneko, M. Ichikawa, T. Kurokawa and M. Takahashi, *et al.*, *Neurosci. Lett.*, 2001, **306**, 57.
53. B. Mead, L. J. Hill, R. J. Blanch, K. Ward, A. Logan and M. Berry, *et al.*, *Cytotherapy*, 2016, **18**, 487.
54. B. Bhatia, H. Jayaram, S. Singhal, M. F. Jones and G. A. Limb, *Exp. Eye Res.*, 2011, **93**, 852.
55. A. Bringmann, T. Pannicke, J. Grosche, M. Francke, P. Wiedemann and S. N. Skatchkov, *et al.*, *Prog. Retinal Eye Res.*, 2006, **25**, 397.
56. E. Newman and A. Reichenbach, *Trends Neurosci.*, 1996, **19**, 307.
57. J. R. Lenkowski and P. A. Raymond, *Prog. Retinal Eye Res.*, 2014, **40**, 94.
58. A. J. Fischer and T. A. Reh, *Nat. Neurosci.*, 2001, **4**, 247.
59. M. O. Karl, S. Hayes, B. R. Nelson, K. Tan, B. Buckingham and T. A. Reh, *Proc. Natl. Acad. Sci. U. S. A.*, 2008, **105**, 19508.

60. S. Ooto, T. Akagi, R. Kageyama, J. Akita, M. Mandai and Y. Honda, *et al.*, *Proc. Natl. Acad. Sci. U. S. A.*, 2004, **101**, 13654.
61. K. Loffler, P. Schafer, M. Volkner, T. Holdt and M. O. Karl, *Glia*, 2015, **63**, 1809.
62. S. Becker, S. Singhal, M. F. Jones, K. Eastlake, P. B. Cottrill and H. Jayaram, *et al.*, *Mol. Vision*, 2013, **19**, 1925.
63. R. B. Brun, *Dev. Biol.*, 1981, **88**, 192.
64. C. L. Cepko, C. P. Austin, X. Yang, M. Alexiades and D. Ezzeddine, *Proc. Natl. Acad. Sci. U. S. A.*, 1996, **93**, 589.
65. F. J. Livesey and C. L. Cepko, *Nat. Rev. Neurosci.*, 2001, **2**, 109.
66. C. E. Holt, T. W. Bertsch, H. M. Ellis and W. A. Harris, *Neuron*, 1988, **1**, 15.
67. M. Andreazzoli, *Birth Defects Res., Part C*, 2009, **87**, 284.
68. M. Xiang, *Cell. Mol. Life Sci.*, 2013, **70**, 2519.
69. T. Marquardt, R. Ashery-Padan, N. Andrejewski, R. Scardigli, F. Guillemot and P. Gruss, *Cell*, 2001, **105**, 43–55.
70. B. R. Nelson, B. Gumuscu, B. H. Hartman and T. A. Reh, *Dev. Neurosci.*, 2006, **28**, 128.
71. M. Ishibashi, K. Moriyoshi, Y. Sasai, K. Shiota, S. Nakanishi and R. Kageyama, *EMBO J.*, 1994, **13**, 1799.
72. K. A. Maurer, A. N. Riesenberg and N. L. Brown, *Development*, 2014, **141**, 3243.
73. J. Hatakeyama and R. Kageyama, *Semin. Cell Dev. Biol.*, 2004, **15**, 83.
74. R. L. Bernardos, S. I. Lentz, M. S. Wolfe and P. A. Raymond, *Dev. Biol.*, 2005, **278**, 381.
75. T. Furukawa, S. Mukherjee, Z. Z. Bao, E. M. Morrow and C. L. Cepko, *Neuron*, 2000, **26**, 383.
76. L. Feng, Z. H. Xie, Q. Ding, X. Xie, R. T. Libby and L. Gan, *Mol. Brain*, 2010, **3**, 36.
77. E. M. Morrow, T. Furukawa, J. E. Lee and C. L. Cepko, *Development*, 1999, **126**, 23.
78. E. M. Morrow, T. Furukawa and C. L. Cepko, *Trends Cell Biol.*, 1998, **8**, 353.
79. T. Nakano, S. Ando, N. Takata, M. Kawada, K. Muguruma and K. Sekiguchi, *et al.*, *Cell Stem Cell*, 2012, **10**, 771.
80. M. Eiraku, N. Takata, H. Ishibashi, M. Kawada, E. Sakakura and S. Okuda, *et al.*, *Nature*, 2011, **472**, 51.
81. J. S. Meyer, R. L. Shearer, E. E. Capowski, L. S. Wright, K. A. Wallace and E. L. McMillan, *et al.*, *Proc. Natl. Acad. Sci. U. S. A.*, 2009, **106**, 16698.
82. J. G. Aparicio, H. Hopp, A. Choi, J. Mandayam Comar, V. C. Liao and N. Harutyunyan, *et al.*, *Exp. Eye Res.*, 2017, **154**, 177.
83. M. Eiraku and Y. Sasai, *Nat. Protoc.*, 2011, 7, 69.
84. M. K. Eldred, M. Charlton-Perkins, L. Muresan and W. A. Harris, *Development*, 2017, **144**, 1097.
85. R. Kaewkhaw, K. D. Kaya, M. Brooks, K. Homma, J. Zou and V. Chaitankar, *et al.*, *Stem Cells*, 2015, **33**, 3504.

86. X. Zhong, C. Gutierrez, T. Xue, C. Hampton, M. N. Vergara and L. H. Cao, *et al.*, *Nat. Commun.*, 2014, **5**, 4047.
87. S. K. Ohlemacher, A. Sridhar, Y. Xiao, A. E. Hochstetler, M. Sarfarazi and T. R. Cummins, *et al.*, *Stem Cells*, 2016, **34**, 1553.
88. J. M. Lawrence, Y. Sauve, D. J. Keegan, P. J. Coffey, L. Hetherington and S. Girman, *et al.*, *Invest. Ophthalmol. Visual Sci.*, 2000, **41**, 518.
89. A. Nishida, M. Takahashi, H. Tanihara, I. Nakano, J. B. Takahashi and A. Mizoguchi, *et al.*, *Invest. Ophthalmol. Visual Sci.*, 2000, **41**, 4268.
90. M. J. Young, J. Ray, S. J. Whiteley, H. Klassen and F. H. Gage, *Mol. Cell Neurosci.*, 2000, **16**, 197.
91. R. E. MacLaren, R. A. Pearson, A. MacNeil, R. H. Douglas, T. E. Salt and M. Akimoto, *et al.*, *Nature*, 2006, **444**, 203.
92. M. S. Divya, V. A. Rasheed, T. Schmidt, S. Lalitha, S. Hattar and J. James, *Front. Cell. Neurosci.*, 2017, **11**, 295.
93. M. Zhang, F. Zhang, J. Sun, Y. Sun, L. Xu and D. Zhang, *et al.*, *Neurosci. Lett.*, 2017, **657**, 62.
94. T. Santos-Ferreira, K. Postel, H. Stutzki, T. Kurth, G. Zeck and M. Ader, *Stem Cells*, 2015, **33**, 79.
95. A. C. Barber, C. Hippert, Y. Duran, E. L. West, J. W. Bainbridge and K. Warre-Cornish, *et al.*, *Proc. Natl. Acad. Sci. U. S. A.*, 2013, **110**, 354.
96. A. Ortin-Martinez, E. L. Tsai, P. E. Nickerson, M. Bergeret, Y. Lu and S. Smiley, *et al.*, *Stem Cells*, 2017, **35**, 932.
97. M. S. Singh, J. Balmer, A. R. Barnard, S. A. Aslam, D. Moralli and C. M. Green, *et al.*, *Nat. Commun.*, 2016, **7**, 13537.
98. T. Santos-Ferreira, S. Llonch, O. Borsch, K. Postel, J. Haas and M. Ader, *Nat. Commun.*, 2016, **7**, 13028.
99. S. Singhal, J. M. Lawrence, T. E. Salt, P. T. Khaw and G. A. Limb, *Exp. Eye Res.*, 2010, **90**, 308.
100. C. Roubeix, D. Godefroy, C. Mias, A. Sapienza, L. Riancho and J. Degardin, *et al.*, *Stem Cell Res. Ther.*, 2015, **6**, 177.
101. N. D. Bull, K. A. Irvine, R. J. Franklin and K. R. Martin, *Invest. Ophthalmol. Visual Sci.*, 2009, **50**, 4244.
102. D. W. Abu-Hassan, X. Li, E. I. Ryan, T. S. Acott and M. J. Kelley, *Stem Cells*, 2015, **33**, 751.
103. C. Dai, P. T. Khaw, Z. Q. Yin, D. Li, G. Raisman and Y. Li, *Trans. Vis. Sci. Technol.*, 2012, **1**, 3.
104. T. V. Johnson and K. R. Martin, *Curr. Opin. Pharmacol.*, 2013, **13**, 78.
105. O. Strauss, *Physiol. Rev.*, 2005, **85**, 845.
106. R. H. Croze and D. O. Clegg, *Dev. Ophthalmol.*, 2014, **53**, 81.
107. K. Bharti, S. S. Miller and H. Arnheiter, *Pigm. Cell Melanoma Res.*, 2011, **24**, 21.
108. A. J. Carr, A. A. Vugler, S. T. Hikita, J. M. Lawrence, C. Gias and L. L. Chen, *et al.*, *PLoS One*, 2009, **4**, e8152.
109. S. D. Schwartz, J. P. Hubschman, G. Heilwell, V. Franco-Cardenas, C. K. Pan and R. M. Ostrick, *et al.*, *Lancet*, 2012, **379**, 713.

110. S. D. Schwartz, G. Tan, H. Hosseini and A. Nagiel, *Invest. Ophthalmol. Visual Sci.*, 2016, **57**, ORSFc1.
111. J. S. Saini, S. Temple and J. H. Stern, *Adv. Exp. Med. Biol.*, 2016, **854**, 557.
112. L. Riolobos, R. K. Hirata, C. J. Turtle, P. R. Wang, G. G. Gornalusse and M. Zavajlevski, *et al.*, *Mol. Ther.*, 2013, **21**, 1232.
113. P. F. Hitchcock and P. A. Raymond, *Zebrafish*, 2004, **1**, 257.
114. R. L. Bernardos, L. K. Barthel, J. R. Meyers and P. A. Raymond, *J. Neurosci.*, 2007, **27**, 7028.
115. F. D. D'Orazi, X. F. Zhao, R. O. Wong and T. Yoshimatsu, *Curr. Biol.*, 2016, **26**, 2268.
116. A. Bringmann, I. Iandiev, T. Pannicke, A. Wurm, M. Hollborn and P. Wiedemann, *et al.*, *Prog. Retinal Eye Res.*, 2009, **28**, 423.
117. K. Eastlake, P. J. Banerjee, A. Angbohang, D. G. Charteris, P. T. Khaw and G. A. Limb, *Glia*, 2016, **64**, 495.
118. C. M. Nelson, K. M. Ackerman, P. O'Hayer, T. J. Bailey, R. A. Gorsuch and D. R. Hyde, *J. Neurosci.*, 2013, **33**, 6524.
119. J. Wan, R. Ramachandran and D. Goldman, *Dev. Cell*, 2012, **22**, 334.
120. A. Angbohang, N. Wu, T. Charalambous, K. Eastlake, Y. Lei and Y. S. Kim, *et al.*, *Stem Cells Dev.*, 2016, **25**, 1.
121. M. L. Vetter, P. F. Hitchcock and AGIwp, *Trans. Vis. Sci. Technol.*, 2017, **6**, 5.
122. X. Xia and I. Ahmad, *Int. J. Stem Cells*, 2016, **9**, 169.
123. M. O. Karl and T. A. Reh, *Trends Mol. Med.*, 2010, **16**, 193.
124. B. V. Fausett, J. D. Gumerson and D. Goldman, *J. Neurosci.*, 2008, **28**, 1109.
125. J. Pollak, M. S. Wilken, Y. Ueki, K. E. Cox, J. M. Sullivan and R. J. Taylor, *et al.*, *Development*, 2013, **140**, 2619.
126. Y. Ueki, M. S. Wilken, K. E. Cox, L. Chipman, N. Jorstad and K. Sternhagen, *et al.*, *Proc. Natl. Acad. Sci. U. S. A.*, 2015, **112**, 13717.
127. N. L. Jorstad, M. S. Wilken, W. N. Grimes, S. G. Wohl, L. S. VandenBosch and T. Yoshimatsu, *et al.*, *Nature*, 2017, **548**, 103.
128. T. Mollick, C. Mohlin and K. Johansson, *Brain Res.*, 2016, **1646**, 522.
129. U. Englund-Johansson, C. Mohlin, I. Liljekvist-Soltic, P. Ekstrom and K. Johansson, *Exp. Eye Res.*, 2010, **90**, 292.
130. T. V. Johnson, N. D. Bull, D. P. Hunt, N. Marina, S. I. Tomarev and K. R. Martin, *Invest. Ophthalmol. Visual Sci.*, 2010, **51**, 2051.
131. A. Muller, T. G. Hauk, M. Leibinger, R. Marienfeld and D. Fischer, *Mol. Cell Neurosci.*, 2009, **41**, 233.
132. V. Pernet and A. Di Polo, *Brain*, 2006, **129**, 1014.
133. P. D. Koeberle and A. K. Ball, *Vis. Res.*, 1998, **38**, 1505.
134. J. Rodger, E. S. Drummond, M. Hellstrom, D. Robertson and A. R. Harvey, *PLoS One*, 2012, **7**, e31061.
135. A. M. Wilson and A. Di Polo, *Gene Ther.*, 2012, **19**, 127.
136. C. Schmeer, G. Straten, S. Kugler, C. Gravel, M. Bahr and S. Isenmann, *Eur. J. Neurosci.*, 2002, **15**, 637.

137. M. P. Zanin, L. N. Pettingill, A. R. Harvey, D. F. Emerich, C. G. Thanos and R. K. Shepherd, *J. Controlled Release*, 2012, **160**, 3.
138. P. A. Sieving, R. C. Caruso, W. Tao, H. R. Coleman, D. J. Thompson and K. R. Fullmer, *et al.*, *Proc. Natl. Acad. Sci. U. S. A.*, 2006, **103**, 3896.
139. H. Chen and A. J. Weber, *Brain Res.*, 2004, **1011**, 99.
140. L. Crigler, R. C. Robey, A. Asawachaicharn, D. Gaupp and D. G. Phinney, *Exp. Neurol.*, 2006, **198**, 54.
141. K. Flachsbarth, K. Kruszewski, G. Jung, W. Jankowiak, K. Riecken and L. Wagenfeld, *et al.*, *Invest. Ophthalmol. Visual Sci.*, 2014, **55**, 7029.
142. H. Levkovitch-Verbin, O. Sadan, S. Vander, M. Rosner, Y. Barhum and E. Melamed, *et al.*, *Invest. Ophthalmol. Visual Sci.*, 2010, **51**, 6394.
143. H. Y. Park, J. H. Kim, H. Sun Kim and C. K. Park, *Brain Res.*, 2012, **1469**, 10.

CHAPTER 13

The Application of Lipid Nanoparticles for Retinal Degenerative Diseases

RAJU V. S. RAJALA*[a,b,c,d], AMMAJI RAJALA[a,d] AND YUHONG WANG[a,d]

[a]Department of Ophthalmology, University of Oklahoma Health Sciences Center, Oklahoma City, OK 73104, USA; [b]Department of Physiology, University of Oklahoma Health Sciences Center, Oklahoma City, OK 73104, USA; [c]Department of Cell Biology, University of Oklahoma Health Sciences Center, Oklahoma City, OK 73104, USA; [d]Dean McGee Eye Institute, Oklahoma City, OK 73104, USA
*E-mail: raju-rajala@ouhsc.edu

13.1 Introduction

Humans have specialized organs for sight, hearing smell, touch, and taste. Our eyes receive light and translate light rays to biological nerve impulses that are carried by the fibers of the nerve cells in the eye to the brain for the perception of visual images. Photoreceptor cells are polarized neurons present in the retina. These cells play an important role in absorbing photons from light and translating it into signals that activate biological events. In short, photoreceptors give us the invaluable gift of vison. In mammalian retina, the photoreceptor cells include rods, cones, and photosensitive

Drug Discovery Series No. 66
Therapies for Retinal Degeneration: Targeting Common Processes
Edited by Enrique J. de la Rosa and Thomas G. Cotter

Published by the Royal Society of Chemistry, www.rsc.org

ganglion cells.[1,2] Rods and cones support sight, whereas photosensitive ganglion cells do not support sight directly, but maintain circadian rhythms and pupillary reflexes.[3–5] Fundamental differences exist between rods and cones, as rods are exceptionally sensitive and are used in dim light, whereas cones support color vision, in which a brighter light is required to harvest a signal. Approximately 120 million rod cells and 6 million cone cells are present in human retina. The number and proportion of rods to cones differs in diurnal or nocturnal animals. In the human visual scheme, there are approximately 2.4–3 million ganglion cells, of which 1–2% are photosensitive. In many retinal degenerative diseases, rod cell death is followed by cone cell death.[6] In retinal degenerative diseases such as retinitis pigmentosa,[7,8] age-related macular degeneration (AMD), diabetic retinopathy,[9–11] and glaucoma,[12,13] the nerve cells in the retina die. Since they cannot be replaced, their loss leads to reduced vision, and often blindness. Over the past 20 years, much knowledge has been generated about the genetic and environmental factors that contribute to retinal degeneration. However, the successful use of this knowledge to successfully treat retinal degeneration has been limited.

Mutations in the genes that are essential for normal retinal cell function and oxidative stress are two causes of retinal cell death. These mutations are present at conception, but the retinal cells often survive for decades before dying. Oxidant stress usually occurs in older cells that have been compromised by age and/or environmental factors, such as smoking, which is the greatest risk factor for AMD other than age. Multiple approaches, including gene therapy,[14–34] administration of antioxidants[35–38] or cytokines,[39–43] tissue transplantation,[44] optogenetics,[45–47] surgical implantation of light-sensitive photocells and prosthesis,[48,49] microRNA administration,[50] and stem cell therapy[51] are being used to treat retinal degenerative diseases. Gene therapy is a fascinating current approach to correct many disease phenotypes. Any drug or gene that is injected systemically must cross the blood–retinal barrier before it enters the eye tissue. Gene therapy is well suited to deliver treatment to eye tissue because of its easy accessibility, relatively low dose, and self-containment; it is also immune-privileged.[52]

The development of effective, non-harmful gene transporters is desirable to target extraneous genetic constituents to specific cell types. These carriers or transporters fall into two categories: viral and non-viral gene delivery systems.[53,54] Several viral vectors in which destructive sequences in the virus have been removed, but the information that is able to transfer genes has been preserved, and designed to transfer genes into cells in a procedure that is commonly termed "gene transduction". Researchers have used many viruses, including retrovirus,[55] adenovirus,[56] herpes simplex virus,[57] adeno-associated virus (AAV),[58] and pox virus,[59] for gene transduction. Highly efficient gene transduction can be achieved with viruses. However, the disadvantages of this approach include the associated biosafety issues, toxicity, induction of immune reaction, and potential recombination of virus that is harmful to the recipient and the environment. Another limitation is the size of the insert that can be cloned into the viral vector. Gene(s) with a size of >5 kb are not suitable for viral-mediated gene delivery.

New approaches are needed to overcome size limitations, achieve safety, and create effective gene therapy. The ideal carrier would accommodate genes with no size restrictions, have high transduction competence, elicit no immune reaction, be affordable, well-tolerated and biodegradable, and have controlled and specific-target tissue expression while preserving the well-being of both the patient and the environment. The existing size limitations and the disadvantages associated with viral gene transduction have inspired many researchers to focus on development of non-viral gene delivery systems. There are excellent reviews of the development and application of nanoparticles as effective delivery vehicles for biomolecules available.[60–63] Those aspects are not discussed in the current review. Instead, we focus primarily on the application of lipid-based nanoparticles.

13.2 Application of LPD for Ocular Diseases

Lipid-based nanoparticles (liposome–protamine–DNA complex; LPD) are derived from liposomes. The name liposome is derived from two Greek words, "lipos", meaning fat, and "soma", meaning body; the liposome is basically a fat body.[64] British hematologist Dr Alec D. Bangham first described liposomes in 1961.[65] Liposomes are circular vesicles with a minimum of one lipid bilayer. These vesicles have long been known to be effective cargo vehicles for transporting biomolecules, nutrients, and therapeutic agents.[66] Disruption of natural membranes by sonication results in the formation of liposomes. Naturally occurring liposomes have bilayers consisting of phospholipids and/or cholesterol[64,67] which are non-toxic and inert under physiological conditions. Liposome size is variable, ranging from 25 nm to 2.5 μm,[68,69] and can be grouped into one of three types: (1) multilamellar vesicles that resemble onions with numerous phospholipid bilayers separated by aqueous layers; (2) large unilamellar vesicles; and (3) small unilamellar vesicles with a single lipid bilayer. Phospholipids, which have a water-attracted head group and a water-repelled tail group, make up the cell membranes.[69] Liposomes can be composed of surfactant components such as dioleoylphosphatidylethanolamine (DOPE), or naturally derived phospholipids with mixed lipid chains, such as egg phosphatidylethanolamine.

The lipid backbone in the liposome used for the preparation of lipid nanoparticles for eye tissues is composed of 1,2-dioleoyl-3-trimethylammonium-propane (DOTAP), DOPE, and cholesterol.[70] Combining the liposome with protamine and nuclear localization and cell-targeting signal peptides promotes efficient gene transduction *in vitro* and *in vivo*[70] (Figure 13.1). Transmission electron microscopy showed that the size of LPD ranges from 80 nm to 120 nm; this size is not affected by the inclusion of a targeting TAT-peptide.[71] Long-term transgene expression up to 5 months has been observed in retinal pigment epithelium cells *in vivo* after sub-retinal

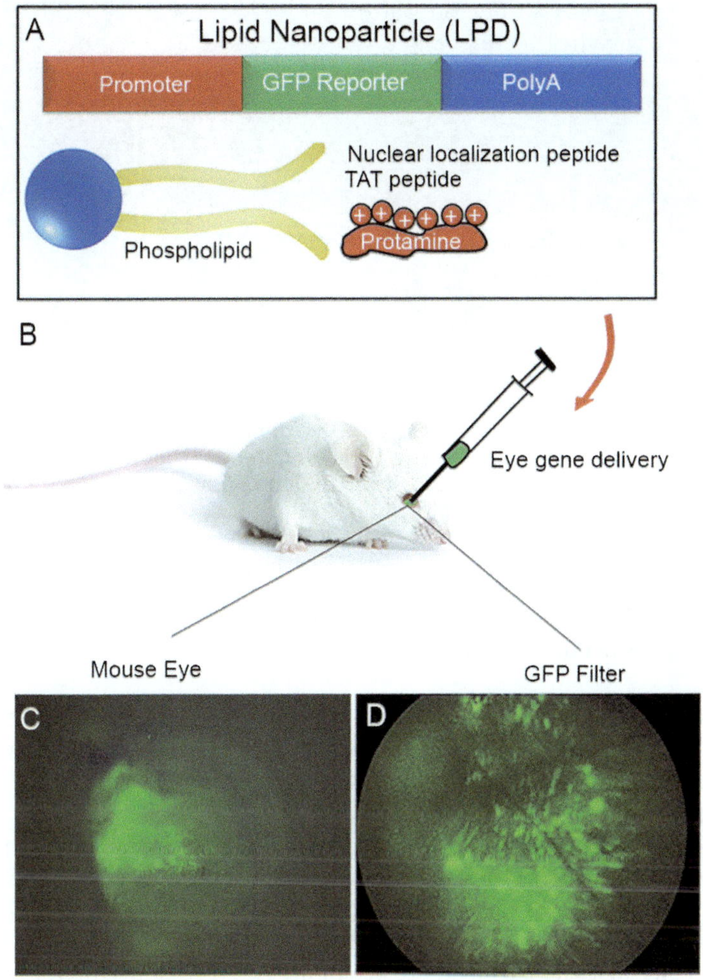

Figure 13.1 Schematic representation of lipid nanoparticle (liposome–protamine–DNA complex; LPD). (A) The LPD used for eye delivery is composed of dioleoylphosphatidylethanolamine (DOPE), 1,2-dioleoyl-3-trimethylammonium-propane (DOTAP), cholesterol, cell-membrane-targeting TAT-peptide (transactivator of transcription), nuclear localization signal peptide from SV40, protamine, and plasmid DNA. (B) Subretinal injection of a plasmid carrying the cDNA-encoding enhanced green fluorescent protein (eGFP) complexed with LPD to mouse eye. 72 h later, eyes were removed and examined for GFP expression under (C) inverted fluorescence microscopy, and by (D) fluorescent fundoscopy.

injection of a reporter plasmid under the control of a cytomegalovirus (CMV) promoter (Figure 13.2).[70] No cytotoxic effects or immune reactions were observed in the retina after LPD treatment.[71] One study showed that systemic administration of LPD with DOTAP, DOPE, and cholesterol delivered genes less efficiently because DOPE interacted with serum proteins.[70]

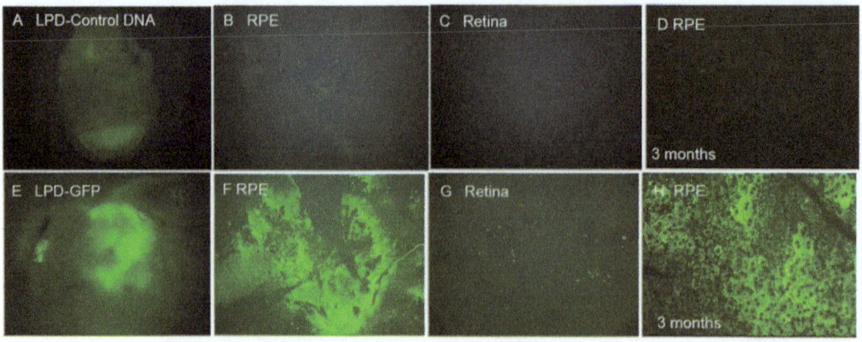

Figure 13.2 *In vivo* transfection by liposome–protamine–DNA complex (LPD). Plasmid DNAs of either control (pcDNA3 vector) or green fluorescent protein (GFP) were complexed with LPD and injected subretinally into BALB/c mice. 72 h later, eyes were removed and examined for (A, E) GFP expression under inverted fluorescence microscopy. Whole (B, F) retinal pigment epithelium (RPE) and (C, G) retinal flat mounts were prepared and examined for GFP expression under inverted fluorescence microscopy. To determine long-term expression, (D, H) whole RPE flat mounts were examined for GFP expression 3 months later. Reprinted with permission from ref. 70. Copyright 2014 American Chemical Society.

Omissions of DOPE overcome this limitation, but result in decreased transgene expression. This is not a problem for ocular tissues, as the administration is local. Thus, lipid nanotechnology is ideal for gene and drug delivery to eye tissue.

Lipid nanoparticle delivery of drugs and genes to the retina has long been desired. The first successful gene therapy with lipid nanoparticles was reported in 2014.[70] Visual chromophore 11-cis-retinal, one form of vitamin A, facilitates visual pigment rhodopsin to capture light.[72] Without the chromophore, we lose the capability to see light. The retinal pigment epithelium 65 (Rpe65) protein regulates the availability of this chromophore; deletion of Rpe65 protein in mice results in a blinding eye disease called Leber's congenital amaurosis.[73,74] LPD-mediated delivery of the Rpe65 gene to mice deficient in Rpe65 protein corrected the blindness in these mice.[70] *Rpe65*$^{-/-}$ mice do not respond to light due to the absence of photobleachable rhodopsin; these mice also undergo cone degeneration at approximately 2 weeks of age, and complete loss of cones occurs by 4 weeks.[75] LPD-mediated delivery of the Rpe65 gene to mice deficient in Rpe65 protein corrected visual function (Figure 13.3A) and preserved the loss of degenerating cones (Figure 13.3B–D).[70] In addition, the study revealed that the efficiency of LPD-mediated gene delivery is comparable to that of AAV and lentiviral gene transfer of Rpe65 gene to Rpe65 knockout mice.[19,30] The researchers reported that LPD administration produced no toxicity or inflammation in the eye. These results are promising, and suggest that LPD can be used safely for gene delivery with clinical applications for the treatment of eye diseases.

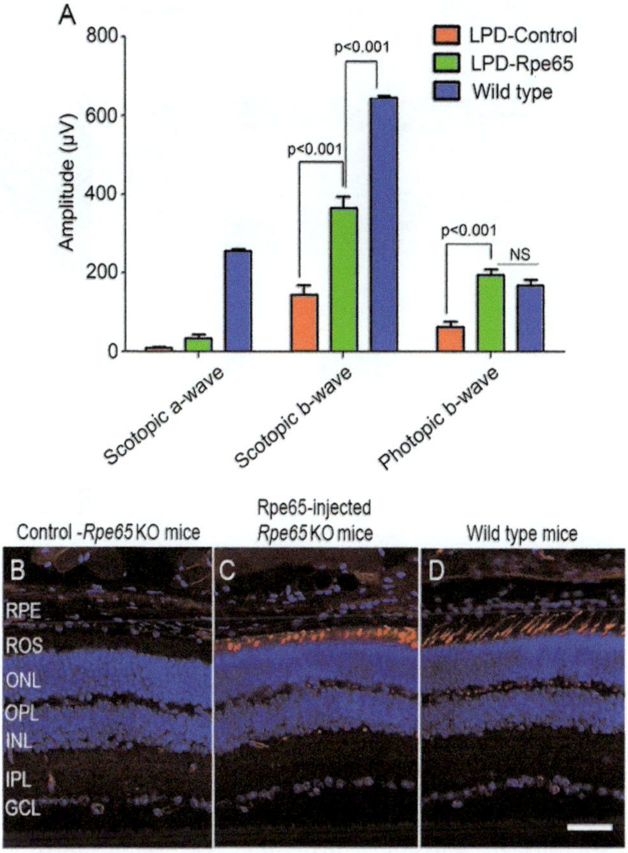

Figure 13.3 Rpe65-complexed liposome–protamine DNA complex (LPD) nanoparticles improve the vision and prevent cone cell loss in *Rpe65* knockout (Rpe65 KO) mice. (A) Scotopic a- and b-wave and photopic b-wave electroretinographic analysis of wild-type and *Rpe65* knockout mice injected with LPD-control DNA (pcDNA3), and LPD-chicken Rpe65. Values are mean ± SEM, $n = 6$, significance between groups $p < 0.001$; NS: not significant. Retinal sections prepared from (B) un-injected and Rpe65-injected (C) *Rpe65* knockout and (D) wild-type mice were subjected to lectin cytochemical analysis using peanut agglutinin. RPE: retinal pigment epithelium; ROS: rod outer segments; ONL: outer nuclear layer; OPL: outer plexiform layer; INL: inner nuclear layer; IPL: inner plexiform layer; GCL: ganglion cell layer. Scale bar = 50 μm. Reprinted with permission from ref. 70. Copyright 2014 American Chemical Society.

13.3 Cell-specific Promoters Enable LPD to Deliver Cargo to Specific Cells of the Retina *In Vivo*

The major limitation of LPD is specificity. Once the LPD is injected *in vivo*, the gene could be delivered anywhere, or everywhere. In many cases, gene expression is cell-specific, and if the same gene is expressed in various cells,

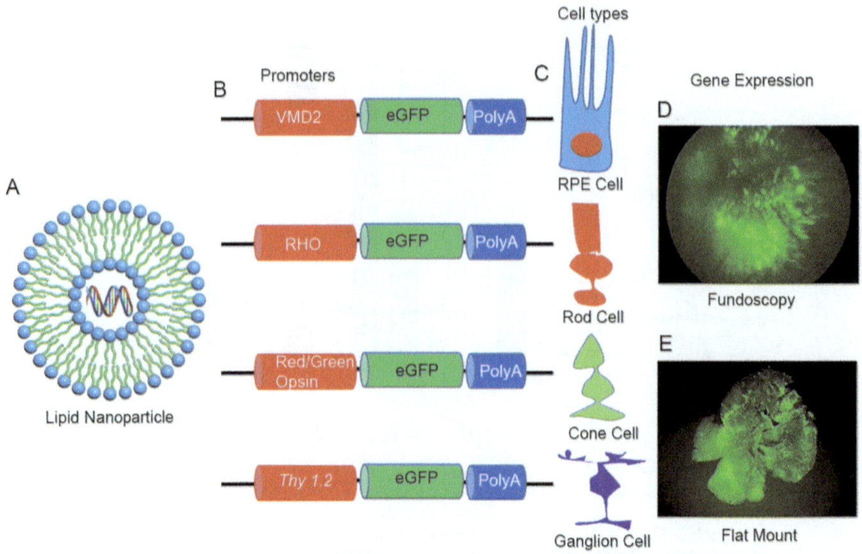

Figure 13.4 Cell-specific delivery of genes to retinal cells. Cell-specific promoters (B) enable liposome–protamine–DNA complex (LPD) (A) to deliver genes specific to retinal pigment epithelium (RPE) cells, rod cells, cone cells, and ganglion cells (C). (D) Fluorescent fundoscopy and (E) retinal flat mounts show the expression of green fluorescent protein (GFP) transgene. Reprinted with permission from ref. 71.

there may be off-target effects. This problem has recently been solved for ocular gene therapy. As a proof of principle, researchers have used retinal cell-specific promoters to target the genes to specific locations of the cell using LPD (Figure 13.4). In retinal degenerative diseases, four types of cells are commonly affected. Targeting genes to those cells provides a window of opportunity to save the dying cells.

Retinal pigment epithelium (RPE) cells provide nutrient support and maintenance to both rod and cone photoreceptor cells. RPE cells perform the daily cellular waste cleanup by acting as the "garbage disposal system" of the eye.[76] Each morning, an amount of cellular waste from the retinal photoreceptor cells is delivered to the RPE cells. This cellular waste is digested by the RPE cells, which recycle the nutrients and vitamin A necessary for normal vision back to the retina cells.[77] As we age, cellular stress arising from lifestyle factors such as smoking, as well as genetic mutations and dysregulation in normal physiological functions of cells, negatively affects the function of the RPE. This leads to retinal and RPE cell death and subsequent loss of vision in diseases such as AMD, diabetic retinopathy, and retinitis pigmentosa. Thus, the relationship between the retina and RPE cells is similar to the buildup of garbage in our environment. If the garbage is not constantly removed and destroyed, the house has a rancid smell and its occupants become more susceptible to disease.

Rod photoreceptors degenerate in retinitis pigmentosa. Retinitis pigmentosa is a group of diseases in which patients experience night blindness. Worldwide, retinitis pigmentosa affects one in every 4000 individuals. If left untreated, patients lose central vision by the age of 60 years. In retinitis pigmentosa, rod degeneration precedes secondary cone degeneration.[6] In humans and rodents, rods constitute approximately 95% of retinal photoreceptor cells, whereas cone photoreceptors constitute only 3–5%.[78,79] Although there are substantially fewer cone photoreceptors than rods, cones are essential for optimal visual perception and color visualization under different light intensities in humans. The most common disorders that initiate cone degeneration are AMD and diabetic retinopathy.[9–11] In retinitis pigmentosa, rods degenerate, then cones degenerate.[6] Rods are directly affected in cone–rod dystrophies.[80,81]

Diabetic retinopathy is a growing worldwide public health problem. Patients with non-proliferative retinopathy exhibit vascular pathologies with impaired vision[82–84] associated with disruption of the neurovascular unit,[11] and dysfunction and death of retinal ganglion cells (RGCs).[85,86] RGC projections carry visual inputs to the brain and regulate critical aspects of vision that are impaired by diabetes, notably contrast sensitivity. Multiple studies have shown that RGC exhibits progressive apoptosis, synaptic protein loss, and axonal degeneration that begin within weeks after the onset of diabetes.[86] Ganglion cells are also affected by glaucoma and traumatic optic neuropathies.[12,86–89] A recent study showed that RPE-cell specific promoter vitelliform macular dystrophy (VMD2) enables LPD to deliver to RPE cells.[71,90] Rod photoreceptor cell-specific promoter,[71,90] rhodopsin, and cone photoreceptor cell-specific red/green cone opsin[71,91] promoters facilitate LPD delivery to specific rod and cone cells, respectively. LPD-mediated delivery to ganglion cells is achieved through a ganglion-cell specific promoter, thymocyte antigen (*Thy1.2*).[71,92] In these studies, green florescent protein expression was monitored as a function of cell specificity under the control of cell-specific promoters.[71] Given the success achieved with cell-specific promoters, LPD may be used to target specific cells with cell-type-specific promoters.

Photoreceptor specificity is also achieved with CAG promoter.[71] CAG promoter contains a CMV early enhancer element, the promoter, the first exon and the first intron of chicken β-actin gene and the splice acceptor of the rabbit β-globin gene.[93,94] The CMV promoter enables the LPD to deliver RPE, but the CAG sequence in the CMV promoter facilitates LPD delivery to photoreceptor cells.[71] This altered cell specificity could be dictated by the existence of a human β-globin intron in the CAG vector.[95] Genes under the control of ubiquitous promoter that are administered intravitreally are able to diffuse to the retina. Intravitreal administration is ideal to target ganglion cells, whereas sub-retinal injection is ideal to target RPE and photoreceptor cells. Thus far, LPD-mediated delivery to eye tissues has been achieved through intravitreal and sub-retinal injections.[71,96] Topical application of LPD has never been attempted, although such a route is ideal for multiple applications. In several studies, the addition of a woodchuck hepatitis virus post-transcriptional regulatory element sequence[97,98] enhanced the expression of transgene, including LPD-mediated delivery to eye tissues.[71]

MicroRNAs (miRNAs) are 17–27 nucleotides long and regulate post-transcriptional mRNA expression, typically by binding to the 3′-untranslated region of the complementary mRNA sequence, resulting in translational repression and gene silencing.[99] Currently, researchers are administering miRNAs as therapeutic agents to silence target genes in many diseases. miRNA 184 has been shown to modulate Wnt signaling through the regulation of frizzled-7 expression in the retina with ischemia-induced neovascularization.[50] Targeting a miRNA 184 mimic with LPD inhibited Wnt signaling in an oxygen-induced retinopathy mouse model.[50] Peroxisome proliferator-activated receptor α (PPARα) is an important receptor that is implicated in diabetic retinopathy and is downregulated in this disease.[100] Fenofibrate is a PPARα agonist that is used to treat diabetes.[100] The mechanism of PPARα downregulation in diabetic retinopathy occurs partly through upregulation of miR-21.[100] Knockdown of mIR-21 prevented PPARα downregulation and decreased oxygen-induced retinopathy.[100] Intravitreal delivery of LPD-mediated miR-21 inhibitor attenuated PPARα downregulation and suppressed retinal inflammation in db/db mice.[100] These findings are promising. In the future, LPD could be used to deliver microRNA in clinical applications for the treatment of human eye diseases.

13.4 Conclusions

Delivering therapeutic agents through a suitable system is desirable; however, the knowledge base in this area remains limited, and such interventions have yet to emerge. Gene therapies using viral vectors have been successfully applied to the treatment of eye diseases, but safety is a major concern for both patients and the environment. Many non-viral vectors, including lipid nanoparticles, have been used to treat eye diseases. Combining the power of lipid nanoparticles with cell-specific promoters enables the delivery of LPD to specific cells of the retina. These approaches may inspire investigators in other fields to couple cell-specific promoters with LPD to directly target genes to specific tissue and cell types. These unique characteristics make LPD a promising delivery system for ocular tissues.

Acknowledgements

This study was supported by grants from the National Institutes of Health (EY00871 and NEI Core grant EY12190) and an unrestricted grant from Research to Prevent Blindness, Inc. to the Department of Ophthalmology. The authors acknowledge Ms Kathy J. Kyler, staff editor, University of Oklahoma Health Sciences Center, for editing this manuscript.

References

1. E. Soucy, Y. Wang, S. Nirenberg, J. Nathans and M. Meister, *Neuron*, 1998, **21**, 481–493.
2. R. G. Foster, I. Provencio, D. Hudson, S. Fiske, G. W. De and M. Menaker, *J. Comp. Physiol. A*, 1991, **169**, 39–50.
3. R. G. Foster, *J. Neuroendocrinol.*, 2004, **16**, 179–180.
4. M. Semo, S. Peirson, D. Lupi, R. J. Lucas, G. Jeffery and R. G. Foster, *Eur. J. Neurosci.*, 2003, **17**, 1793–1801.
5. S. Hattar, H. W. Liao, M. Takao, D. M. Berson and K. W. Yau, *Science*, 2002, **295**, 1065–1070.
6. C. Punzo, W. Xiong and C. L. Cepko, *J. Biol. Chem.*, 2012, **287**, 1642–1648.
7. A. H. Milam, Z. Y. Li and R. N. Fariss, *Prog. Retinal Eye Res.*, 1998, **17**, 175–205.
8. E. L. Berson, *Proc. Natl. Acad. Sci. U. S. A.*, 1996, **93**, 4526–4528.
9. R. Adler, C. Curcio, D. Hicks, D. Price and F. Wong, *Mol. Vision*, 1999, **5**, 31.
10. E. M. Stone, T. A. Braun, S. R. Russell, M. H. Kuehn, A. J. Lotery, P. A. Moore, C. G. Eastman, T. L. Casavant and V. C. Sheffield, *N. Engl. J. Med.*, 2004, **351**, 346–353.
11. D. A. Antonetti, R. Klein and T. W. Gardner, *N. Engl. J. Med.*, 2012, **366**, 1227–1239.
12. H. A. Quigley, *Lancet*, 2011, **377**, 1367–1377.
13. T. M. Nork, *Trans. Am. Ophthalmol. Soc.*, 2000, **98**, 331–363.
14. G. M. Acland, G. D. Aguirre, J. Ray, Q. Zhang, T. S. Aleman, A. V. Cideciyan, S. E. Pearce-Kelling, V. Anand, Y. Zeng, A. M. Maguire, S. G. Jacobson, W. W. Hauswirth and J. Bennett, *Nat. Genet.*, 2001, **28**, 92–95.
15. J. Adijanto and M. I. Naash, *Eur. J. Pharm. Biopharm.*, 2015, **95**, 353–367.
16. R. R. Ali, *Hum. Gene Ther.*, 2012, **23**, 337–339.
17. J. W. Bainbridge, A. J. Smith, S. S. Barker, S. Robbie, R. Henderson, K. Balaggan, A. Viswanathan, G. E. Holder, A. Stockman, N. Tyler, S. Petersen-Jones, S. S. Bhattacharya, A. J. Thrasher, F. W. Fitzke, B. J. Carter, G. S. Rubin, A. T. Moore and R. R. Ali, *N. Engl. J. Med.*, 2008, **358**, 2231–2239.
18. W. A. Beltran, A. V. Cideciyan, A. S. Lewin, S. Iwabe, H. Khanna, A. Sumaroka, V. A. Chiodo, D. S. Fajardo, A. J. Roman, W. T. Deng, M. Swider, T. S. Aleman, S. L. Boye, S. Genini, A. Swaroop, W. W. Hauswirth, S. G. Jacobson and G. D. Aguirre, *Proc. Natl. Acad. Sci. U. S. A.*, 2012, **109**, 2132–2137.
19. A. P. Bemelmans, C. Kostic, S. V. Crippa, W. W. Hauswirth, J. Lem, F. L. Munier, M. W. Seeliger, A. Wenzel and Y. Arsenijevic, *PLoS Med.*, 2006, **3**, e347.
20. A. P. Bemelmans, C. Kostic, D. Hornfeld, M. Jaquet, S. V. Crippa, W. W. Hauswirth, J. Lem, Z. Wang, D. E. Schorderet, F. L. Munier, A. Wenzel and Y. Arsenijevic, *Adv. Exp. Med. Biol.*, 2006, **572**, 247–253.
21. L. C. Byrne, D. Dalkara, G. Luna, S. K. Fisher, E. Clerin, J. A. Sahel, T. Leveillard and J. G. Flannery, *J. Clin. Invest.*, 2015, **125**, 105–116.

22. X. Cai, S. M. Conley, Z. Nash, S. J. Fliesler, M. J. Cooper and M. I. Naash, *FASEB J.*, 2010, **24**, 1178–1191.
23. Y. Chen, G. Moiseyev, Y. Takahashi and J. X. Ma, *Invest. Ophthalmol. Visual Sci.*, 2006, **47**, 1177–1184.
24. D. Dalkara, L. C. Byrne, R. R. Klimczak, M. Visel, L. Yin, W. H. Merigan, J. G. Flannery and D. V. Schaffer, *Sci. Transl. Med.*, 2013, **5**, 189ra176.
25. Z. Han, S. M. Conley, R. S. Makkia, M. J. Cooper and M. I. Naash, *J. Clin. Invest.*, 2012, **122**, 3221–3226.
26. Z. Han, S. M. Conley and M. I. Naash, *Invest. Ophthalmol. Visual Sci.*, 2011, **52**, 3051–3059.
27. Z. Han, S. M. Conley, R. Makkia, J. Guo, M. J. Cooper and M. I. Naash, *PLoS One*, 2012, **7**, e52189.
28. Z. Han, M. J. Banworth, R. Makkia, S. M. Conley, M. R. Al-Ubaidi, M. J. Cooper and M. I. Naash, *FASEB J.*, 2015, **29**, 2535–2544.
29. A. M. Komaromy, J. J. Alexander, J. S. Rowlan, M. M. Garcia, V. A. Chiodo, A. Kaya, J. C. Tanaka, G. M. Acland, W. W. Hauswirth and G. D. Aguirre, *Hum. Mol. Genet.*, 2010, **19**, 2581–2593.
30. C. M. Lai, M. J. Yu, M. Brankov, N. L. Barnett, X. Zhou, T. M. Redmond, K. Narfstrom and P. E. Rakoczy, *Genet. Vaccines Ther.*, 2004, **2**, 3.
31. A. Koirala, S. M. Conley and M. I. Naash, *Biomaterials*, 2013, **34**, 7158–7167.
32. J. Kong, S. R. Kim, K. Binley, I. Pata, K. Doi, J. Mannik, J. Zernant-Rajang, O. Kan, S. Iqball, S. Naylor, J. R. Sparrow, P. Gouras and R. Allikmets, *Gene Ther.*, 2008, **15**, 1311–1320.
33. X. Li, W. Li, X. Dai, F. Kong, Q. Zheng, X. Zhou, F. Lu, B. Chang, B. Rohrer, W. W. Hauswirth, J. Qu and J. J. Pang, *Invest. Ophthalmol. Visual Sci.*, 2011, **52**, 7–15.
34. J. Tosi, J. Sancho-Pelluz, R. J. Davis, C. W. Hsu, K. V. Wolpert, J. D. Sengillo, C. S. Lin and S. H. Tsang, *Exp. Biol. Med.*, 2011, **236**, 1211–1217.
35. A. J. Chucair, N. P. Rotstein, J. P. Sangiovanni, A. During, E. Y. Chew and L. E. Politi, *Invest. Ophthalmol. Visual Sci.*, 2007, **48**, 5168–5177.
36. D. Joshi, J. Field, J. Murphy, M. Abdelrahim, H. Schonherr, J. R. Sparrow, G. Ellestad, K. Nakanishi and A. Zask, *J. Nat. Prod.*, 2013, **76**, 450–454.
37. R. A. Kowluru and Q. Zhong, *Invest. Ophthalmol. Visual Sci.*, 2011, **52**, 8665–8671.
38. K. Komeima, B. S. Rogers, L. Lu and P. A. Campochiaro, *Proc. Natl. Acad. Sci. U. S. A.*, 2006, **103**, 11300–11305.
39. Y. Ueki, Y. Z. Le, S. Chollangi, W. Muller and J. D. Ash, *Proc. Natl. Acad. Sci. U. S. A.*, 2009, **106**, 21389–21394.
40. S. Burgi, M. Samardzija and C. Grimm, *Mol. Vision*, 2009, **15**, 1631–1637.
41. S. Chollangi, J. Wang, A. Martin, J. Quinn and J. D. Ash, *Neurobiol. Dis.*, 2009, **34**, 535–544.
42. Y. Ueki, J. Wang, S. Chollangi and J. D. Ash, *J. Neurochem.*, 2008, **105**, 784–796.
43. S. Joly, C. Lange, M. Thiersch, M. Samardzija and C. Grimm, *J. Neurosci.*, 2008, **28**, 13765–13774.

44. N. D. Radtke, R. B. Aramant, H. M. Petry, P. T. Green, D. J. Pidwell and M. J. Seiler, *Am. J. Ophthalmol.*, 2008, **146**, 172–182.
45. V. Busskamp, J. Duebel, D. Balya, M. Fradot, T. J. Viney, S. Siegert, A. C. Groner, E. Cabuy, V. Forster, M. Seeliger, M. Biel, P. Humphries, M. Paques, S. Mohand-Said, D. Trono, K. Deisseroth, J. A. Sahel, S. Picaud and B. Roska, *Science*, 2010, **329**, 413–417.
46. V. Busskamp, S. Picaud, J. A. Sahel and B. Roska, *Gene Ther.*, 2012, **19**, 169–175.
47. P. S. Lagali, D. Balya, G. B. Awatramani, T. A. Munch, D. S. Kim, V. Busskamp, C. L. Cepko and B. Roska, *Nat. Neurosci.*, 2008, **11**, 667–675.
48. G. Dagnelie, *Curr. Opin. Neurol.*, 2012, **25**, 67–75.
49. D. Yanai, J. D. Weiland, M. Mahadevappa, R. J. Greenberg, I. Fine and M. S. Humayun, *Am. J. Ophthalmol.*, 2007, **143**, 820–827.
50. Y. Takahashi, Q. Chen, R. V. Rajala and J. Ma, *FEBS Lett.*, 2015, **589**, 1143–1149.
51. B. Mead and B. A. Scheven, *Neural Regener. Res.*, 2015, **10**, 371–373.
52. J. Y. Niederkorn, *J. Invest. Dermatol. Symp. Proc.*, 2003, **8**, 168–172.
53. N. Nayerossadat, T. Maedeh and P. A. Ali, *Adv. Biomed. Res.*, 2012, **1**, 27.
54. S. Chira, C. S. Jackson, I. Oprea, F. Ozturk, M. S. Pepper, I. Diaconu, C. Braicu, L. Z. Raduly, G. A. Calin and I. Berindan-Neagoe, *Oncotarget*, 2015, **6**, 30675–30703.
55. D. G. Miller, M. A. Adam and A. D. Miller, *Mol. Cell. Biol.*, 1990, **10**, 4239–4242.
56. T. Ragot, N. Vincent, P. Chafey, E. Vigne, H. Gilgenkrantz, D. Couton, J. Cartaud, P. Briand, J. C. Kaplan and M. Perricaudet, *Nature*, 1993, **361**, 647–650.
57. M. Toda, R. L. Martuza and S. D. Rabkin, *Mol. Ther.*, 2000, **2**, 324–329.
58. I. E. Alexander, D. W. Russell, A. M. Spence and A. D. Miller, *Hum. Gene Ther.*, 1996, 7, 841–850.
59. E. Paoletti, *Proc. Natl. Acad. Sci. U. S. A.*, 1996, **93**, 11349–11353.
60. F. Li, J. Lu, X. Kong, T. Hyeon and D. Ling, *Adv. Mater.*, 2017, **29**, 1605897.
61. E. Keles, Y. Song, D. Du, W. J. Dong and Y. Lin, *Biomater. Sci.*, 2016, **4**, 1291–1309.
62. A. Mangraviti, D. Gullotti, B. Tyler and H. Brem, *J. Controlled Release*, 2016, **240**, 443–453.
63. M. Moritz and M. Geszke-Moritz, *Adv. Clin. Exp. Med.*, 2015, **24**, 749–758.
64. A. D. Bangham and R. W. Horne, *J. Mol. Biol.*, 1964, **8**, 660–668.
65. A. D. Bangham, *Nature*, 1961, **192**, 1197–1198.
66. G. Gregoriadis and B. E. Ryman, *Biochem. J.*, 1971, **124**, 58P.
67. V. P. Torchilin, *Adv. Drug Delivery Rev.*, 2006, **58**, 1532–1555.
68. S. Bai, V. Gupta and F. Ahsan, *Eur. J. Pharm. Sci.*, 2009, **38**, 165–171.
69. A. Sharma and U. S. Sharma, *Int. J. Pharm.*, 1997, **154**, 123–140.
70. A. Rajala, Y. Wang, Y. Zhu, M. Ranjo-Bishop, J. X. Ma, C. Mao and R. V. Rajala, *Nano Lett.*, 2014, **14**, 5257–5263.
71. Y. Wang, A. Rajala, B. Cao, M. Ranjo-Bishop, M. P. Agbaga, C. Mao and R. V. Rajala, *Theranostics*, 2016, **6**, 1514–1527.

72. T. M. Redmond, S. Yu, E. Lee, D. Bok, D. Hamasaki, N. Chen, P. Goletz, J. X. Ma, R. K. Crouch and K. Pfeifer, *Nat. Genet.*, 1998, **20**, 344–351.
73. F. P. Cremers, J. A. van den Hurk and A. I. den Hollander, *Hum. Mol. Genet.*, 2002, **11**, 1169–1176.
74. A. I. den Hollander, R. Roepman, R. K. Koenekoop and F. P. Cremers, *Prog. Retinal Eye Res.*, 2008, **27**, 391–419.
75. S. L. Znoiko, B. Rohrer, K. Lu, H. R. Lohr, R. K. Crouch and J. X. Ma, *Invest. Ophthalmol. Visual Sci.*, 2005, **46**, 1473–1479.
76. D. A. Ferrington, D. Sinha and K. Kaarniranta, *Prog. Retinal Eye Res.*, 2016, **51**, 69–89.
77. W. C. Gordon, E. B. Rodriguez de Turco and N. G. Bazan, *Curr. Eye Res.*, 1992, **11**, 73–83.
78. L. D. Carter-Dawson and M. M. LaVail, *J. Comp. Neurol.*, 1979, **188**, 245–262.
79. L. D. Carter-Dawson, M. M. LaVail and R. L. Sidman, *Invest. Ophthalmol. Visual Sci.*, 1978, **17**, 489–498.
80. C. P. Hamel, *Orphanet J. Rare Dis.*, 2007, **2**, 7.
81. U. Kellner and M. H. Foerster, *Ger. J. Ophthalmol.*, 1993, **2**, 170–177.
82. M. A. Di Leo, S. Caputo, B. Falsini, V. Porciatti, A. Minnella, A. V. Greco and G. Ghirlanda, *Diabetes Care*, 1992, **15**, 620–625.
83. G. R. Jackson, I. U. Scott, D. A. Quillen, L. E. Walter and T. W. Gardner, *Br. J. Ophthalmol.*, 2012, **96**, 699–703.
84. M. Parravano, F. Oddone, D. Mineo, M. Centofanti, P. Borboni, R. Lauro, L. Tanga and G. Manni, *Br. J. Ophthalmol.*, 2008, **92**, 1656–1660.
85. A. J. Barber, E. Lieth, S. A. Khin, D. A. Antonetti, A. G. Buchanan and T. W. Gardner, *J. Clin. Invest.*, 1998, **102**, 783–791.
86. T. S. Kern and A. J. Barber, *J. Physiol.*, 2008, **586**, 4401–4408.
87. I. Soto, M. E. Pease, J. L. Son, X. Shi, H. A. Quigley and N. Marsh-Armstrong, *Invest. Ophthalmol. Visual Sci.*, 2011, **52**, 434–441.
88. Y. C. Tham, X. Li, T. Y. Wong, H. A. Quigley, T. Aung and C. Y. Cheng, *Ophthalmology*, 2014, **121**, 2081–2090.
89. W. Wasinska-Borowiec, K. A. Aghdam, J. M. Saari and A. Grzybowski, *Curr. Pharm. Des.*, 2017, **23**, 586–595.
90. N. Esumi, Y. Oshima, Y. Li, P. A. Campochiaro and D. J. Zack, *J. Biol. Chem.*, 2004, **279**, 19064–19073.
91. Y. Wang, J. P. Macke, S. L. Merbs, D. J. Zack, B. Klaunberg, J. Bennett, J. Gearhart and J. Nathans, *Neuron*, 1992, **9**, 429–440.
92. G. Feng, R. H. Mellor, M. Bernstein, C. Keller-Peck, Q. T. Nguyen, M. Wallace, J. M. Nerbonne, J. W. Lichtman and J. R. Sanes, *Neuron*, 2000, **28**, 41–51.
93. H. Niwa, K. Yamamura and J. Miyazaki, *Gene*, 1991, **108**, 193–199.
94. J. Miyazaki, S. Takaki, K. Araki, F. Tashiro, A. Tominaga, K. Takatsu and K. Yamamura, *Gene*, 1989, **79**, 269–277.
95. W. Xiong, A. E. MacColl Garfinkel, Y. Li, L. I. Benowitz and C. L. Cepko, *J. Clin. Invest.*, 2015, **125**, 1433–1445.
96. Y. Wang, A. Rajala and R. V. Rajala, *J. Funct. Biomater.*, 2015, **6**, 379–394.

97. J. C. Paterna, T. Moccetti, A. Mura, J. Feldon and H. Bueler, *Gene Ther.*, 2000, **7**, 1304–1311.
98. J. E. Donello, J. E. Loeb and T. J. Hope, *J. Virol.*, 1998, **72**, 5085–5092.
99. S. L. Ameres and P. D. Zamore, *Nat. Rev. Mol. Cell Biol.*, 2013, **14**, 475–488.
100. Q. Chen, F. Qiu, K. Zhou, H. G. Matlock, Y. Takahashi, R. V. S. Rajala, Y. Yang, E. Moran and J. X. Ma, *Diabetes*, 2017, **66**, 1671–1682.

CHAPTER 14

Biologicals in Retinal Therapy

EMMA CONNOLLY AND SARAH L. DOYLE*

Dept. Clinical Medicine, School of Medicine, Trinity College Dublin, Dublin 2 and The National Children's Research Centre, Our Lady's Children's Hospital Crumlin, Dublin 12, Ireland
*E-mail: sarah.doyle@tcd.ie

14.1 Introduction

Biological products, or biologics, are medical products that are produced by a variety of natural biological or "living" sources such as humans, animals, microorganisms and in some cases plants. Biologics can have diagnostic qualities or they can be used to prevent or treat disease. Biologics encompass a wide and varying range of products that can generally be categorised into subcategories, such as vaccines, blood or blood components, allergenic extracts, human cells and tissues, monoclonal antibodies, recombinant therapeutic proteins, peptide inhibitors, cell therapies and gene therapies. Biologics are the fastest growing segment of the pharmaceutical market, tripling in value from $36 billion to $163 billion between 2001 and 2012, thanks to their high specificity for their drug targets. Apart from tissue and cellular therapies, most biologics are very complex molecules that target specific disease-causing genes and proteins, a process termed molecular targeting, the

Drug Discovery Series No. 66
Therapies for Retinal Degeneration: Targeting Common Processes
Edited by Enrique J. de la Rosa and Thomas G. Cotter

Published by the Royal Society of Chemistry, www.rsc.org

theory being to specifically attack the molecular pathways that cause disease without disrupting normal functions in our cells and tissues. For this reason, it is most often the latest cutting-edge research at the genetic and molecular level that allows for the discovery and development of biological therapeutics by providing a greater knowledge of how disease processes are initiated. Over the past decade biologics have accounted for one-third of all new medicine approvals, and the use of biologics for retinal disease is a major success story.

The retina is the part of the eye that is responsible for receiving and processing light to convert into neural signals and relay to the brain for visual activity. Retinal disease can occur in many forms and encompasses >250 inherited forms in addition to those that are acquired *via* circumstance and genetic risk, such as age-related macular degeneration (AMD), diabetic retinopathy, diabetic macular edema (DME), retinal vascular disease, posterior inflammatory uveitis and retinopathy of prematurity (ROP). Together, AMD and diabetic eye diseases are the most common diseases of the retina, accounting for 89% of all cases of retinal disease. Despite the different underlying triggers of disease, most cause visual disruption and many have common pathologies, for example wet AMD, diabetic retinopathy, DME and ROP all exhibit unwanted neovascularisation and/or leakage of plasma and plasma proteins from the blood vessels that supply nutrients to the eye, whereas both infectious and non-infectious uveitis, diabetic retinopathy and DME all share symptoms of intraocular inflammation. It is estimated that 296 million people suffer from some type of retinal disease globally, with a small percentage suffering from more than one disorder. The highest rates of retinal disease are in Western Europe, the US and Japan, where the elderly account for large proportions of the population. However, rates of retinal disease are climbing in developing countries such as China and India in line with increasing life expectancies and incidence of diabetes. Retinitis pigmentosa is the leading cause of hereditary blindness in developed countries with a prevalence of ~1.5 million people throughout the world. There is currently no curative treatment for retinitis pigmentosa, which is designated a rare and orphan disease in the United States and Europe. Retinitis pigmentosa causes progressive loss of vision due to degeneration of the rod photoreceptors resulting in the loss of peripheral vision followed by degeneration of the cone photoreceptors resulting in the loss of central vision. Biological therapy in the form of gene therapy has major potential for the treatment of inherited retinopathies. Almost 300 inherited retinal disease-causing gene loci have been mapped and there is concerted research focused on developing gene therapy as a means of treating individuals with retinitis pigmentosa. In fact, the first gene therapy product to be approved by the United States Food and Drug Administration (FDA) has been developed for the treatment of Leber hereditary optic neuropathy. However, this chapter focuses on biologics for treatment of neovascular and inflammatory retinal diseases, which in the main consist of monoclonal antibodies and recombinant proteins.

14.2 Biologics for Treatment of Neovascular Retinal Disease

The formation of new blood vessels, termed angiogenesis, is essential for growth and development and requires the sprouting, migration and proliferation of endothelial cells to form capillary tubes and new blood vessels (neovascularisation). However, there are many instances where neovascularisation can have pathologic consequences, and this is particularly relevant in the eye. The leading causes of blindness in the developed world result from several disorders that have pathologic ocular neovascularisation as the common pathway that leads to vision loss. These retinal diseases are exudative or wet AMD, diabetic retinopathy, ROP, retinal vein occlusions and ocular tumours. In these diseases, the sprouting of new vessels or the leakage of blood products from existing compromised vessels can result in significant loss of vision and scarring.

14.2.1 Age-related Macular Degeneration

AMD is the leading cause of central vision loss in the over-50's, the projected number of people with AMD in 2020 is 196 million, increasing to 288 million in 2040.[1] AMD is a progressive disease categorised into early, intermediate and late disease stages. Early AMD is generally asymptomatic and is diagnosed upon detection of drusen deposits in fundus photographs. Drusen are lipid/protein deposits that lie subjacent to the retinal pigment epithelium (RPE) in the macula, disrupting the laminar structure of the retina; as AMD progresses these deposits can result in distortion of vision. Approximately 20% of individuals diagnosed with early AMD will go on to develop late-stage disease. Of those who progress to late stage, 70% will develop neovascular AMD, also known as exudative or wet AMD, and the remainder will develop geographic atrophy. In geographic atrophy the RPE deteriorates and can no longer perform its functions sufficiently, many of which are vital for the maintenance of the visual cycle and include supporting the overlying photoreceptors. At any time one RPE cell is in contact with ~50 photoreceptor cells; atrophy of the RPE is followed by death of the photoreceptors it supports, resulting in central vision loss. Neovascular AMD occurs when the choroidal vasculature that nourishes the outer retina becomes leaky and begins to proliferate into Bruch's membrane. These choroidal neovascular (CNV) capillaries can breach the outer blood–brain barrier, comprised of the RPE, and reach the delicate neural retina, causing instant central vision loss.

The advent of biologics that target the key molecular switch for the angiogenic process, vascular endothelial growth factor (VEGF), has transformed the treatment of neovascular disease and in particular wet AMD. Prior to 2006, the outlook for patients with wet AMD was fairly bleak, as diagnosis meant certain vision loss. In wet AMD, growth of the CNV membranes is aggressive and they quickly proliferate under the fovea; furthermore, there is a high risk of developing CNV growth in both eyes. Laser photocoagulation therapy and subsequently photodynamic laser therapy (PDT) were shown to be effective in a percentage of patients with CNV; however, they were not without risk and were known to enlarge the creepage of scar tissue towards the fovea and to damage the healthy choroidal vasculature over time, respectively.[2]

14.2.2 Diabetic Retinopathy

Diabetic retinopathy is the leading cause of legal blindness in working-age adults in the Western world. The systemic changes associated with the elevated blood glucose and metabolic syndrome symptomatic of diabetes lead to end-organ damage in the eye and especially in the retinal microvasculature. The earliest manifestation of diabetic retinopathy is in the form of pericyte loss leading to instability of the retinal capillaries, formation of capillary microaneurysms, development of ischemic zones within the retina, capillary leakage and subsequent retinal exudation and edema. The tissue ischemia leads to stabilisation of the transcription factor hypoxia inducible factor-1, which acts to induce gene expression of VEGF. VEGF encourages retinal neovascularisation characterised by the growth of new retinal and pre-retinal fronds. For decades, proliferative diabetic retinopathy was treated with laser photocoagulation, specifically scatter laser surgery, sometimes called panretinal laser surgery or panretinal photocoagulation. Treatment involves making 1000–2000 tiny laser burns in areas of the retina away from the macula. These laser burns are intended to cause abnormal blood vessels to shrink. Although laser photocoagulation was the mainstay of treatment since the early 1980s and numerous clinical trials demonstrated that it can preserve central vision, scatter laser surgery can cause some loss of peripheral, color and night vision and it is recognised that not all patients respond well to treatment.[2] The accelerating rise in incident diabetes is a major public health concern across the Western world, as it is estimated that after 20 years–60% of individuals with type 2 diabetes will develop retinopathy and almost all individuals with type 1 diabetes will develop retinopathy. Increasing evidence from recent trials (Table 14.1) has shown that anti-VEGF treatment is not only effective for treating diffuse macular edema, but is also effective for slowing progression of diabetic retinopathy, including proliferative diabetic retinopathy, so anti-VEGF therapy is increasingly used as a first-line treatment for proliferative diabetic retinopathy and has become the new gold-standard treatment for diabetic macular edema.

14.2.3 Retinopathy of Prematurity

ROP was first described in the 1940s, when disease symptoms were linked to premature births. It is one of the leading cause of childhood blindness, and can present in varying degrees from mild abnormal vessel growth requiring no treatment to vision loss caused by partial or complete detachment of the

Table 14.1 Large cohort studies using anti-vascular endothelial growth factor (VEGF) therapies.[a]

Treatment	Study	Disease	Manufacturer	Summary	Phase	Years
Aflibercept (Eylea)	VISTA/VIVID	DME	Bayer/Regeneron	RCT of intravitreal aflibercept treatment for patients with DME	Phase 3	2011–2014
	COPERNICUS/ GALILEO	Vein occlusion	Bayer/Regeneron	RCT of aflibercept treatments in macular edema secondary to retinal vein occlusion.	Phase 3	2009–2012
Ranibizumab (Lucentis)	RISE/RIDE	DME	Genentech	RCT of 0.3 mg or 0.5 mg intravitreal ranibizumab for the treatment of DME[58]	Phase 3	2007–2012
	BRAVO	BRVO	Genentech	RCT of 0.3 mg and 0.5 mg intravitreal injections of ranibizumab for treatment of macular edema secondary to BRVO[59]	Phase 3	2007–2009
	CRUISE	CRVO	Genentech	Evaluating the safety of 0.5 mg ranibizumab treatments in macular edema secondary to CRVO[59]	Phase 3	2007–2009
	ANCHOR/ MARINA	nvAMD	Genentech	RCT of 0.3 mg or 0.5 mg intravitreal ranibizumab for nvAMD[22,60]	Phase 3	2003–2006
Bevacizumab (Avastin)	BEAT-ROP	ROP	Genentech	RCT comparing intravitreal bevacizumab monotherapy with conventional laser therapy[16]	Phase 2	2008–2020
Pegaptanib (Macugen)	VISION	nvAMD	Eyetech/Pfizer	RCT of intravitreal pegaptanib 0.3 mg, 1.0 mg and 3.0 mg for nvAMD[61]	Phase 4	
Comparison studies						
Bevacizumab/ ranibizumab	CATT/IVAN	nvAMD		A multicentre trial comparing bevacizumab to ranibizumab treatment on a fixed or variable schedule[9,12]		2008–2015
Aflibercept/ ranibizumab	VIEW 1/2	nvAMD		RCT comparing safety and efficacy of intravitreal aflibercept to ranibizumab in nvAMD[62]		2007–2011
Aflibercept/ ranibizumab/ bevacizumab	DRCR	DME		RCT evaluating the effectiveness of three anti-VEGF treatments for DME		2012–2013

[a]BRVO, branch retinal vein occlusion; CRVO, central retinal vein occlusion; DME, diabetic macular edema; nvAMD, neovascular age-related macular degeneration; RCT, randomised controlled trial; ROP, retinopathy of prematurity.

retina. Risk factors for this potentially blinding condition include low birthweight (<1500 g) and delivery before 31 weeks, which can interrupt the latter stages of eye development, leading to abnormal vessel growth. During the third trimester of gestation and for up to a month after birth the vasculature of the eye grows from the central optic nerve out towards the periphery of the retina. In those born prematurely, this process is interrupted resulting in incomplete vascular development and areas of peripheral avascular retina. Studies looking at supplemental oxygen therapy for neonates elucidated the two-phase pathology of the disease, where hyperoxia enhances phase one, the interruption of vessel development, leading to an avascular anterior region of the retina which become metabolically active and hypoxic leading to the second phase of disease, the upregulation of pro-angiogenic VEGF. The accumulation of VEGF in the vitreous induces neovascularisation at the boundary between the vascular posterior and avascular anterior regions of the retina. These new vessels are fragile and leaky, resulting in the formation of fibrous scars which pull on the retina causing detachment and vision loss. Since the 1980s cryo-therapy or laser therapy has been used to treat those at stage 3 ROP, the point where abnormal vessels begin growing towards the centre of the eye. This involves ablation of the non-vascularised, VEGF-producing anterior segment of the retina. While studies have shown that this treatment can reduce the rate of progression and retinal detachment these therapies are highly invasive, impair peripheral vision and often cause clinically significant myopia. In the advent of anti-VEGF therapies being used to modulate neovascular growth in adult retinal diseases, trials began to assess whether these treatments can inhibit hypoxia-induced VEGF and slow progression from early ROP.

14.2.4 Anti-VEGF Therapies

The new era for the treatment of retinal disease with biologics began in 2006 when the FDA approved Lucentis (ranibizumab) for the treatment of wet AMD. Ranibizumab is a monoclonal antibody designed specifically to target and block VEGF-A. The use of anti-VEGF biologics for anti-angiogenic therapy has transformed the treatment of wet AMD and diabetic retinopathy with improved visual outcomes over laser therapies, and has been so widely successful that anti-VEGF's are now the gold standard of care. VEGF has been identified as the critical regulator of pathologic neovascularisation and the role of VEGF in ocular neovascular-related diseases has been characterized in both experimental and translational studies. Elevated levels of VEGF are closely correlated with the pathogenesis of several blinding eye diseases. Of the four major isoforms, $VEGF_{165}$ seems to be the predominant isoform in these processes. Despite there being many molecules that promote angiogenesis, few will also promote permeability.[3] VEGF is the master regulator of both angiogenesis and vascular permeability.[4,5] The *de novo* formation of capillary blood vessels is a dynamic, multistep process that requires endothelial cells to switch from a quiescent state to an angiogenic phenotype.

"Sprouting" angiogenesis originates from pre-existing vasculature and initially encompasses an increase in vasopermeability, due to the disruption of endothelial cell–cell adhesions; this allows for extravasation of plasma proteins that function as a temporary scaffold for migrating endothelial cells. Matrix metalloproteases break down the basement membrane, enabling endothelial cell migration, followed by endothelial cell proliferation and further migration of the newly formed endothelial cells to form a solid endothelial cord prior to lumen formation, pericyte coverage and capillary stabilisation.[6] It is this VEGF-driven neovascularization and increased vascular permeability that leads to many of the complications described earlier for neovascular-related retinal disease, and as such it is not surprising that the first biologics trialed for neovascular retinal disease were geared towards modifying the levels of VEGF in the eye to inhibit pathologic angiogenesis. Three anti-VEGF biological agents have FDA approval, and are used in the treatment of retinal neovascular disease; they are ranibizumab, bevacizumab and aflibercept. Large cohort studies using anti-VEGFs for retinal disease are illustrated in Table 14.1.

14.2.4.1 Ranibizumab

The most common monoclonal antibody based therapeutics are humanised or fully human antibodies. Humanised antibodies replace only the antigen binding site from the human antibody with that of the specific mouse region. In December 2005, Genentech submitted a biologics license application to the FDA for the use of ranibizumab (tradename Lucentis; Roche/Genentech) in the treatment of neovascular wet AMD based on 1-year clinical efficacy and safety data from the two pivotal phase 3 trials, ANCHOR and MARINA, and the phase I-II FOCUS trial.[7] In 2006, ranibizumab was approved for therapeutic use in wet AMD. Ranibizumab is a recombinant humanised monoclonal antibody (rhuMAb) fragment produced in *Escherichia coli* cells by standard recombinant DNA technology and is targeted against human VEGF-A. Ranibizumab was originally derived from a murine monoclonal antibody (muMAb A4.6.1), which was produced at Genentech using hybridomas generated from mice immunised with the predominant 165-residue form of recombinant human VEGF (rhuVEGF$_{165}$) conjugated with keyhole limpet hemocyanin. Ranibizumab is the monoclonal antibody fragment (Fab) moiety of a rhuMAb accomplished by site-directed mutagenesis of a human immunoglobulin G1 (IgG1) framework (95%) with murine complementarity-determining regions (5%); the molecular weight of ranibizumab is ~48 kDa (23 kDa and 25 kDa for the light and heavy chains, respectively). As a Fab fragment, ranibizumab does not contain the Fc region that is involved in antibody-mediated effector functions. Ranibizumab binds with high affinity to VEGF-A isoforms such as VEGF$_{121}$, VEGF$_{165}$ and their biologically active proteolytic cleavage product VEGF$_{110}$, and dissociates slowly; it is likely that once the ranibizumab–VEGF complex is formed, it is very stable. The binding of ranibizumab to VEGF-A effectively "sponges-up" the available VEGF-A, preventing the interaction of VEGF-A with its receptors VEGFR-1 and VEGFR-2 on the surface of endothelial cells, disabling the cell signaling pathways downstream of VEGF-A. Receptor activation normally induces tyrosine auto-phosphorylation and the subsequent series of signal transduction events elicit mitogenic and pro-survival activity signals for the vascular endothelial cells.

Since there is a very low or undetectable expression of VEGF receptors in most normal tissues (with exception of renal glomeruli) but a significant upregulation in the vasculature in retinal disease, the neutralisation of VEGF by ranibizumab through local intravitreal injection provides the rationale for specific inhibition of the retinal vascular angiogenesis and thereby inhibition of lesion growth and leakage. The ANCHOR and MARINA studies – large, multicentre, randomised clinical trials – showed that a fixed monthly dosing regimen of ranibizumab gave extremely good visual outcomes in all subtypes of wet AMD. The clear benefits of intravitreal anti-VEGF drugs over PDT laser led to a rapid transition from PDT to the introduction of ranibizumab after 2007. PDT is now mainly only used for polypoidal vasculopathy and chronic central serous chorioretinopathy. Based on data from the RIDE and RISE phase 3 clinical trials ranibizumab was approved by the FDA for treatment of individuals with diabetic retinopathy with DME in 2015, and has recently been approved to treat all forms of diabetic retinopathy, with or without DME, in spring of 2017 in response to the Diabetic Retinopathy Clinical Research Network's (www.DRCR.net) Protocol S study. In addition, ranibizumab is FDA-approved for macular edema after retinal vein occlusion.

14.2.4.2 Bevacizumab

Bevacizumab, (trade name Avastin; Roche/Genentech), is a recombinant humanised form of a murine monoclonal IgG1 antibody containing human constant region sequences and murine light and heavy chain complementarity determining region sequences (93% human, 7% murine sequences) that selectively binds with high affinity to all isoforms of human VEGF-A and neutralises VEGF's biological activity through a steric blocking of the binding of VEGF to its receptors VEGFR-1 and VEGFR-2 on the surface of endothelial cells, in a similar fashion to ranibizumab. In fact, ranibizumab is a Fab created from the same parent mouse antibody as bevacizumab. However, bevacizumab is a full-length IgG1 isotype antibody composed of two identical light chains (214 amino acid residues) and two heavy chains (453 residues) with a total molecular weight of 149 kDa, and is therefore substantially larger than ranibizumab. Bevacizumab is approved for use as a therapy for cancer patients, as there is significant upregulation of VEGF in the vasculature of many tumours; therefore, the neutralisation of VEGF by bevacizumab provides the rationale for relative specific inhibition of the tumour angiogenesis and consequently inhibition of tumour growth and metastasis. Due to its use as an anti-VEGF agent in the treatment of cancer, bevacizumab is packaged in large volumes and is relatively inexpensive; for this reason many ophthalmologists use bevacizumab off-label as an intravitreal treatment for neovascular retinal disease for those without the financial means to access ranibizumab or aflibercept. The difference in the cost of dosing is hugely significant; bevacizumab costs ~$42 a dose compared with ranibizumab at a cost of more than $1500 a dose. When translated into economic terms as an example of the savings to be made in 2008, bevacizumab cost Medicare only $20 million for ~480 000 injections, while ranibizumab cost Medicare $537 million for only 337 000 injections.[8] In an effort to establish whether bevacizumab used off-label provided equivalent effects on visual acuity when administered according to the same schedule, the Comparison of Age-related Macular Degeneration Treatment Trials (CATT) research group initiated a multicentre, single-blind, non-inferiority trial, in which 1208 randomly assigned patients with neovascular AMD received intravitreal injections of either ranibizumab or bevacizumab on either a monthly schedule or as needed with monthly evaluation. The primary outcome was the mean change in visual acuity at 1 year. This clinical trial was sponsored by the National Eye Institute, and the initial results were published in the *New England Journal of Medicine* in May 2011.[9] The trial concluded that the two drugs "had equivalent effects on visual acuity when administered according to the same schedule". The results of several subsequent head-to-head trials of the two anti-VEGF treatments later reinforced CATT's findings. The two therapies performed equally at restoring visual acuity according to a 2012 meta-analysis and in the IVAN trial, alone and in the investigators' meta-analysis, pooling their own results with CATT's.[10–12] A 2012 meta-analysis focused specifically on safety issues concluded that the rates of several adverse events were more common with bevacizumab, although the absolute rates of ocular serious adverse events were low with both therapies: ocular adverse events were about 2.8 times as frequent with bevacizumab than with ranibizumab, and the "proportion of patients with serious infections and gastrointestinal disorders was also higher".[13]

As with AMD, off-label use of bevacizumab for diabetic retinopathy followed ranibizumab's approval as a therapy for diabetic retinopathy. In a similar fashion to the CATT study, the DRCR network, also sponsored by the National Eye Institute, reported the results of a comparative effectiveness randomised clinical trial evaluating the three anti-VEGF agents ranibizumab, bevacizumab and aflibercept for treatment of DME.[14] On average, all three biological agents led to improved visual acuity in eyes with DME involving the centre of the retina. Interestingly, worse visual acuity when initiating therapy was associated with greater visual acuity benefit of aflibercept over bevacizumab or ranibizumab 1 year later; however, for eyes with better levels of baseline visual acuity, little difference between anti-VEGF agents was reported. Later, a 2017 systematic review update found that while ranibizumab and bevacizumab provide similar functional outcomes in diabetic macular edema, there is low-certainty evidence suggesting that ranibizumab is more effective in reducing central retinal thickness than bevacizumab.[15]

In terms of ROP, the BEAT-ROP study compared an intravitreal bevacizumab monotherapy to conventional laser therapy and demonstrated a reduced rate of ROP recurrence in the monotherapy group; however, zonal location of disease determined the effectiveness of treatment.[16] Whereas bevacizumab is effective in treating zone 1 disease, closest to the optic nerve, ranibizumab is effective in treating zone 2 disease, the area between zone 1 and the nasal ora serrata. While these therapies are promising, questions on their safety and long-term effects have arisen, and currently there are clinical trials underway investigating the optimal dose and number and timing of treatments required, as well as looking at the systemic effects of these anti-VEGF agents in premature infants.

14.2.4.3 Aflibercept

Aflibercept (tradename Eylea; Regeneron) gained FDA approval for treatment of neovascular AMD in 2011. Within a year, aflibercept had taken 50% of the market share of anti-VEGF biologics, with the other 50% attributable to Genentech. In contrast to the humanised monoclonal antibody-based VEGF-binding strategy used by ranibizumab and bevacizumab, aflibercept is a VEGF "trap". As noted earlier, VEGF binds two related receptor tyrosine kinases, named Flt-1 (VEGFR-1), and KDR/Flk-1 (VEGFR-2). Over the years, various studies had found that one of the most effective ways to block the VEGF signaling pathway was to prevent VEGF from binding to its normal receptors by administering decoy VEGF receptors. Regeneron decided to harness this "trapping" mechanism as a method to block VEGF signaling in retinal disease. At the time, the highest-affinity VEGF blocker described was a soluble decoy receptor created by fusing the first three Ig domains of VEGFR1 to the constant region (Fc portion) of human IgG1, resulting in a forced homodimer that has picomolar binding affinity.[17] However, despite its high affinity, the VEGFR1-Fc was not a feasible clinical candidate because of its poor pharmacokinetic profile, in addition to toxicological side effects that were not seen with the VEGFR2-Fc.[18] Interestingly, while VEGFR1 binds to VEGF with substantially higher affinity, most of the biologic effects of VEGF seem to be mediated *via* VEGFR2. By determining the requirements to maintain high affinity while extending *in vivo* half-life, Regeneron engineered a potent high-affinity VEGF blocker [VEGF-TRAP(R1R2)] that has prolonged *in vivo* pharmacokinetics and pharmacodynamics, lacks non-specific toxicities and could effectively suppress the growth and vascularisation of a number of different types of tumours *in vivo*. VEGF-TRAP(R1R2) is a 110 kDa fusion protein, which combines ligand binding elements taken from the extracellular domains of VEGF receptors 1 and 2 fused to the Fc portion of IgG1.[17] By fusing the extracellular protein sequences that make up the second IgG-binding domain of the VEGFR-1 receptor and the third IgG domain of the VEGFR-2 receptor to the Fc segment of a human IgG backbone, in a manner similar to the rheumatoid arthritis drug etanercept, developers have created a chimeric protein with a very high VEGF binding affinity ($K_d \approx 1$ pM) that binds all isomers of the VEGF-A family, and although the clinical significance of this is not yet known, it also binds VEGF-B and placental growth factor. Subcutaneous injections or a single intravitreous injection of VEGF-TRAP(R1R2) strongly suppressed choroidal neovascularisation in mice with laser-induced rupture of Bruch's membrane; this led to clinical trials to assess the efficacy of this VEGF-trap in neovascular AMD.[19,20] Aflibercept was granted approval for use with neovascular AMD on the basis of two concurrent AMD trials: the VIEW 1 trial, which enrolled 1217 patients in North American centers, and the VIEW 2 trial, which enrolled 1240 patients in South American, European, Asian and Australian centers. The clinical superiority of ranibizumab over both observation and photodynamic therapy was well documented in both the MARINA (7.2 letters *vs.* −10.4 letters) and ANCHOR (11.3 letters *vs.* −9.5 letters) studies, thus establishing ranibizumab as the standard against which all subsequent drugs are compared. All investigational arms for aflibercept reached their primary endpoint; non-inferiority for maintenance of vision (≤15 letters of vision loss) compared to ranibizumab (94% for ranibizumab arms and 95–96% for all VTE arms).[21,22]

Aflibercept's intermediate size at 110 kD (compared to 48 kD for ranibizumab and 148 kD for bevacizumab) and potent binding capacity for VEGF create a 1 month intravitreal binding activity that exceeds that of both ranibizumab and bevacizumab.[23] Both ranibizumab and bevacizumab need monthly dosing by intravitreal injection. Efforts to stretch the ranibizumab dosing interval to 3 months (PIER study) resulted in forfeiting previous vision gains.[24] Despite this, the majority of studies use treat-and-extend or treat-and-observe strategies to minimise both the number of injections and office visits. The first indication that aflibercept could be dosed less frequently came from the phase 2 CLEAR-IT 2 study where patients required, on average, only two injections between the 12 week loading period and the 52 week termination visit.[25] The phase 3 trials showed that injections of aflibercept every 8 weeks delivered comparable letter gains to monthly ranibizumab (VIEW 1 and VIEW 2), demonstrating that patients may not require monthly evaluations, thereby decreasing the growing burden of AMD patients on the healthcare system, while simultaneously relieving patients of the need for monthly physician visits.

14.2.4.4 Disadvantages of Anti-VEGF Therapy in Retinal Disease

Introduction of anti-angiogenic biologics such as ranibizumab, bevacizumab and aflibercept to the ophthalmology field has represented a true paradigm shift in the treatment of patients. In particular, these treatments have revolutionised AMD therapy, stabilising the areas of CNV and delaying further neovascularisation. With anti-VEGF therapy, more than 90% of AMD patients avoid moderate vision loss, and approximately one-third achieve vision of 20/40 or better.[21] However, there is no end stage to treatment, with patients requiring regular intra-ocular injections to sustain efficacy. In addition, a significant portion of patients with neovascular retinal disease are refractory or become refractory to treatments, and anti-VEGFs have not been as successful for those with diabetic retinopathy. Furthermore, anti-VEGFs do not address the underlying cause of retinal neovascular disease and there is substantial evidence that VEGF plays a crucial role in cellular homeostasis as a neurotrophic and cell survival factor.[26] Certainly, for AMD, longer-term studies suggest that the degenerative process continues to develop even when neovascularisation is controlled. Results from longer-term studies revealing vision outcomes at 7 years demonstrate fluid or at least degenerative cysts in 72%, and, perhaps most importantly, macular atrophy by fundus autofluorescence in almost all patients (up to 98.2%).[27] Indeed, a recent report raises questions regarding the efficacy of anti-VEGF treatments beyond 2 years.[28] In addition, there have been reports of RPE tears after administration of VEGF-modulating therapies, with an incidence as high as 17%, suggesting that neutralisation of VEGF may have adverse effects.[29,30] RPE tears may occur as a part of the natural progression of AMD; however, in most cases, the tears arise within 6 weeks of the first injection, lending to the suspicion that the anti-VEGF therapy may be causative. Despite these concerns about long-term treatment, anti-VEGF biologics remain the gold standard of care available to clinicians for the treatment of neovascular retinal disease today.

14.2.5 New Horizons in Biologics for Neovascular Retinal Disease

The ceiling effect and lack of an end stage of anti-VEGF treatment has focused the research and pharmaceutical community to develop other drugs to complement anti-VEGF therapy with the hope that targeting alternative mechanisms of neovascularisation and edema in addition to blockade of VEGF will improve visual outcome. Table 14.2 illustrates the alternative biological therapies that are being trialed for the treatment of neovascular retinal disease. A number of monoclonal antibody studies targeting the platelet-derived growth factor (PDGF) pathway have been terminated, due to the fact that they did not meet primary endpoints. A phase 2 study with REGN2176-3 was terminated as no additional efficacy was observed over aflibercept alone, while perhaps most disappointingly, researchers studying Fovista (pegpleranib), another anti-PDGF that had reached phase 3 clinical trials announced in August 2017 that the addition of 1.5 mg of Fovista to an Eylea or Avastin regimen did not result in benefit as measured by the mean change in visual acuity at the 12 month time point. THR-317 is an anti-placental growth factor recombinant monoclonal antibody.[31] Placental growth factor is a VEGF homolog that regulates the angiogenic switch in disease. RG7716 is a bispecific humanised IgG1 monoclonal antibody that blocks both angiopoietin 2 (Ang2) and VEGFA. Ang2 and VEGF act together to regulate vascular permeability *in vitro*.[32] A similar fully humanised monoclonal antibody REGN910 developed by Regeneron also targets Ang2. Both RG7716 and REGN910 are in phase 2 clinical trial for both AMD and DME.

Table 14.2 Other monoclonal antibodies in clinical trials for neovascular age-related macular degeneration (nvAMD)/diabetic macular edema (DME).[a]

Treatment	Disease	Manufacturer	Mechanism of action	Phase	ClinicalTrials.gov identifier
THR-317	DME	Thrombogenic	Anti-PlGF recombinant monoclonal antibody	Phase 2	NCT03071068
RO6867461 (RG7716)	DME, nvAMD	Roche	Bispecific antibody blocking simultaneous Ang-2 and VEGF-A	Phase 2	NCT03038880
REGN910 (nesvacumab)	DME, nvAMD	Regeneron	Fully human IgG1 monoclonal antibody that specifically binds and inactivates the Tie2 receptor ligand Ang-2 with high affinity, but shows no binding to Ang-1	Phase 2	NCT02713204
DS7080a	nvAMD	Daiichi Sankya, Inc.	Monoclonal antibody	Phase 1	NCT02530918
TK001 (sevacizumab)	nvAMD	Jiangsu T-Mab Biopharma Co., Ltd	Recombinant humanised monoclonal antibody	Phase 1	NCT03021785
Brolucizumab (formerly RTH-258 and ESBA1008)	nvAMD	Alcon	Humanised, single-chain antibody fragment that is much smaller than commercially available anti-VEGF agents	Phase 2	NCT02507388
REGN2176-3	nvAMD	Regeneron	Combination antibody to PDGFR-â coformulated with aflibercept	Phase 2 terminated	NCT02418754
iSONEP (LT1009)	nvAMD	Lpath, Inc.	Humanised monoclonal antibody against S1P (systemic administration)	Phase 2	NCT01414153
Fovista	AMD	Ophthotech	Administered in combination with an anti-VEGF agent (ranibizumab)	Phase 3 terminated	NCT01944839

[a]DME, diabetic macular edema; nvAMD, neovascular Age-related Macular Degeneration; PDGFR-β, beta-type Platelet-derived growth factor receptor; PlGF: placental growth factor; S1P, Sphingosine-1-phosphate; VEGF: vascular endothelial growth factor.

14.3 Biologics for Non-infectious Inflammatory Retinal Disease

In the same way that ocular neovascularisation is a common pathway that leads to vision loss in several disorders that cause blindness, inflammation has been implicated as one of the pivotal mediators of retinal injury in the pathogenesis of ocular disease. An overactive inflammatory response is clear in diseases including diabetic retinopathy and DME, two potential outcomes of the metabolic disease diabetes, and in non-infectious uveitis. The role of the inflammatory response in AMD is less clear, especially in the cases of early dry AMD and geographic atrophy. While there is certainly dysfunction in mediators of the inflammatory system, there are arguments made for both a lack of inflammatory function and for overt inflammatory responses in the pathobiology of AMD.[33] There are several biologics targeting specific molecules involved in inflammatory processes. They mainly fall into either the classification of monoclonal antibody therapy or recombinant protein therapies; those of note for use in retinal disease are listed in Table 14.3.

14.3.1 Anti-TNFα Therapies

Tumour necrosis factor-α (TNFα) is a key cytokine involved in the pathogenesis of many inflammatory diseases. TNFα is a multifunctional cytokine; it activates T-cells and macrophages by increasing the expression of endothelial adhesion markers, and through inducing the expression of pro-inflammatory cytokines TNFα further amplifies the inflammatory response. In addition, TNFα has been shown to induce the secretion of VEGFA and VEGFC from RPE cells and choroidal fibroblasts, implicating it as a mediator of inflammation-inducible edema and neovascularisation.[34] Animal models have indicated that TNFα plays a deleterious role in the retina, with pharmacological blockade leading to reduced retinal cell death in a rat model of diabetic retinopathy.[35,36] Several TNF-blocking antibodies have been approved for use in inflammatory diseases such as rheumatoid arthritis, psoriasis and ankylosing spondylitis. The currently available biological TNFα inhibitors are etanercept, infliximab, adalimumab, golimumab and certolizumab. Etanercept (tradename Enbrel; Amgen) is a recombinant dimeric protein generated by fusion of the ligand-binding portion of the human type 2 TNF receptor linked to the Fc portion of human IgG1; in this way etanercept mimics the inhibitory effects of naturally occurring soluble TNF receptors. Infliximab (tradename Remicade; Janssen) is a mouse–human chimeric monoclonal antibody with human IgG1 and a mouse variable fragment that neutralises TNFα by high-affinity binding to both soluble and transmembrane forms of TNFα, preventing the binding of TNFα to its receptors. Adalimumab (tradename Humira; Abbvie) is a fully human monoclonal antibody that binds all forms of TNFα. Golimumab (tradename Simponi; Merck Sharp and Dohme) is a fully human IgG1 monoclonal antibody against TNFα, while certolizumab (tradename Cimzia; UCB) is a PEGylated humanised monoclonal antibody

Table 14.3 Anti-inflammatory biological agents.[a]

Treatment	Description	Targeted disease	Manufacturer	Current status	Years	Clinical Trials.gov identifier
CCR 2/5	Chemokine receptor agonist	DME	Pfizer	Phase 2	2013–2015	NCT01994291
Daclizumab	Anti-CD25	nvAMD	Abbvie	Phase 2	2006–2010	NCT00304954
Infliximab	Anti-TNF	nvAMD	Janssen	Phase 2	2006–2010	NCT00304954
Adalimumab	Anti-TNF	nvAMD	Abbvie	Phase 2	2010–2011	NCT01136252
Canakinumab	Anti-IL-1α	Diabetic retinopathy/ DME	Novartis			NCT01589029

[a]DME, diabetic macular edema; IL, interleukin; nvAMD, neovascular Age-related Macular Degeneration; TNF, tumour necrosis factor.

that combines the fragment antigen-binding Fab of the TNF antibody with polyethylene glycol (PEG). Infliximab is delivered intravenously, whereas the others are injected subcutaneously.

14.3.1.1 *Anti-TNFα for Non-infectious Uveitis*

Non-infectious intermediate, posterior and pan uveitis are antigen-specific CD4 T-cell mediated autoimmune diseases. TNFα represents one of the most important amplifying factors in this disease and is found at high concentrations both in the aqueous humour and the serum of uveitis patients. Given the established inflammatory pathways of uveitis, anti-inflammatory biological therapeutics have provided an attractive potential avenue for treatment, especially in cases that are refractory to conventional medicines such as glucocorticoid or anti-metabolite therapies. Indeed, the systemic use of anti-TNFα appears to be an appropriate "off-label" alternative to traditional immunosuppressive treatments for non-infectious uveitis. In particular, systemic infliximab has been shown to be effective as a treatment for both anterior and posterior uveitis; biological use for uveitis has been reviewed recently by Posarelli *et al.*[37]

14.3.1.2 *Disadvantages of Systemic Anti-TNFα Therapy*

There are a number of reported risks for systemic anti-TNFα therapy; these include fatal blood disorders, secondary infections, reactivation of latent infections, in particular *Tuberculosis mycobacterium*, tumorigenesis, drug-induced lupus and demyelinating central nervous system disorders. Due to these potentially severe systemic side effects, local immunomodulation is a desirable approach, leading ophthalmologists to investigate alternative routes of administration. Intravitreal injections of anti-TNFα, in a similar manner to intravitreal injections of anti-VEGFs, would have the advantage of bypassing the severe side effects of systemic anti-TNFα administration while simultaneously achieving a higher local concentration within the eye.

14.3.1.3 *Anti-TNFα Utility for Diabetic Retinopathy, DME and AMD*

In 2005, Markomichelakis *et al.* reported three cases of CNV regression secondary to AMD in patients under systemic administration of infliximab for the treatment of inflammatory arthritis, exciting the AMD community.[38] However, a prospective, randomised, pilot study of systemic immunosuppression with infliximab in 2010 for patients receiving anti-VEGFs as standard of care, but with recurrent exudation from neovascular lesions, demonstrated no advantage with anti-TNFα treatment.[39] Due to a number of small pilot studies, the use of intravitreal anti-TNFα, rather than systemic anti-TNFα, also gained early interest for the treatment of refractory DME and neovascular AMD. Following the apparent utility of systemic infliximab for AMD reported by Markomichelakis *et al.*, Theodossiadis *et al.* treated three patients with exudative AMD, unresponsive to ranibizumab, with two injections of 2 mg of infliximab given 2 months apart.[40] They reported visual improvement in all three cases with no safety concerns. However, soon afterwards, Giganti *et al.* reported highly contrasting results, having found that intravitreal infliximab at a dose of 0.5 mg in two eyes of DME and in two cases of neovascular AMD was associated with anatomical, electroretinographic and microperimetric worsening, and three of these cases developed intraocular inflammation.[41] Further to these reported adverse events Arias *et al.* conducted a prospective interventional trial of four eyes with exudative AMD refractory to anti-VEGF agents and reported that intravitreal infliximab, at a dose of 2 mg, resulted in half of their patients developing severe intraocular inflammatory reactions; further infliximab did not provide any visual or anatomical benefit.[42] The most comprehensive publication on the use of intravitreal infliximab in both DME and exudative AMD was reported by the Pan-American Collaborative Retina Study Group. In their series, eyes with refractory DME received a single intravitreal injection of infliximab 1.0 mg/0.05 mL (15 eyes) or 2.0 mg/0.10 mL (19 eyes). No significant anatomical or functional benefit was evidenced. In addition, 42% of eyes injected with the higher dose developed a severe uveitis, and three of these eyes (37.5%) required pars plana vitrectomy.[43] Similar results were found in 26 neovascular AMD eyes with partial responses to anti-VEGFs. No functional gains were seen, and in addition, 37.5% of patients developed a severe uveitis that resolved with intensive topical steroid therapy.[44] In contrast to the adverse events reported with the use of intravitreal infliximab, no ocular side effects were reported by the Pan-American Collaborative Retina Study Group following intravitreal injections of adalimumab. However, intravitreal injections of adalimumab (2.0 mg/0.08 mL) did not provide any anatomical or functional benefit in five cases of DME non-responsive to conventional therapies and four cases of neovascular AMD. The potential utility of combination intravitreal adalimumab with bevacizumab was analysed in a short-term pilot study and found that a clinically relevant response was achieved in patients with macular edema of various etiologies, including partially responsive neovascular AMD. However, overall, the several concerning reports describing severe ocular adverse events and lack of definitive benefit in visual acuity has curbed the initial interest in off-label intravitreal use of anti-TNFα for retinal disease.

14.3.2 Other Anti-inflammatory Biologics in Retinal Disease

Table 14.3 lists other monoclonal antibodies, in addition to infliximab and adalimumab, that have been trialed for therapeutic use for either DME or neovascular AMD. Daclizumab (tradename Zinbryta; Biogen) is a humanised antibody targeting the α-portion of the interleukin (IL)-2 (CD25) receptor, which is found highly expressed on activated T-cells. Several studies have suggested that daclizumab can stabilise visual acuity in patients with non-infectious intermediate and posterior uveitis, with no adverse events reported.[45–47] A small randomised pilot study investigated whether systemic immune suppression with daclizumab would similarly improve visual acuity and retinal thickness in patients with recurrent choroidal neovascularisation associated with AMD. However, this study did not demonstrate a definitive beneficial effect of daclizumab for treating AMD for these primary endpoints, although the authors suggest that the daclizumab group approach significance for a reduction in the anti-VEGF injection rate. IL-1β, like TNFα, is a potently inflammatory cytokine with pleiotropic effects. IL-1β pathology is highly associated with auto-inflammatory diseases such as cryopyrin-associated periodic syndromes. Canakinumab (tradename Ilaris; Novartis), a human IgG monoclonal antibody targeting IL-1β is approved for use for numerous auto-inflammatory diseases. Clinical studies investigating the blockade of IL-1β have shown improvement in glycemic control of patients with type 2 diabetes, likely due in part to inhibition of IL-1β-induced pancreatic β-cell death. Furthermore, IL-1β is detectable in the vitreous of proliferative diabetic retinopathy. These observations have led to studies evaluating the effect of systemic IL-1β blockade using canakinumab in proliferative diabetic retinopathy. Systemic administration showed stabilisation but no regression of retinal neovascularisation in patients with proliferative diabetic retinopathy; however, reduced leakage from neovascular lesions and reduced macular edema was observed indicating that the use of canakinumab for DME warrants further study.[48] The CANTOS (Canakinumab Anti-inflammatory Thrombosis Outcomes Study) trial has enrolled more than 10 000 patients over the past 6 years and will likely address canakinumab's utility for diabetic retinopathy and DME over the coming years.

14.4 Biological Agents for Treatment of Geographic Atrophy End Stage of AMD

The role of the complement system in the pathogenesis of AMD has been studied extensively over the past decade. There are a variety of lines of evidence to support the role of the complement system in the pathogenesis of AMD. Firstly, various complement components, including C3, C5b-9, CFB and CFH have been detected in drusen as well as in AMD lesions.[49,50] In addition, increased plasma levels of C3a, C3d, Bb, and C5a have been observed in AMD patients, suggesting increased local and systemic complement activation in AMD.[51–53] Secondly, genetic evidence suggests that the complement system, and in particular the alternative pathway, may be dysregulated in AMD patients, with polymorphisms in a number of complement genes (CFH,

CFB, C2, SERPING1 and C3) increasing the risk of AMD.[54,55] Furthermore, overexpression of C3 in animal models has been shown to induce many of the pathologies associated with AMD, whereas inhibition of the complement activation pathways can suppress CNV development and photoreceptor cell death in animal models. Despite convincing evidence that complement activation contributes to retinal pathology in dry AMD, up until very recently the results from clinical trials had been disappointing and most trials have ended early as a result of minimal improvement in visual acuity or reduction in disease progression at time of interim reporting (see Table 14.4). However, in August 2017 Apellis pharmaceuticals announced that its C3 inhibitor (APL-2), met its primary endpoint in a phase 2 clinical trial (FILLY) in patients with geographic atrophy-associated AMD. APL-2 was administered as an intravitreal injection in the study eye monthly or every other month for 12 months, followed by 6 months of monitoring after the end of treatment. At 12 months, APL-2, administered monthly *via* intravitreal injection, showed a 29% reduction in the rate of geographic atrophy lesion growth compared to sham. With alternate-month administration, a 20% reduction was observed. Additionally, a greater effect was observed during the second 6 months of the study, with a reduction in growth rate of 47% with monthly administration, and a reduction of 33% with alternate-month administration. APL-2 is a cyclic peptide conjugated to a PEG polymer that binds specifically to C3 and C3b, effectively blocking all three pathways of complement activation (classical, lectin and alternative). Peptides are a different class of biological molecule, and compared with larger protein antibodies they can penetrate further into the tissue. It is possible that this property of APL-2 has contributed to what has set it apart from the previous complement antibodies that have failed to progress through clinical trials. Another promising biologic targeting the complement pathway for treatment of dry AMD is lampalizumab (Genentech/Roche). Lampalizumab is an antigen-binding fragment derived from a humanised monoclonal antibody to factor D. A phase 3 trial of lampalizumab, a factor D inhibitor, is currently underway. MAHALO, the phase 2 study of the drug, demonstrated that there was a 20.4% reduction in mean change from baseline in geographic atrophy area in patients treated with monthly intravitreal injections. Along with complement factors, amyloid-β is also found in drusen deposits and is associated with dry AMD.[56,57] Based on this, GSK trialed GSK933776, a humanised mouse IgG1 monoclonal antibody directed against the *N*-terminus of the Aβ peptide in a phase 2 trial for treatment of dry AMD. However, although safe and well-tolerated, the phase 2 study investigating intravenous infusions failed to meet its primary endpoint of slowing the rate of geographic atrophy enlargement. Intravenous RN6G (PF-4382923; Pfizer) a humanised monoclonal antibody targeted against Aβ40 and Aβ4B42 which binds to amyloid that accumulates in drusen was also terminated after phase 2 trials, having failed to slow geographic atrophy.

Table 14.4 Biologics in clinical trials for treatment of geographic atrophy (GA).[a]

Treatment	Targeted disease	Manufacturer	Mechanism of action	Phase	Years	ClinicalTrials.gov identifier
GSK933776	GA AMD	GlaxoSmithKline	Monoclonal antibody against Aβ	Phase 2	2011–2016	NCT01342926
Lampalizumab	GA AMD	Roche	Monoclonal antibody against CFD	Phase 3	2014–2017	NCT02247479
Zimura	GA AMD	Ophthotech	Anti-C5 aptamer	Phase 2/3 recruiting	2016–2018	NCT02686658
LFG316	GA AMD	Novartis	Anti-C5	Phase 2	2012–2015	NCT01527500
Clg561	GA AMD	Alcon	Anti-properdin	Phase 2	2015–2017	NCT02515942
Eculizumab	GA AMD	Alexion	C5 inhibitor	Phase 2	2009–2012	NCT00935883
APL-2	GA AMD	Apellis Pharma	Peptide inhibitor of C3	Phase 2	2015–2017	NCT02503332
RN6G	GA AMD	Pfizer	Monoclonal antibody against Aβ	Phase 2 (terminated)	2012–2013	NCT01577381

[a]Aβ, Amyloid Beta; AMD, age-related macular degeneration; CFD, complement factor D; GA, Geographic atrophy.

14.5 Conclusions

There can be no doubt that the advent of biologics has revolutionised the medical landscape for retinal disease. In particular, monoclonal humanised and fully human antibodies have been life-changing for many millions of people suffering with various neovascular and inflammatory retinal conditions. Biologics are providing vision-preserving effects that greatly enhance the quality of life and reduce institutionalisation rates of those with retinal disease. It will be exciting to see how the new biologics coming along the pipeline through clinical trial complement and add value to those already in use as standard of care over the coming years.

References

1. W. L. Wong, X. Su, X. Li, C. M. Cheung, R. Klein and C. Y. Cheng, *et al.*, Global prevalence of age-related macular degeneration and disease burden projection for 2020 and 2040: a systematic review and meta-analysis, *Lancet Glob. Health*, 2014, **2**(2), e106–e116.
2. J. M. Gibson, 25th RCOphth Congress, President's Session paper: 25 years of progress in medical retina, *Eye*, 2014, **28**(9), 1041–1052.
3. G. Thurston, J. S. Rudge, E. Ioffe, H. Zhou, L. Ross and S. D. Croll, *et al.*, Angiopoietin-1 protects the adult vasculature against plasma leakage, *Nat. Med.*, 2000, **6**(4), 460–463.
4. J. W. Miller, J. Le Couter, E. C. Strauss and N. Ferrara, Vascular endothelial growth factor a in intraocular vascular disease, *Ophthalmology*, 2013, **120**(1), 106–114.
5. J. S. Penn, A. Madan, R. B. Caldwell, M. Bartoli, R. W. Caldwell and M. E. Hartnett, Vascular endothelial growth factor in eye disease, *Prog. Retinal Eye Res.*, 2008, **27**(4), 331–371.
6. W. Risau, Mechanisms of angiogenesis, *Nature*, 1997, **386**(6626), 671–674.
7. Ranabizumab SD. Available from http://www.ema.europa.eu/docs/en_GB/document_library/EPAR_-_Scientific_Discussion/human/000715/WC500043550.pdf.
8. A. Tufail, P. J. Patel, C. Egan, P. Hykin, L. da Cruz and Z. Gregor, *et al.*, Bevacizumab for neovascular age related macular degeneration (ABC Trial): multicentre randomised double masked study, *BMJ*, 2010, **340**, c2459.
9. C. R. Group, D. F. Martin, M. G. Maguire, G. S. Ying, J. E. Grunwald and S. L. Fine, *et al.*, Ranibizumab and bevacizumab for neovascular age-related macular degeneration, *N. Engl. J. Med.*, 2011, **364**(20), 1897–1908.
10. M. L. Subramanian, G. Abedi, S. Ness, E. Ahmed, M. Fenberg and M. K. Daly, *et al.*, Bevacizumab vs ranibizumab for age-related macular degeneration: 1-year outcomes of a prospective, double-masked randomised clinical trial, *Eye*, 2010, **24**(11), 1708–1715.

11. S. Jiang, C. Park and J. C. Barner, Ranibizumab for age-related macular degeneration: a meta-analysis of dose effects and comparison with no anti-VEGF treatment and bevacizumab, *J. Clin. Pharm. Ther.*, 2014, **39**(3), 234–239.
12. U. Chakravarthy, S. P. Harding, C. A. Rogers, S. M. Downes, A. J. Lotery and L. A. Culliford, *et al.*, Alternative treatments to inhibit VEGF in age-related choroidal neovascularisation: 2-year findings of the IVAN randomised controlled trial, *Lancet*, 2013, **382**(9900), 1258–1267.
13. C. Schmucker, C. Ehlken, H. T. Agostini, G. Antes, G. Ruecker and M. Lelgemann, *et al.*, A safety review and meta-analyses of bevacizumab and ranibizumab: off-label versus goldstandard, *PLoS One*, 2012, **7**(8), e42701.
14. J. A. Wells, 3rd, A. R. Glassman and L. M. Jampol, Targeting the effect of VEGF in diabetic macular edema, *N. Engl. J. Med.*, 2015, **373**(5), 481–482.
15. G. Virgili, M. Parravano, J. R. Evans, I. Gordon and E. Lucenteforte, Anti-vascular endothelial growth factor for diabetic macular oedema: a network meta-analysis, *Cochrane Database Syst. Rev.*, 2017, **6**, CD007419.
16. H. A. Mintz-Hittner, K. A. Kennedy, A. Z. Chuang and B.-R. C. Group, Efficacy of intravitreal bevacizumab for stage 3+ retinopathy of prematurity, *N. Engl. J. Med.*, 2011, **364**(7), 603–615.
17. J. Holash, S. Davis, N. Papadopoulos, S. D. Croll, L. Ho and M. Russell, *et al.*, VEGF-Trap: a VEGF blocker with potent antitumor effects, *Proc. Natl. Acad. Sci. U. S. A.*, 2002, **99**(17), 11393–11398.
18. C. J. Kuo, F. Farnebo, E. Y. Yu, R. Christofferson, R. A. Swearingen and R. Carter, *et al.*, Comparative evaluation of the antitumor activity of antiangiogenic proteins delivered by gene transfer, *Proc. Natl. Acad. Sci. U. S. A.*, 2001, **98**(8), 4605–4610.
19. Y. Saishin, Y. Saishin, K. Takahashi, R. Lima e Silva, D. Hylton and J. S. Rudge, *et al.*, VEGF-TRAP(R1R2) suppresses choroidal neovascularization and VEGF-induced breakdown of the blood-retinal barrier, *J. Cell. Physiol.*, 2003, **195**(2), 241–248.
20. Y. Saishin, Y. Saishin, K. Takahashi, M. Melia, S. A. Vinores and P. A. Campochiaro, Inhibition of protein kinase C decreases prostaglandin-induced breakdown of the blood-retinal barrier, *J. Cell. Physiol.*, 2003, **195**(2), 210–219.
21. P. J. Rosenfeld, D. M. Brown, J. S. Heier, D. S. Boyer, P. K. Kaiser and C. Y. Chung, *et al.*, Ranibizumab for neovascular age-related macular degeneration, *N. Engl. J. Med.*, 2006, **355**(14), 1419–1431.
22. D. M. Brown, P. K. Kaiser, M. Michels, G. Soubrane, J. S. Heier and R. Y. Kim, *et al.*, Ranibizumab versus verteporfin for neovascular age-related macular degeneration, *N. Engl. J. Med.*, 2006, **355**(14), 1432–1444.
23. M. W. Stewart and P. J. Rosenfeld, Predicted biological activity of intravitreal VEGF Trap, *Br. J. Ophthalmol.*, 2008, **92**(5), 667–668.

24. C. D. Regillo, D. M. Brown, P. Abraham, H. Yue, T. Ianchulev and S. Schneider, *et al.*, Randomized, double-masked, sham-controlled trial of ranibizumab for neovascular age-related macular degeneration: PIER study year 1, *Am. J. Ophthalmol.*, 2008, **145**(2), 239–248.
25. J. S. Heier, D. Boyer, Q. D. Nguyen, D. Marcus, D. B. Roth and G. Yancopoulos, *et al.*, The 1-year results of CLEAR-IT 2, a phase 2 study of vascular endothelial growth factor trap-eye dosed as-needed after 12-week fixed dosing, *Ophthalmology*, 2011, **118**(6), 1098–1106.
26. M. Saint-Geniez, A. S. Maharaj, T. E. Walshe, B. A. Tucker, E. Sekiyama and T. Kurihara, *et al.*, Endogenous VEGF is required for visual function: evidence for a survival role on muller cells and photoreceptors, *PLoS One*, 2008, **3**(11), e3554.
27. R. B. Bhisitkul, T. S. Mendes, S. Rofagha, W. Enanoria, D. S. Boyer and S. R. Sadda, *et al.*, Macular atrophy progression and 7-year vision outcomes in subjects from the ANCHOR, MARINA, and HORIZON studies: the SEVEN-UP study, *Am. J. Ophthalmol.*, 2015, **159**(5), 915–924 e2.
28. Y. Tao, T. Libondi and J. B. Jonas, Long-term follow-up after multiple intravitreal bevacizumab injections for exudative age-related macular degeneration, *J. Ocul. Pharmacol. Ther.*, 2010, **26**(1), 79–83.
29. B. Yeh and S. Ferrucci, Retinal pigment epithelium tears after bevacizumab injection, *Optometry*, 2011, **82**(3), 152–157.
30. M. Nicolo, D. Ghiglione and G. Calabria, Retinal pigment epithelial tear following intravitreal injection of bevacizumab (Avastin), *Eur. J. Ophthalmol.*, 2006, **16**(5), 770–773.
31. C. Fischer, B. Jonckx, M. Mazzone, S. Zacchigna, S. Loges and L. Pattarini, *et al.*, Anti-PlGF inhibits growth of VEGF(R)-inhibitor-resistant tumors without affecting healthy vessels, *Cell*, 2007, **131**(3), 463–475.
32. S. Rangasamy, R. Srinivasan, J. Maestas, P. G. McGuire and A. Das, A potential role for angiopoietin 2 in the regulation of the blood-retinal barrier in diabetic retinopathy, *Invest. Ophthalmol. Visual Sci.*, 2011, **52**(6), 3784–3791.
33. T. A. Ferguson and R. S. Apte, Angiogenesis in eye disease: immunity gained or immunity lost? *Semin. Immunopathol.*, 2008, **30**(2), 111–119.
34. C. N. Nagineni, V. K. Kommineni, A. William, B. Detrick and J. J. Hooks, Regulation of VEGF expression in human retinal cells by cytokines: implications for the role of inflammation in age-related macular degeneration, *J. Cell. Physiol.*, 2012, **227**(1), 116–126.
35. P. G. Theodossiadis, N. N. Markomichelakis and P. P. Sfikakis, Tumor necrosis factor antagonists: preliminary evidence for an emerging approach in the treatment of ocular inflammation, *Retina*, 2007, **27**(4), 399–413.
36. A. M. Joussen, S. Doehmen, M. L. Le, K. Koizumi, S. Radetzky and T. U. Krohne, *et al.*, TNF-alpha mediated apoptosis plays an important role in the development of early diabetic retinopathy and long-term histopathological alterations, *Mol. Vision*, 2009, **15**, 1418–1428.

37. C. Posarelli, I. Arapi, M. Figus and P. Neri, Biologic agents in inflammatory eye disease, *J. Ophthalmic Vision Res.*, 2011, **6**(4), 309–316.
38. N. N. Markomichelakis, P. G. Theodossiadis and P. P. Sfikakis, Regression of neovascular age-related macular degeneration following infliximab therapy, *Am. J. Ophthalmol.*, 2005, **139**(3), 537–540.
39. R. B. Nussenblatt, G. Byrnes, H. N. Sen, S. Yeh, L. Faia and C. Meyerle, *et al.*, A randomized pilot study of systemic immunosuppression in the treatment of age-related macular degeneration with choroidal neovascularization, *Retina*, 2010, **30**(10), 1579–1587.
40. P. G. Theodossiadis, V. S. Liarakos, P. P. Sfikakis, I. A. Vergados and G. P. Theodossiadis, Intravitreal administration of the anti-tumor necrosis factor agent infliximab for neovascular age-related macular degeneration, *Am. J. Ophthalmol.*, 2009, **147**(5), 825–830, 30 e1.
41. M. Giganti, P. M. Beer, N. Lemanski, C. Hartman, J. Schartman and N. Falk, Adverse events after intravitreal infliximab (Remicade), *Retina*, 2010, **30**(1), 71–80.
42. L. Arias, J. M. Caminal, M. B. Badia, M. J. Rubio, J. Catala and O. Pujol, Intravitreal infliximab in patients with macular degeneration who are nonresponders to antivascular endothelial growth factor therapy, *Retina*, 2010, **30**(10), 1601–1608.
43. L. Wu, E. Hernandez-Bogantes, J. A. Roca, J. F. Arevalo, K. Barraza and A. F. Lasave, Intravitreal tumor necrosis factor inhibitors in the treatment of refractory diabetic macular edema: a pilot study from the Pan-American Collaborative Retina Study Group, *Retina*, 2011, **31**(2), 298–303.
44. L. Wu, J. F. Arevalo, E. Hernandez-Bogantes, C. V. Regatieri, J. A. Roca and M. E. Farah, Intravitreal tumor necrosis factor-alpha inhibitors for neovascular age-related macular degeneration suboptimally responsive to antivascular endothelial growth factor agents: a pilot study from the Pan American Collaborative Retina Study Group, *J. Ocul. Pharmacol. Ther.*, 2013, **29**(3), 366–371.
45. R. B. Nussenblatt, E. Fortin, R. Schiffman, L. Rizzo, J. Smith and P. Van Veldhuisen, *et al.*, Treatment of noninfectious intermediate and posterior uveitis with the humanized anti-Tac mAb: a phase I/II clinical trial, *Proc. Natl. Acad. Sci. U. S. A.*, 1999, **96**(13), 7462–7466.
46. R. B. Nussenblatt, J. S. Peterson, C. S. Foster, N. A. Rao, R. F. See and E. Letko, *et al.*, Initial evaluation of subcutaneous daclizumab treatments for noninfectious uveitis: a multicenter noncomparative interventional case series, *Ophthalmology*, 2005, **112**(5), 764–770.
47. M. L. Hernandez Garfella, M. Diaz Llopis, D. Salom Alonso and E. Cervera Taulet, [Recurrent uveitis and therapy with monoclonal antibody (daclizumab)], *Arch. Soc. Esp. Oftalmol.*, 2004, **79**(12), 593–598.
48. M. Stahel, M. Becker, N. Graf and S. Michels, SYSTEMIC INTERLEUKIN 1beta INHIBITION IN PROLIFERATIVE DIABETIC RETINOPATHY: a prospective open-label study using canakinumab, *Retina*, 2016, **36**(2), 385–391.

49. D. H. Anderson, R. F. Mullins, G. S. Hageman and L. V. Johnson, A role for local inflammation in the formation of drusen in the aging eye, *Am. J. Ophthalmol.*, 2002, **134**(3), 411–431.
50. D. H. Anderson, M. J. Radeke, N. B. Gallo, E. A. Chapin, P. T. Johnson and C. R. Curletti, *et al.*, The pivotal role of the complement system in aging and age-related macular degeneration: hypothesis re-visited, *Prog. Retinal Eye Res.*, 2010, **29**(2), 95–112.
51. R. Reynolds, M. E. Hartnett, J. P. Atkinson, P. C. Giclas, B. Rosner and J. M. Seddon, Plasma complement components and activation fragments: associations with age-related macular degeneration genotypes and phenotypes, *Invest. Ophthalmol. Visual Sci.*, 2009, **50**(12), 5818–5827.
52. H. P. Scholl, P. Charbel Issa, M. Walier, S. Janzer, B. Pollok-Kopp and F. Borncke, *et al.*, Systemic complement activation in age-related macular degeneration, *PLoS One*, 2008, **3**(7), e2593.
53. J. Lechner, M. Chen, R. E. Hogg, L. Toth, G. Silvestri and U. Chakravarthy, *et al.*, Higher plasma levels of complement C3a, C4a and C5a increase the risk of subretinal fibrosis in neovascular age-related macular degeneration: complement activation in AMD, *Immun. Ageing*, 2016, **13**, 4.
54. V. Cipriani, H. T. Leung, V. Plagnol, C. Bunce, J. C. Khan and H. Shahid, *et al.*, Genome-wide association study of age-related macular degeneration identifies associated variants in the TNXB-FKBPL-NOTCH4 region of chromosome 6p21.3, *Hum. Mol. Genet.*, 2012, **21**(18), 4138–4150.
55. S. Katta, I. Kaur and S. Chakrabarti, The molecular genetic basis of age-related macular degeneration: an overview, *J. Genet.*, 2009, **88**(4), 425–449.
56. T. Dentchev, A. H. Milam, V. M. Lee, J. Q. Trojanowski and J. L. Dunaief, Amyloid-beta is found in drusen from some age-related macular degeneration retinas, but not in drusen from normal retinas, *Mol. Vision*, 2003, **9**, 184–190.
57. A. Ning, J. Cui, E. To, K. H. Ashe and J. Matsubara, Amyloid-beta deposits lead to retinal degeneration in a mouse model of Alzheimer disease, *Invest. Ophthalmol. Visual Sci.*, 2008, **49**(11), 5136–5143.
58. Q. D. Nguyen, D. M. Brown, D. M. Marcus, D. S. Boyer, S. Patel and L. Feiner, *et al.*, Ranibizumab for diabetic macular edema: results from 2 phase III randomized trials: RISE and RIDE, *Ophthalmology*, 2012, **119**(4), 789–801.
59. P. A. Keane and S. R. Sadda, Retinal vein occlusion and macular edema – critical evaluation of the clinical value of ranibizumab, *Clin. Ophthalmol.*, 2011, **5**, 771–781.
60. P. J. Rosenfeld, R. M. Rich and G. A. Lalwani, Ranibizumab: Phase III clinical trial results, *Ophthalmol. Clin. North Am.*, 2006, **19**(3), 361–372.

61. E. S. Gragoudas, A. P. Adamis, E. T. Cunningham, Jr., M. Feinsod, D. R. Guyer and Group VISiONCT, Pegaptanib for neovascular age-related macular degeneration, *N. Engl. J. Med.*, 2004, **351**(27), 2805–2816.
62. J. S. Heier, D. M. Brown, V. Chong, J. F. Korobelnik, P. K. Kaiser and Q. D. Nguyen, *et al.*, Intravitreal aflibercept (VEGF trap-eye) in wet age-related macular degeneration, *Ophthalmology*, 2012, **119**(12), 2537–2548.

Subject Index